Springer Proceedings in Mathematics & Statistics

Volume 515

This book series features volumes composed of selected contributions from workshops and conferences in all areas of current research in mathematics and statistics, including data science, operations research and optimization. In addition to an overall evaluation of the interest, scientific quality, and timeliness of each proposal at the hands of the publisher, individual contributions are all refereed to the high quality standards of leading journals in the field. Thus, this series provides the research community with well-edited, authoritative reports on developments in the most exciting areas of mathematical and statistical research today.

Andreas Ott · Wolfgang Reichel ·
John Alasdair Warwicker
Editors

Applications of Mathematics in Sciences, Engineering, and Economics

MathSEE Symposium, Karlsruhe, September 27–29, 2023

Springer

Editors
Andreas Ott
Institute for Mathematics
Heidelberg University
Heidelberg, Germany

Wolfgang Reichel
Department of Mathematics
Karlsruhe Institute of Technology
Karlsruhe, Germany

John Alasdair Warwicker
Institute for Operations Research
Karlsruhe Institute of Technology
Karlsruhe, Germany

ISSN 2194-1009 ISSN 2194-1017 (electronic)
Springer Proceedings in Mathematics & Statistics
ISBN 978-3-032-01278-4 ISBN 978-3-032-01279-1 (eBook)
https://doi.org/10.1007/978-3-032-01279-1

Mathematics Subject Classification: 14-02, 35-02, 35R30, 45Q, 65K10, 62-02, 65-02, 55-02, 34-02

This Springer imprint is published by the registered company Springer Nature Switzerland AG
The registered company address is: Gewerbestrasse 11, 6330 Cham, Switzerland

Preface

This proceedings volume is an outcome of the 2nd Symposium on Applications of Mathematical Sciences held at the KIT Center MathSEE at Karlsruhe Institute of Technology (KIT), Germany, on September 27–29, 2023. The symposium featured eight plenary lectures and several contributed sessions on the applications of mathematical methods along the following tracks:

- Mathematical Structures: Shapes, Geometry, Number Theory and Algebra
- Mathematical Modeling, Differential Equations, Numerics and Simulation
- Inverse Problems and Optimization
- Stochastic Modeling, Statistical Data Analysis and Forecasting.

The symposium also included a panel discussion on the topic of "Interdisciplinary Mathematical Research: Challenges and Opportunities" followed by a symposium dinner for registered participants. On the first day of the symposium, a poster session took place with a competition for the "Best Poster Award".

MathSEE (Mathematics in Sciences, Engineering, and Economics) is one of nine KIT centers, which are the largest organizational units of KIT combining central cross-divisional research activities and representing the strategic research fields. MathSEE views mathematical research as an indispensable component in research and progress in natural sciences, innovation, society, and for the economy. Its central mission is to foster interdisciplinary mathematical research. Scientists working in mathematical theory and in applications are cooperating in joint research projects. Ranging from young researchers to experienced mentors they are pooling their competencies and passion for mathematical research together in MathSEE. The center lives by the motto of playing a leading role in interdisciplinary mathematical research in sciences, engineering, and economics, and to enrich science, the economy, and society.

For the second time, MathSEE has held and sponsored the symposium so that attendance was free of charge. Thanks go to the scientific team consisting of Punit Gandhi, Martin Krejca, Annette Möller, Marc-Oliver Pohle, Bastian Rieck, Matthias Schlottbom, Bernadette Jana Stolz-Pretzer, and Alexandria Volkening for having designed the attractive program and selected the excellent speakers. Thanks also go to the organizational team Sebastian Lerch, Andreas Ott, Julian Quinting, Björn de Rijk, Aarti Singh, John Alasdair Warwicker, and their numerous helpers for having made the symposium a unique and enjoyable event worth remembering

All participants of the symposium were offered to send in contributions to this proceedings volume. Following a thorough peer-review process, 14 contributions were finally accepted. The volume aims at all scientists interested in the interplay between mathematical methods and applications. The topics of the contributed papers reflect very well the unique spirit of the symposium and they may serve as an introductory guideline to the specific problems in the respective fields. We are very grateful to the referees for having carefully checked the papers and cordially thank them for their very valuable suggestions for improvement. We also thank all authors for having submitted their papers to this volume and thus sharing their latest research insights with a large community. Finally, our sincere thanks go to Springer for their support in compiling and publishing this volume.

Heidelberg, Germany — Andreas Ott
Karlsruhe, Germany — Wolfgang Reichel
Karlsruhe, Germany — John Alasdair Warwicker
April 2025

Contents

1 **Robust Capture Pursuit Based on Differential Games** 1
Valery Y. Glizer and Vladimir Turetsky

2 **Simplicial Fermat–Weber Problems Are Dual to Minimum Cost Flow Problems** 33
Andrei Comăneci

3 **Forecasting Lightning Probabilities Derived from the Lightning Potential Index Using Neural Networks** 43
Manuel Baumgartner, Guido Schröder, and Cristina Primo

4 **Numerical Approaches to Modelling of Lava Dome Growth** 67
Natalya Zeinalova, Alik Ismail-Zadeh, Igor Tsepelev, Oleg Melnik, and Frank Schilling

5 **Enabling Microscale Simulations for Long-Term Aging of Li-Ion Batteries** 81
Falco Schneider and Jochen Zausch

6 **Compression and Simulation of Large Insurance Portfolios with New Business** 105
Maximilian Diehl and Jörn Sass

7 **A PAC-Bayes Oracle Inequality for Sparse Neural Networks** 131
Maximilian F. Steffen and Mathias Trabs

8 **Comparison of Model Output Statistics and Neural Networks to Postprocess Wind Gusts** 153
Cristina Primo, Benedikt Schulz, Sebastian Lerch, and Reinhold Hess

9 **AutoEncoding Cognitive Maps of Non-Trivial Topology** 181
M. Beketov, K. Sorokin, M. Subbotin, V. Sotskov, K. Anokhin, and A. Ayzenberg

10 Residual Based Error Estimator for Chemical-Mechanically Coupled Battery Active Particles 197
Raphael Schoof, Lennart Flür, Florian Tuschner, and Willy Dörfler

11 Analysis of Kerr Frequency Combs in Microresonators with Two-Photon Absorption and Free-Carriers 221
Lukas Bengel and Wolfgang Reichel

12 The Reversible Simulator a Data-Driven Approach for Solving Forward and Inverse Problems 243
Tristan van Leeuwen

13 The Effect of Protest Management Strategies on Protesting Activity and Social Tension: A Mathematical Perspective 261
Christina Wuyan Wang, Timothy Wessler, P. Jeffrey Brantingham, and Nancy Rodriguez

14 Inferring a Cell Structure on the Space of Cyclooctane Conformations 303
Ulrich Bauer and Fabian Lenzen

Chapter 1
Robust Capture Pursuit Based on Differential Games

Valery Y. Glizer and Vladimir Turetsky

Abstract A problem of pursuit with a capture of a maneuvering evader is considered. The notions of a pursuer's robust capture control and its capture zone are introduced. Based on solutions of auxiliary zero-sum differential games, two versions of the pursuer's robust capture control are designed and the corresponding capture zones are constructed. An illustrative example of a pursuit-evasion engagement between two flying vehicles is presented.

Keywords Pursuit-evasion engagement · Differential game · Robust capture pursuit

1.1 Introduction

The problem of pursuit of a maneuvering evader has a considerable interest in theory and applications (see e.g. [2, 4, 11, 13, 15, 17–19]). This pursuit problem can be formulated in two different versions. If the evader's behaviour is predictable, we can formulate an optimal control problem for the pursuer (see e.g. [3, 9]). Without such information, we should design an admissible pursuer's control providing an acceptable outcome of the pursuit against any admissible evader's behaviour, i.e., robustly with respect to the evader's behaviour. Its best outcome is the capture where the distance between the pursuer and the evader (the miss distance) becomes zero in a given time against any admissible evader's behavior. To reach this outcome, two approaches are mainly used in the literature. In the first one the robust pursuit problem is formulated as a zero-sum differential game with the miss distance as the cost function and the bounded controls of the players. The pursuer minimizes the cost function, while the evader maximizes it. If the pursuer's optimal feedback control

V. Y. Glizer (✉)
The Galilee Research Center for Applied Mathematics, Karmiel, Israel
e-mail: valgl120@gmail.com; val1202023@outlook.com

V. Turetsky
Braude College of Engineering, Karmiel, Israel
e-mail: turetsky1@braude.ac.il

A. Ott et al. (eds.), *Applications of Mathematics in Sciences, Engineering, and Economics*, Springer Proceedings in Mathematics & Statistics 515,
https://doi.org/10.1007/978-3-032-01279-1_1

provides zero upper value of the game from some set of initial game's positions then this control guarantees the capture from any of these positions against any admissible evader's behavior. This control and the corresponding set of the initial game's positions are called a robust capture control and a capture zone. The capture zone obtained by the first approach is maximal possible, while the robust capture control is as a rule of a bang-bang structure yielding a chattering which is extremely undesirable in real-life applications (for more details see [17, 18]). The second approach consists in a preliminary design of a pursuer's feedback control with desirable properties (continuity/smoothness/linearity) with respect to the state variable. Then, due to proper additional conditions imposed on this control, its capture zone is constructed. The robust capture control of the second approach does not lead to chattering, however, its capture zone is not maximal possible. The first approach can be found in [12, 15, 16], while the second approach can be found in [7, 18, 19]. In the present paper, a combination of these two approaches is proposed allowing a design of an essentially novel pursuer's feedback control which, being continuous, has the capture zone coinciding with the maximal possible capture zone of the first approach. This robust capture control does not lead to chattering. Its design and the construction of its capture zone are based on solutions of two auxiliary zero-sum differential games. Along with this control, a bang-bang robust capture control also is designed. We apply in the paper the following main notations and notions: **1)** the superscript T denotes the transposition of a vector $x \in \mathbb{R}^n$, (x^T); **2)** I_n denotes the $n \times n$ identity matrix; **3)** $\text{clo}\big(\mathcal{D}\big)$ denotes the closure of a set $\mathcal{D} \in \mathbb{R}^n$ and $\text{int}\big(\mathcal{D}\big)$ denotes the set of all its interior points; **4)** we call a function $\varphi(x) : \mathbb{R}^n \to \mathbb{R}$ piecewise continuous in $\mathbb{R}^n$ if $\varphi(x)$ is bounded in $\mathbb{R}^n$ and there exists a finite number of nonintersecting open sets $\mathcal{D}_i$, $(i = 1, \ldots, K)$, such that $\bigcup_{i=1}^{K} \text{clo}\big(\mathcal{D}_i\big) = \mathbb{R}^n$ and $\varphi(x)$ is continuous in each set $\mathcal{D}_i$, $(i = 1, \ldots, K)$; **5)** for a continuous function $\theta(t) : [t_1, t_2] \to \mathbb{R}$, having a finite number of distinct zeros in $[t_1, t_2]$, $\text{sign}\big(\theta(t)\big) = 1$ if $\theta(t) > 0$ and $\text{sign}\big(\theta(t)\big) = -1$ if $\theta(t) \leq 0$.

1.2 Problem Formulation of the Robust Capture Pursuit

In this section, we first present the mathematical model of a planar engagement between two participants (the pursuer and the evader). This model is based on the participants' trajectories in the initial collision course and on the assumption that, during the entire engagement, the trajectories of the participants remain in a small vicinity of the corresponding initial collision course trajectories (see Sect. 1.2.1). Then, in Sect. 1.2.2, we present main definitions of the robust capture pursuit and formulate the pursuit-evasion game with bounded players' controls.

1.2.1 Mathematical Model of Planar Pursuit-Evasion Engagement

The mathematical model of the engagement is constructed subject to the following assumptions:

(A1) the engagement is between two participants called the pursuer and the evader;
(A2) the engagement takes place in a plane;
(A3) the pursuer and the evader are considered as points in this plane;
(A4) the origin of the right-handed orthogonal Cartesian coordinate system in the aforementioned plane coincides with the pursuer's initial position, the X-axis coincides with the pursuer's initial line-of-sight and it is directed from the pursuer's initial position to the evader's initial position;
(A5) the velocity magnitudes (speeds) of both participants are known continuously differentiable positive functions of time;
(A6) the trajectories of both participants are close to their initial collision course trajectories.

To derive the mathematical model of the pursuit-evasion engagement, we should construct the participants' trajectories in their initial collision course.

1.2.1.1 Trajectories of the Pursuer and the Evader in the Initial Collision Course

The geometry of this course is presented in Fig. 1.1.

In this figure, the points P_0 and E_0 are the initial positions of the pursuer and the evader, respectively. The initial velocity vectors of the pursuer and the evader are denoted by $\mathbf{V}_{P0} = \mathbf{V}_P(t_0)$ and $\mathbf{V}_{E0} = \mathbf{V}_E(t_0)$, where t_0 is an initial time instant of the engagement. Here and in what follows, $\mathbf{V}_P(t)$ and $\mathbf{V}_E(t)$ are current velocity vectors of the pursuer and the evader, while the magnitudes of these vectors are denoted by $V_P(t)$ and $V_E(t)$, respectively. By ϕ_{P0} and ϕ_{E0}, we denote the initial aspect angles

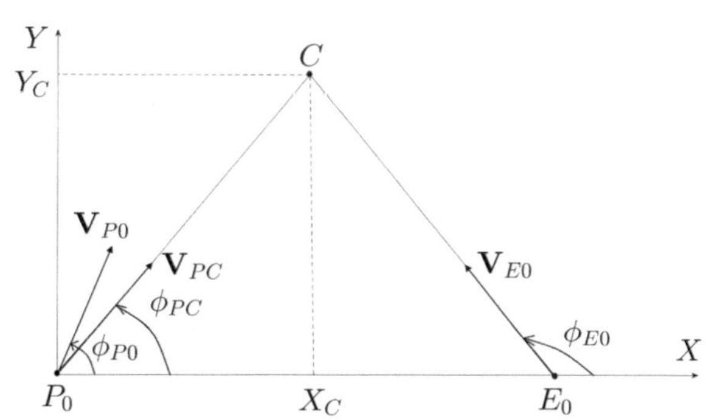

Fig. 1.1 Geometry of the initial collision course

of the pursuer and the evader, i.e., the angles between the corresponding velocity vectors ($\mathbf{V}_{P0}$ and $\mathbf{V}_{E0}$) and the X-axis. The vector $\mathbf{V}_{PC}$ shows the direction of the pursuer's motion in the initial collision course, and ϕ_{PC} is the angle between this vector and the X-axis. The collision point in the initial collision course is denoted by $C = C\big(X_C, Y_C\big)$. By $R_0 > 0$, we denote the initial range between the pursuer and the evader, i.e., $R_0 = |E_0 P_0|$.

Due to Fig. 1.1, the trajectories of the pursuer and the evader in the initial collision course are the segments of the straight lines P_0C and E_0C, respectively. The motions of the pursuer and the evader along these trajectories are in the directions of $\mathbf{V}_{PC}$ and $\mathbf{V}_{E0}$, respectively, but with the speeds $V_P(t)$ and $V_E(t)$. In what follows, based on these (virtual) trajectories, the mathematical model of the real kinematics of the engagement between the pursuer and the evader is derived. Also, it should be noted that the pursuer, being aware of the evader's angle $\phi_{E0} \in [0, 2\pi)$ and assuming the evader's motion in the direction of $\mathbf{V}_{E0}$, chooses its own angle ϕ_{PC} (and, therefore, the direction of $\mathbf{V}_{PC}$) in the initial collision course preferably such that the collision will happen as soon as possible. However, another choice of the angle ϕ_{PC}, providing the collision, also can be acceptable from the practical viewpoint. Let $t_C > t_0$ be the collision time-instant in the initial collision course, i.e., the time instant when the pursuer and the evader, moving along their initial collision course trajectories, reach simultaneously the collision point $C\big(X_C, Y_C\big)$. Thus, using Fig. 1.1 and the aforementioned notations and notion of the initial collision course, as well as the equalities $\cos(\pi - \phi_{E0}) = -\cos(\phi_{E0})$ and $\sin(\pi - \phi_{E0}) = \sin(\phi_{E0})$, we directly obtain the following equations:

$$\int_{t_0}^{t_C} V_P(t)dt\cos(\phi_{PC}) = R_0 + \int_{t_0}^{t_C} V_E(t)dt\cos(\phi_{E0}),$$
$$\int_{t_0}^{t_C} V_P(t)dt\sin(\phi_{PC}) = \int_{t_0}^{t_C} V_E(t)dt\sin(\phi_{E0}). \tag{1.1}$$

The equations in (1.1) constitute the system of two nonlinear algebraic equations with respect to two unknowns, ϕ_{PC} and t_C. Using the equality $\cos^2(\phi) + \sin^2(\phi) = 1$ and the inequality $V_P(t) > 0$, $t \geq t_0$, the system (1.1) is reduced to the nonlinear algebraic equation with respect to t_C

$$\int_{t_0}^{t_C} V_P(t)dt = \left[R_0^2 + 2R_0\int_{t_0}^{t_C} V_E(t)dt\cos(\phi_{E0}) + \left(\int_{t_0}^{t_C} V_E(t)dt\right)^2\right]^{1/2}. \tag{1.2}$$

We assume that this equation has at least one positive solution, and let $t_C = t_{C,min} > t_0$ be the smallest positive solution of (1.2). Substituting $t_C = t_{C,min}$ into (1.1), we obtain from the resulting equations

$$\cos(\phi_{PC}) = \left(R_0 + \int_{t_0}^{t_{C,min}} V_E(t)dt\cos(\phi_{E0})\right) \Big/ \int_{t_0}^{t_{C,min}} V_P(t)dt,$$

$$\sin(\phi_{PC}) = \left(\int_{t_0}^{t_{C,min}} V_E(t)dt\sin(\phi_{E0})\right) \Big/ \int_{t_0}^{t_{C,min}} V_P(t)dt. \quad (1.3)$$

These expressions, being feasible, determine the unique angle ϕ_{PC} satisfying the inequality $0 \leq \phi_{PC} < 2\pi$. The solution $(\phi_{PC}, t_{C,min})$ of (1.1), along with the angle ϕ_{E0}, uniquely determines the trajectories of the pursuer and the evader in the initial collision course. Note that the system (1.1) is derived and its solution is obtained in the case of the initial collision course presented in Fig. 1.1. Other cases of the initial collision course can be treated similarly.

1.2.1.2 Mathematical Model of the Pursuit-Evasion Kinematics in a Small Vicinity of the Initial Collision Course

First, we derive the mathematical model of the kinematics separately for each of the engagement participants. We start with the pursuer's kinematics. Due to the assumption (A6), the current pursuer's position $P(t) = \big(x_P(t), y_P(t)\big), t \in [t_0, t_C]$ is close to the corresponding point $P_C(t) = \big(x_{PC}(t), y_{PC}(t)\big), t \in [t_0, t_C]$ on its initial collision course trajectory. More precisely, we assume that the distance between these two points is negligible, i.e., $P(t) \approx P_C(t)$. Therefore, in the construction of the mathematical model of the pursuer's kinematics, we set that $P(t) = P_C(t)$ (see Fig. 1.2). In this figure, $\alpha_P(t) = \phi_P(t) - \phi_{PC}$, while $\phi_P(t), t \in [t_0, t_C]$ is the current pursuer's aspect angle, i.e., the angle between the current pursuer's velocity vector $\mathbf{V}_P(t)$ and the X-axis. By $\mathbf{a}_P(t), t \in [t_0, t_C]$, we denote the pursuer's current lateral acceleration vector. This vector is perpendicular to the current pursuer's velocity vector. By $a_P(t)$, we denote the magnitude of the vector $\mathbf{a}_P(t)$ taken with the sign "+" if $\mathbf{a}_P(t)$ rotates the vector $\mathbf{V}_P(t)$ counterclockwise, and taken with the sign "−"

Fig. 1.2 Geometry of the engagement along the initial collision course

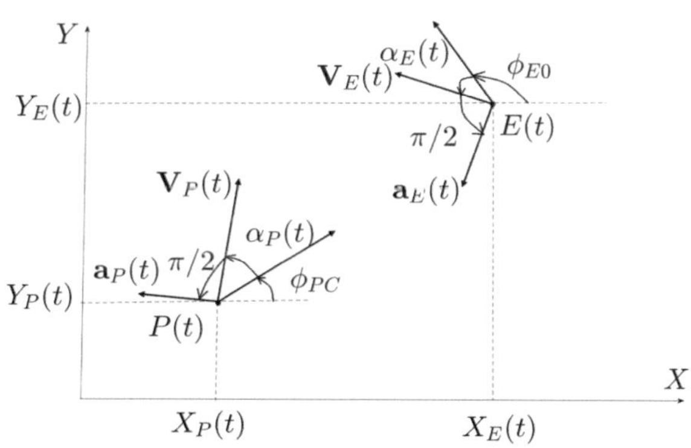

if $\mathbf{a}_P(t)$ rotates the vector $\mathbf{V}_P(t)$ clockwise. Using Fig. 1.2 and the aforementioned notations, we directly obtain the following equations:

$$\frac{dx_P(t)}{dt} = V_P(t)\cos\big(\phi_{PC} + \alpha_P(t)\big), \quad t \in [t_0, t_C], \tag{1.4}$$

$$\frac{dy_P(t)}{dt} = V_P(t)\sin\big(\phi_{PC} + \alpha_P(t)\big), \quad t \in [t_0, t_C], \tag{1.5}$$

$$\frac{d\alpha_P(t)}{dt} = \frac{a_P(t)}{V_P(t)}, \quad t \in [t_0, t_C]. \tag{1.6}$$

Differentiation of the Eq. (1.5) and use of the Eq. (1.6) yield

$$\begin{aligned}\frac{d^2y_P(t)}{dt^2} = {} & \frac{dV_P(t)}{dt}\sin\big(\phi_{PC} + \alpha_P(t)\big) \\ & + a_P(t)\cos\big(\phi_{PC} + \alpha_P(t)\big), \ t \in [t_0, t_C].\end{aligned} \tag{1.7}$$

Thus, the Eqs. (1.4)–(1.6), along with (1.7), describe the pursuer's kinematics in a small vicinity of its initial collision course trajectory. Proceed to the derivation of the evader's kinematics (see Fig. 1.2). In this figure, $E(t) = \big(x_E(t), y_E(t)\big)$, $t \in [t_0, t_C]$ is the current evader's position, $\alpha_E(t) = \phi_E(t) - \phi_{E0}$, where $\phi_E(t)$, $t \in [t_0, t_C]$ is the angle between the current evader's velocity vector $\mathbf{V}_E(t)$ and the X-axis. By $\mathbf{a}_E(t)$, $t \in [t_0, t_C]$, we denote the evader's current lateral acceleration vector. This vector is perpendicular to the current evader's velocity vector. By $a_E(t)$, we denote the magnitude of the vector $\mathbf{a}_E(t)$ taken with the sign "+" if $\mathbf{a}_E(t)$ rotates the vector $\mathbf{V}_E(t)$ counterclockwise, and taken with the sign "−" if $\mathbf{a}_E(t)$ rotates the vector $\mathbf{V}_E(t)$ clockwise. Similarly to the derivation of the pursuer's kinematics, we set that $E(t) = E_C(t)$, $t \in [t_0, t_C]$, where $E_C(t) = \big(x_{EC}(t), y_{EC}(t)\big)$, $t \in [t_0, t_C]$ is the current evader position on its initial collision course trajectory. Using these notations, we obtain the following equations for the evader's kinematics in a small vicinity of its initial collision course trajectory:

$$\frac{dx_E(t)}{dt} = V_E(t)\cos\big(\phi_{E0} + \alpha_E(t)\big), \quad t \in [t_0, t_C], \tag{1.8}$$

$$\frac{dy_E(t)}{dt} = V_E(t)\sin\big(\phi_{E0} + \alpha_E(t)\big), \quad t \in [t_0, t_C], \tag{1.9}$$

$$\frac{d\alpha_E(t)}{dt} = \frac{a_E(t)}{V_E(t)}, \quad t \in [t_0, t_C], \tag{1.10}$$

$$\frac{d^2 y_E(t)}{dt^2} = \frac{dV_E(t)}{dt}\sin\big(\phi_{E0} + \alpha_E(t)\big) + a_E(t)\cos\big(\phi_{E0} + \alpha_E(t)\big),\ t \in [t_0, t_C]. \tag{1.11}$$

The sets (1.4)–(1.7) and (1.8)–(1.11) are sets of nonlinear differential equations, because of nonlinear dependence of their right-hand sides on $\alpha_P(t)$ and $\alpha_E(t)$. Using the assumption (A6), we can linearize these sets. Due to (A6), the absolute values of the angles $\alpha_P(t)$ and $\alpha_E(t)$, $t \in [t_0, t_C]$ are small, yielding

$$\begin{array}{ll} \cos\big(\phi_{iC} + \alpha_i(t)\big) \approx \cos(\phi_{iC}), & t \in [t_0, t_C], \\ \sin\big(\phi_{iC} + \alpha_i(t)\big) \approx \sin(\phi_{iC}) + \cos(\phi_{iC})\alpha_i(t), & t \in [t_0, t_C], \end{array} \quad i = P, E. \tag{1.12}$$

The accuracies of the approximations in (1.12) for cosine and sine functions are $O\Big(\big|\alpha_i(t)\sin\big(\phi_{iC} + \alpha_i(t)/2\big)\big|\Big)$ and $O\Big(\big|\alpha_i^2(t)\sin\big(\phi_{iC} + \varphi(\alpha_i(t))\big)\big|\Big)$, $t \in [t_0, t_C]$, $(i = P, E)$, where $\varphi(\alpha_i(t))$, $(i = P, E)$ is the angle satisfying the equalities

$$\cos\big(\varphi(\alpha_i(t))\big) = \frac{1}{\sqrt{1 + \alpha_i^2(t)/9}},\ \sin\big(\varphi(\alpha_i(t))\big) = \frac{\alpha_i(t)/3}{\sqrt{1 + \alpha_i^2(t)/9}},\ t \in [t_0, t_C]$$

and the inequality $0 \le \varphi(\alpha_i(t)) < 2\pi$, $i = P, E$, $t \in [t_0, t_C]$.

The approximations in (1.12) allow us to linearize the sets (1.4)–(1.7) and (1.8)–(1.11) as:

$$\frac{dx_P(t)}{dt} = V_P(t)\cos(\phi_{PC}), \quad t \in [t_0, t_C], \tag{1.13}$$

$$\frac{dy_P(t)}{dt} = V_P(t)\big(\sin(\phi_{PC}) + \cos(\phi_{PC})\alpha_P(t)\big), \quad t \in [t_0, t_C], \tag{1.14}$$

$$\frac{d\alpha_P(t)}{dt} = \frac{a_P(t)}{V_P(t)}, \quad t \in [t_0, t_C], \tag{1.15}$$

$$\frac{d^2 y_P(t)}{dt^2} = \frac{dV_P(t)}{dt}\big(\sin(\phi_{PC}) + \cos(\phi_{PC})\alpha_P(t)\big) + a_P(t)\cos(\phi_{PC}),\ t \in [t_0, t_C], \tag{1.16}$$

and

$$\frac{dx_E(t)}{dt} = V_E(t)\cos(\phi_{E0}), \quad t \in [t_0, t_C], \tag{1.17}$$

$$\frac{dy_E(t)}{dt} = V_E(t)\big(\sin(\phi_{E0}) + \cos(\phi_{E0})\alpha_E(t)\big), \quad t \in [t_0, t_C], \tag{1.18}$$

$$\frac{d\alpha_E(t)}{dt} = \frac{a_E(t)}{V_E(t)}, \qquad t \in [t_0, t_C], \tag{1.19}$$

$$\begin{aligned}\frac{d^2 y_E(t)}{dt^2} = &\frac{dV_E(t)}{dt}\big(\sin(\phi_{E0}) + \cos(\phi_{E0})\alpha_E(t)\big)\\ &+ a_E(t)\cos(\phi_{E0}), \ t \in [t_0, t_C].\end{aligned} \tag{1.20}$$

Now, based on the pursuer's and evader's linearized kinematics models (1.13)–(1.16) and (1.17)–(1.20), let us derive the model of the relative engagement's kinematics of the participants. We start with the relative motion of the pursuer and the evader with respect to the X-axis. Taking into account that $x_P(t_0) = 0$, $x_E(t_0) = R_0$ and integrating the differential Eqs. (1.13) and (1.17) from $t = t_0$ to $t = t_C$, we directly obtain

$$x_P(t_C) = \int_{t_0}^{t_C} V_P(t)dt\cos(\phi_{PC}), \ \ x_E(t_C) = R_0 + \int_{t_0}^{t_C} V_E(t)dt\cos(\phi_{E0}). \tag{1.21}$$

The equations in (1.21), along with the equation (1.1), yield the equality $x_P(t_C) = x_E(t_C)$, meaning that the X-coordinates of the pursuer and the evader meet each other at the time instant $t = t_C$. This circumstance allows us to choose the final time-instance of the engagement t_f as t_C, i.e.,

$$t_f = t_C. \tag{1.22}$$

To derive the model of the relative motion of the pursuer and the evader with respect to the Y-axis, let us introduce the following new state variables:

$$\begin{aligned}&x_1(t) = y_E(t) - y_P(t), \ \ x_2(t) = \tfrac{dy_E(t)}{dt} - \tfrac{dy_P(t)}{dt}, \ \ x_3(t) = a_P(t),\\ &x_4(t) = a_E(t), \ \ x_5(t) = \alpha_P(t), \ \ x_6(t) = \alpha_E(t).\end{aligned} \tag{1.23}$$

These state variables are the ones of the relative motion of the pursuer and the evader with respect to the Y-axis. Using the equations (1.14)–(1.16), (1.18)–(1.20) and (1.22), we derive the equations of this motion in the form

$$\frac{dx_1(t)}{dt} = x_2(t), \qquad t \in [t_0, t_f], \tag{1.24}$$

$$\begin{aligned}\frac{dx_2(t)}{dt} = &-\cos(\phi_{PC})x_3(t) + \cos(\phi_{E0})x_4(t)\\ &-\big(dV_P(t)/dt\big)\cos(\phi_{PC})x_5(t) + \big(dV_E(t)/dt\big)\cos(\phi_{E0})x_6(t)\\ &-\big(dV_P(t)/dt\big)\sin(\phi_{PC}) + \big(dV_E(t)/dt\big)\sin(\phi_{E0}), \quad t \in [t_0, t_f],\end{aligned} \tag{1.25}$$

$$\frac{dx_5(t)}{dt} = \frac{x_3(t)}{V_P(t)}, \quad t \in [t_0, t_f], \tag{1.26}$$

$$\frac{dx_6(t)}{dt} = \frac{x_4(t)}{V_E(t)}, \quad t \in [t_0, t_f]. \tag{1.27}$$

Let us derive the initial conditions for the set (1.24)–(1.27). Taking into account the assumption (A4), as well as the equalities $\phi_E(t_0) = \phi_{E0}$, $\phi_P(t_0) = \phi_{P0}$ and the equations (1.14) and (1.18), we obtain

$$\begin{aligned} x_1(t_0) = 0, \quad & x_2(t_0) = x_{20}(\phi_{PC}, \phi_{E0}) \stackrel{\triangle}{=} V_E(t_0)\sin(\phi_{E0}) \\ & -V_P(t_0)\big(\sin(\phi_{PC}) + \cos(\phi_{PC})\alpha_{P0}(\phi_{PC})\big), \\ x_5(t_0) = \phi_{P0} - \phi_{PC} & \stackrel{\triangle}{=} \alpha_{P0}(\phi_{PC}), \quad x_6(t_0) = 0. \end{aligned} \tag{1.28}$$

Note that the set (1.24)–(1.27) contains six state variables $x_j(t)$, $(j = 1, \ldots, 6)$, but only four differential equations for $x_1(t)$, $x_2(t)$, $x_5(t)$, $x_6(t)$. The differential equations for $x_3(t)$ and $x_4(t)$ will be introduced in the next subsection.

Remark 1.1 As it was aforementioned, the linear relative motion of the pursuer and the evader with respect to the X-axis yields the unavoidable coincidence of their X-coordinates at $t = t_f$. However, it is not so for their linear relative motion with respect to the Y-axis. Therefore, the pursuer and the evader focus only on their relative motion with respect to the Y-axis, having opposite aims. The pursuer aims to decrease the value $|x_1(t_f)|$ as much as possible, while the evader aims to increase $|x_1(t_f)|$ as much as possible. We call $|x_1(t_f)|$ the *miss distance in the engagement* (or, briefly, the *miss distance*). The best outcome of the engagement from the pursuer's viewpoint is zero miss distance. This outcome has a considerable interest in theory and applications.

1.2.1.3 Complete Mathematical Model of the Relative Motion Between the Pursuer and the Evader

To complete the kinematics model of the engagement between the pursuer and the evader, given by the equations (1.24)–(1.28), we should determine a status of the variables $x_3(t)$ and $x_4(t)$, $t \in [t_0, t_f]$. Namely, we should decide whether the pursuer (the evader) directly disposes of the variable $x_3(t)$ (the variable $x_4(t)$), or the corresponding variable is subject to some differential equation. In this section and in the subsequent sections of the paper, we focus on the case where each of the variables $x_3(t)$ and $x_4(t)$ is subject to a first-order linear differential equation. This case, called the first-order control dynamics of the pursuer/evader, is of a considerable interest in the theory of pursuit-evasion problems and its numerous applications (see e.g. [11, 12, 15, 17–19]). Thus, the differential equations of the first-order control dynamics for the pursuer and the evader are, respectively,

$$\frac{dx_3(t)}{dt} = \frac{a_{P,c}(t) - x_3(t)}{\tau_P}, \quad t \in [t_0, t_f], \quad x_3(t_0) = x_{30} \stackrel{\triangle}{=} a_P(t_0), \tag{1.29}$$

$$\frac{dx_4(t)}{dt} = \frac{a_{E,c}(t) - x_4(t)}{\tau_E}, \quad t \in [t_0, t_f], \quad x_4(t_0) = x_{40} \stackrel{\triangle}{=} a_E(t_0), \tag{1.30}$$

where $a_{P,c}(t)$ and $a_{E,c}(t), t \in [t_0, t_f]$ are the lateral acceleration commands (controls) of the pursuer and the evader, respectively; τ_P and τ_E are the time constants of the pursuer's and the evader's control dynamics.

The lateral acceleration commands are bounded as: $\left|a_{P,c}(t)\right| \le a_{P,c}^{\max}(t)$, $\left|a_{E,c}(t)\right| \le a_{E,c}^{\max}(t)$, $t \in [t_0, t_f]$, where $a_{P,c}^{\max}(t)$ and $a_{E,c}^{\max}(t)$ are positive continuous functions known to both participants of the engagement.

Let us introduce new controls $u(t)$ and $v(t)$ by normalizing $a_{P,c}(t)$ and $a_{E,c}(t)$, respectively, in the equations (1.29) and (1.30):

$$u(t) = a_{P,c}(t)/a_{P,c}^{\max}(t), \quad v(t) = a_{E,c}(t)/a_{E,c}^{\max}(t), \quad t \in [t_0, t_f]. \tag{1.31}$$

Thus, the new controls are bounded as:

$$|u(t)| \le 1, \quad |v(t)| \le 1, \quad t \in [t_0, t_f]. \tag{1.32}$$

Now, let us consider the following matrix and vectors for $t \in [t_0, t_f]$:

$$A(t, \phi_{PC}, \phi_{E0}) = \begin{pmatrix} 0 & 1 & 0 & 0 & 0 & 0 \\ 0 & 0 & -\cos(\phi_{PC}) & \cos(\phi_{E0}) & -\frac{dV_P(t)}{dt}\cos(\phi_{PC}) & \frac{dV_E(t)}{dt}\cos(\phi_{E0}) \\ 0 & 0 & -\frac{1}{\tau_P} & 0 & 0 & 0 \\ 0 & 0 & 0 & -\frac{1}{\tau_E} & 0 & 0 \\ 0 & 0 & \frac{1}{V_P(t)} & 0 & 0 & 0 \\ 0 & 0 & 0 & \frac{1}{V_E(t)} & 0 & 0 \end{pmatrix}, \tag{1.33}$$

$$x(t) = \begin{pmatrix} x_1(t) \\ x_2(t) \\ x_3(t) \\ x_4(t) \\ x_5(t) \\ x_6(t) \end{pmatrix}, \quad b_u(t) = \begin{pmatrix} 0 \\ 0 \\ \frac{a_{P,c}^{\max}(t)}{\tau_P} \\ 0 \\ 0 \\ 0 \end{pmatrix}, \quad c_v(t) = \begin{pmatrix} 0 \\ 0 \\ 0 \\ \frac{a_{E,c}^{\max}(t)}{\tau_E} \\ 0 \\ 0 \end{pmatrix}, \tag{1.34}$$

$$f(t, \phi_{PC}, \phi_{E0}) = \begin{pmatrix} 0 \\ f_2(t, \phi_{PC}, \phi_{E0}) \\ 0 \\ 0 \\ 0 \\ 0 \end{pmatrix}, \quad x_0(\phi_{PC}, \phi_{E0}) = \begin{pmatrix} 0 \\ x_{20}(\phi_{PC}, \phi_{E0}) \\ x_{30} \\ x_{40} \\ \alpha_{P0}(\phi_{PC}) \\ 0 \end{pmatrix}, \tag{1.35}$$

where $f_2(t, \phi_{PC}, \phi_{E0}) = -\left(dV_P(t)/dt\right)\sin(\phi_{PC}) + \left(dV_E(t)/dt\right)\sin(\phi_{E0})$.

Using the Eqs. (1.24)–(1.28), (1.29)–(1.31), (1.33)–(1.35), we obtain the complete linear mathematical model of the relative motion between the pursuer and the evader in the vector form

$$\frac{dx(t)}{dt} = A(t, \phi_{PC}, \phi_{E0})x(t) + b_u(t)u(t) + c_v(t)v(t) + f(t, \phi_{PC}, \phi_{E0}),$$
$$t \in [t_0, t_f], \quad x(t_0) = x_0(\phi_{PC}, \phi_{E0}). \tag{1.36}$$

This system is subject to the control constraints (1.32).

1.2.2 Robust Capture Pursuit

1.2.2.1 Main Definitions

Due to Remark 1.1, the aim of the pursuer in the engagement (1.36), (1.32) is to decrease the miss distance $|x_1(t_f)|$ by a proper choice of the control u. The best result of this aim is $x_1(t_f) = 0$, which is called the *capture*. It is extremely important for the pursuer that the capture would be achieved against any possible behaviour of the evader, i.e., robustly with respect to any admissible evader's control v which is unknown to the pursuer. Such behaviour of the pursuer is called the *robust pursuit with capture* (or briefly, the *robust capture pursuit*). It is clear that not any pursuer's control u and not any starting engagement position $(\bar{t}, \bar{x}) \in [t_0, t_f) \times \mathbb{R}^6$ are suitable for the robust pursuit. Based on this observation, we present the following definitions.

Definition 1.1 For any given $\bar{t} \in [t_0, t_f)$, the evader's control $v(t) : [\bar{t}, t_f] \to \mathbb{R}$ is called admissible if it satisfies the constraint $|v(t)| \le 1$, $t \in [\bar{t}, t_f]$.

Definition 1.2 Pursuer's control $u = u(t, x) : [t_0, t_f] \times \mathbb{R}^6 \to \mathbb{R}$ is called admissible if, for any given point $(\bar{t}, \bar{x}) \in [t_0, t_f) \times \mathbb{R}^6$ and any admissible evader's control $v(t)$, the following conditions are fulfilled: (i) the differential equation in (1.36) with $u(t, x)$ instead of $u(t)$ has a solution $x(t) = x\big(t; \bar{t}, \bar{x}, v(\cdot)\big)$, $t \in [\bar{t}, t_f)$, satisfying the initial condition $x(\bar{t}) = \bar{x}$; (ii) for any such solution $x(t)$, there exists the finite limit $x(t_f) \stackrel{\triangle}{=} \lim_{t \to t_f - 0} x(t)$.

Remark 1.2 It should be noted that we understand a solution (a continuous one) of the initial-value problem, mentioned in Definition 1.2, in the sense of a particular case of the Krasovskii's constructive motion [14], i.e., as a limit of a convergent sequence of piecewise-linear Eiler's functions, associated with this initial-value problem. In this particular case, in contrast with [14], we do not replace the given initial state's position with a convergent sequence of initial positions, as well as we do not replace the evader's control with a convergent sequence of such controls.

Definition 1.3 Let $(\bar{t}, \bar{x})$ be any given point of the set $[t_0, t_f) \times \mathbb{R}^6$. An admissible pursuer's control $u = u(t, x)$ is called a *robust capture control* from the point $(\bar{t}, \bar{x})$ if, for all admissible evader's controls $v(t)$, the following conditions are fulfilled: (a) the first component $x_1(t)$ of any solution $x(t) = x\big(t; \bar{t}, \bar{x}, v(\cdot)\big)$, mentioned in Definition 1.2, becomes zero at $t = t_f$, i.e., $x_1(t_f) = 0$; (b) the time realization $u(t) \stackrel{\triangle}{=} u\big(t, x(t)\big)$ of the control $u(t, x)$ along the solution $x(t)$ satisfies the inequality $|u(t)| \leq 1$, $t \in [\bar{t}, t_f]$.

Definition 1.4 Let $u = u(t, x)$ be any given admissible pursuer's control. The set $C_x\big(u(t, x)\big)$ of all points $(\bar{t}, \bar{x}) \in [t_0, t_f) \times \mathbb{R}^6$, from which $u(t, x)$ is the robust capture control, is called a *capture zone* of this pursuer's control. If there exist a point $\bar{x} \in \mathbb{R}^6$ such that $(t_0, \bar{x}) \in C_x\big(u(t, x)\big)$, then the capture zone $C_x\big(u(t, x)\big)$ is called *t-complete*.

In the present paper, the design of a robust capture control and the construction of its capture zone are based on the differential games approach.

1.2.2.2 Pursuit-Evasion Game with Bounded Players' Controls

Consider the differential game consisting of the differential system (1.36), the control constraints (1.32) and the cost function

$$J_x = J_x\big(u(\cdot), v(\cdot)\big) = |d^T x(t_f)|, \quad d = \mathrm{col}(1, 0, 0, 0, 0, 0). \tag{1.37}$$

The pursuer disposes of the control $u(\cdot)$, while the evader disposes of the control $v(\cdot)$. The pursuer aims to minimize the cost function (1.37) by a proper choice of $u(\cdot)$, while the evader aims to maximize this cost function by a proper choice of $v(\cdot)$. We assume the following information pattern in the considered pursuit-evasion game: the pursuer and the evader are aware of all the game's data, presenting in (1.36), (1.32), (1.37), and of the current game's position $\big(t, x(t)\big)$. In what follows of the paper, we denote this game as PEGBC (the Pursuit-Evasion Game with Bounded Controls). Due to the Eq. (1.23) and Remark 1.1, the cost function (1.37) represents the miss distance. Hence, in the PEGBC, the pursuer aims to minimize the miss distance, while the evader tries to maximize this distance. Based on the PEGBC feedback solution, we will design two versions of the robust capture control and construct the corresponding capture zones, which is the objective of the paper.

1.3 Auxiliary Results

In this section, two auxiliary results are presented. Namely, in Sect. 1.3.1, a saddle-point feedback solution to a zero-sum differential game with linear dynamics and bounded controls is derived based on the transformation of the original game to a scalar zero-sum differential game. In Sect. 1.3.2, a scalar zero-sum linear-quadratic differential game with cheap controls is considered. A saddle-point feedback solution of this game, uniform in the cheap control parameter, is derived.

1.3.1 Feedback Solution of a Zero-Sum Differential Game with Linear Dynamics and Bounded Controls

1.3.1.1 Original Game Formulation

Consider the differential game with the dynamics system

$$\frac{dx(t)}{dt} = A(t)x(t) + b(t)u(t) + c(t)v(t) + f(t),$$
$$t \in [t_0, t_f], \quad x(t_0) = x_0, \tag{1.38}$$

where $x(t) \in \mathbb{R}^n$ is a state vector; $u(t)$ and $v(t)$ are scalar controls of the players; t_0 and t_f, $(t_0 < t_f)$ are initial and terminal time instants; the matrix-valued function $A(t)$ and the vector-valued functions $b(t)$, $c(t)$, $f(t)$ are given continuous ones for $t \in [t_0, t_f]$; $x_0 \in \mathbb{R}^n$ is a given initial state vector.

The controls of the players satisfy the constraints

$$|u(t)| \leq 1, \quad |v(t)| \leq 1, \quad t \in [t_0, t_f]. \tag{1.39}$$

In the space $\mathbb{R}^n$, we consider the hyperplane $\mathcal{D} = \{x \in \mathbb{R}^n : d^T x + d_0 = 0\}$, where $d \in \mathbb{R}^n$ is a given non-zero vector, d_0 is a given scalar value. The player with the control $u(t)$ (the minimizer) aims to minimize the distance $\mathrm{Dist}\big(\mathcal{D}, x(t_f)\big)$ between the terminal state vector $x(t_f)$ and the hyperplane $\mathcal{D}$, while the player with the control $v(t)$ (the maximizer) tends to maximize $\mathrm{Dist}\big(\mathcal{D}, x(t_f)\big)$, where $\mathrm{Dist}\big(\mathcal{D}, x(t_f)\big) \triangleq |d^T x(t_f) + d_0|/\|d\|$. Since the norm $\|d\| > 0$, the cost function, evaluating the performance of the minimizer and the maximizer, can be chosen as:

$$J_x = J_x\big(u(\cdot), v(\cdot)\big) = |d^T x(t_f) + d_0|. \tag{1.40}$$

As the information pattern in this game, we assume that both players are aware of all the game's data, presenting in (1.38)–(1.40), and of the current game's position $\big(t, x(t)\big)$. The system (1.38), the control constraints (1.39) and the cost function (1.40), along with the aims of the players and the information pattern, constitute a zero-sum differential game. We call this differential game the Vector-valued Game with Bounded Controls (VGBC).

1.3.1.2 Main Definitions

Let $\mathcal{G}^x$ be the set of all scalar functions $g(t, x)$, $(t, x) \in [t_0, t_f] \times \mathbb{R}^n$, measurable with respect to $t \in [t_0, t_f]$ for any $x \in \mathbb{R}^n$, piecewise continuous with respect to $x \in \mathbb{R}^n$ for any $t \in [t_0, t_f]$, and satisfying the inequality $|g(t, x)| \leq 1\ \forall (t, x) \in [t_0, t_f] \times \mathbb{R}^n$. Consider a pair of functions $\big(u(t, x), v(t, x)\big) \in \mathcal{G}^x \times \mathcal{G}^x$ and substitute $u(t, x)$ and

$v(t, x)$ into the system (1.38) instead of $u(t)$ and $v(t)$, respectively. This substitution yields the system

$$\frac{dx(t)}{dt} = A(t)x(t) + b(t)u\big(t, x(t)\big) + c(t)v\big(t, x(t)\big) + f(t),$$
$$t \in [t_0, t_f], \quad x(t_0) = x_0. \tag{1.41}$$

Remark 1.3 Since at least one of the functions, $u(t, x)$ or $v(t, x)$, may be discontinuous with respect to $x \in \mathbb{R}^n$ for some $t \in [t_0, t_f]$, the classical Caratheodory notion of a differential equation's solution may not be applicable to (1.41). To overcome this obstacle, we use the results of [14] (Sect. 6). Due to these results, the solution of (1.41) is understood in the sense of the Krasovskii's constructive motion. Namely, the solution (the continuous one) of (1.41) is defined as a limit of a convergent sequence of piecewise-linear Eiler's vector-valued functions, associated with this system. Due to the results of [14], such a sequence exists for any initial position (t_0, x_0) and any pair of functions $\big(u(t, x), v(t, x)\big) \in \mathcal{G}^x \times \mathcal{G}^x$ yielding the existence of the aforementioned solution to (1.41). However this solution is not in general unique, because of possible existence of several convergent sequences of piecewise-linear Eiler's vector-valued functions associated with (1.41) and having different limits. The non-uniqueness of solution to the system (1.41) can yield the non-uniqueness of the value of the VGBC cost function $J_x\big(u(\cdot), v(\cdot)\big)$ (see the equation (1.40)), which is unfeasible. Taking into account all the aforementioned in this remark, we make the following definition.

Definition 1.5 A pair of functions $\big(u(t, x), v(t, x)\big) \in \mathcal{G}^x \times \mathcal{G}^x$ is called an admissible pair of players' controls in the VGBC if, for any given point $(\bar{t}, \bar{x}) \in [t_0, t_f) \times \mathbb{R}^n$, all the solutions $x(t) = x\big(t; u(\cdot), v(\cdot), \bar{t}, \bar{x}\big)$ of the differential equation in (1.41) with the initial condition $x(\bar{t}) = \bar{x}$ yield the same value $|d^T x\big(t_f; u(\cdot), v(\cdot), \bar{t}, \bar{x}\big) + d_0|$. The set of all the admissible pairs of players' feedback controls in the VGBC is denoted as $\mathcal{F}^x$.

Remark 1.4 Due to Definition 1.5 and the equation (1.40), for any starting game's position, any pair of the players' feedback controls $\big(u(t, x), v(t, x)\big) \in \mathcal{F}^x$ generates a single value of the cost function in the VGBC. Therefore, the cost function of the VGBC and the entire game are feasible in the set $\mathcal{F}^x$.

For a given $u(t, x) \in \mathcal{G}^x$, consider the set $\mathcal{H}_v^x\big(u(t, x)\big) \stackrel{\triangle}{=} \big\{v(t, x) \in \mathcal{G}^x : \big(u(t, x), v(t, x)\big) \in \mathcal{F}^x\big\}$, and let $\mathcal{K}_u^x \stackrel{\triangle}{=} \big\{u(t, x) \in \mathcal{G}^x : \mathcal{H}_v^x\big(u(t, x)\big) \neq \emptyset\big\}$. Similarly, for a given $v(t, x) \in \mathcal{G}^x$, consider the set $\mathcal{H}_u^x\big(v(t, x)\big) \stackrel{\triangle}{=} \big\{u(t, x) \in \mathcal{G}^x : \big(u(t, x), v(t, x)\big) \in \mathcal{F}^x\big\}$, and let $\mathcal{K}_v^x \stackrel{\triangle}{=} \big\{v(t, x) \in \mathcal{G}^x : \mathcal{H}_u^x\big(v(t, x)\big) \neq \emptyset\big\}$.

Definition 1.6 The minimizer's control $u^0(\cdot) \in \mathcal{K}_u^x$ is called optimal in the VGBC, if for any starting game's position $(\bar{t}, \bar{x}) \in [t_0, t_f) \times \mathbb{R}^n$ and for any $u(\cdot) \in \mathcal{K}_u^x$: $\sup_{v(\cdot) \in \mathcal{H}_v^x(u^0(\cdot))} J_x\big(u^0(\cdot), v(\cdot)\big) \leq \sup_{v(\cdot) \in \mathcal{H}_v^x(u(\cdot))} J_x\big(u(\cdot), v(\cdot)\big)$. The value $J_{xu}(\bar{t}, \bar{x}) = \sup_{v(\cdot) \in \mathcal{H}_v^x(u^0(\cdot))} J_x\big(u^0(\cdot), v(\cdot)\big)$ is called the VGBC upper value.

Definition 1.7 The maximizer's control $v^0(\cdot) \in \mathcal{K}_v^x$ is called optimal in the VGBC, if for any starting game's position $(\bar{t}, \bar{x}) \in [t_0, t_f) \times \mathbb{R}^n$ and for any $v(\cdot) \in \mathcal{K}_v^x$: $\inf_{u(\cdot)\in\mathcal{H}_u^x(v^0(\cdot))} J_x\big(u(\cdot), v^0(\cdot)\big) \geq \inf_{u(\cdot)\in\mathcal{H}_u^x(v(\cdot))} J_x\big(u(\cdot), v(\cdot)\big)$. The value $J_{xl}(\bar{t}, \bar{x}) = \inf_{u(\cdot)\in\mathcal{H}_u^x(v^0(\cdot))} J_x\big(u(\cdot), v^0(\cdot)\big)$ is called the VGBC lower value.

Definition 1.8 If $\big(u^0(\cdot), v^0(\cdot)\big) \in \mathcal{F}^x$ and $J_{xu}(\bar{t}, \bar{x}) = J_{xl}(\bar{t}, \bar{x}) \stackrel{\triangle}{=} J_x^0(\bar{t}, \bar{x})\ \forall(\bar{t}, \bar{x}) \in [t_0, t_f) \times \mathbb{R}^n$, then $J_x^0(\bar{t}, \bar{x})$ is called the VGBC value and the pair $\big(u^0(\cdot), v^0(\cdot)\big)$ is called the VGBC saddle point. Any solution $x^0(t) = x^0(t; \bar{t}, \bar{x})$, $t \in [\bar{t}, t_f]$ of the differential equation in (1.41) for $\big(u(t,x), v(t,x)\big) = \big(u^0(t,x), v^0(t,x)\big)$, satisfying the initial condition $x^0(\bar{t}) = \bar{x}$, is called a saddle-point trajectory of the VGBC.

1.3.1.3 Scalar Differential Game

Let the $n \times n$-matrix $\Phi(t)$ be the unique solution of the terminal-value problem

$$\frac{d\Phi(t)}{dt} = -\Phi(t)A(t), \quad t \in [t_0, t_f], \quad \Phi(t_f) = I_n.$$

Using $\Phi(t)$, we construct the scalar function of two variables

$$\zeta(t,x) = d^T\left(\Phi(t)x + \int_t^{t_f} \Phi(s)f(s)ds\right) + d_0, \quad (t,x) \in [t_0, t_f] \times \mathbb{R}^n, \quad (1.42)$$

and make the following transformation of the state variable $x(t)$ in the system (1.38): $z(t) = \zeta\big(t, x(t)\big)$, where $z(t)$ is a new scalar state variable. For any given $u(t)$ and $v(t)$ satisfying (1.39), this transformation converts the system (1.38) to the scalar system

$$\begin{aligned} &\tfrac{dz(t)}{dt} = h_u(t)u(t) + h_v(t)v(t), \quad t \in [t_0, t_f], \quad z(t_0) = z_0, \\ &h_u(t) = d^T\Phi(t)b(t), \quad h_v(t) = d^T\Phi(t)c(t), \quad t \in [t_0, t_f], \end{aligned} \quad (1.43)$$

where the initial and terminal state values are

$$z_0 = d^T\left(\Phi(t_0)x_0 + \int_{t_0}^{t_f} \Phi(s)f(s)ds\right) + d_0, \quad z_f \stackrel{\triangle}{=} z(t_f) = d^T x(t_f) + d_0. \quad (1.44)$$

Due to the equation (1.44), we can rewrite the cost function (1.40) by $z(t)$ as:

$$J_z = J_z\big(u(\cdot), v(\cdot)\big) = |z(t_f)|. \quad (1.45)$$

The scalar differential system (1.43), the control constraints (1.39) and the cost function (1.45) constitute a scalar differential game with bounded controls (SGBC). In this game (like in the VGBC), the player with the control $u(t)$ (the minimizer) tends to minimize the cost function (1.45), while the player with the control $v(t)$ (the maximizer) aims to maximize this cost function. Thus, like the VGBC, the SGBC is a zero-sum differential game. However, being scalar, the SGBC is much simpler than the VGBC. The feedback solution of the VGBC is obtained based on the feedback solution of the SGBC.

Let Q^z be the set of all scalar functions $q(t,z)$, $(t,z) \in [t_0, t_f] \times \mathbb{R}$, measurable with respect to $t \in [t_0, t_f]$ for any $z \in \mathbb{R}$, piecewise continuous with respect to $z \in \mathbb{R}$ for any $t \in [t_0, t_f]$, and satisfying the inequality $|q(t,z)| \le 1$, $\forall (t,z) \in [t_0, t_f] \times \mathbb{R}$.

Remark 1.5 Note that the SGBC can be considered as a particular case of the VGBC. Therefore, the definitions of the admissible pair of players' controls $\big(u_{sg}(t,z), v_{sg}(t,z)\big) \in Q^z \times Q^z$ for the SGBC, the optimal minimizer's (maximizer's) control $u^0_{sg}(t,z)$ $(v^0_{sg}(t,z))$, the upper (lower) value $J_{zu}(\bar{t},\bar{z})$ $(J_{zl}(\bar{t},\bar{z}))$, $(\bar{t},\bar{z}) \in [t_0, t_f) \times \mathbb{R}$, the game value $J^0_z(\bar{t},\bar{z})$, the saddle point $\big(u^0_{sg}(t,z), v^0_{sg}(t,z)\big)$, and the saddle-point trajectory $z^0(t) = z^0(t;\bar{t},\bar{x})$ are similar to the corresponding definitions for the VGBC. Therefore, saving the space, we do not present these definitions for the SGBC in details.

1.3.1.4 Saddle-Point Solutions of the SGBC and of VGBC

In this subsection, we present the results of the work [6] in a brief form.

In what follows, we assume:

AI. The functions $h_u(t)$ and $h_v(t)$ have no more than a finite number of distinct zeros in the interval $[t_0, t_f]$.

AII. The function $R(t) \triangleq |h_v(t)| - |h_u(t)|$ (the determining function) has no more than a finite number of distinct zeros in the interval (t_0, t_f) and $R(t)$ changes its sign in this interval $N_R \ge 0$ times.

Let us consider the set $\mathcal{S}_{[t_0,t_f]} \triangleq \big\{(t,z) : t \in [t_0, t_f], z \in \mathbb{R}\big\}$. This set is called the space of the SGBC. The saddle-point solution in feedback controls depends on a proper decomposition of $\mathcal{S}_{[t_0,t_f]}$ into two non-intersecting subsets called the regular region $\mathcal{R}_r$ and the singular region $\mathcal{R}_s$. Let us briefly describe how each of these regions is defined and constructed. To do this, we consider the behaviour of the trajectories of the terminal-value problem

$$\frac{dz(t)}{dt} = R(t)\mathrm{sign}(z_f), \quad t \in [t_0, t_f], \quad z(t_f) = z_f. \tag{1.46}$$

For any given $z_f \neq 0$, let us integrate the differential equation in (1.46) from $t = t_f$ in the backward time. This integration yields the following trajectory:

$$\widetilde{z}(t) = \widetilde{z}(t; t_f, z_f) = \left(|z_f| + \int_{t_f}^{t} R(\sigma)d\sigma\right)\mathrm{sign}(z_f), \; t \in [\widetilde{t}, t_f] \subseteq [t_0, t_f], \tag{1.47}$$

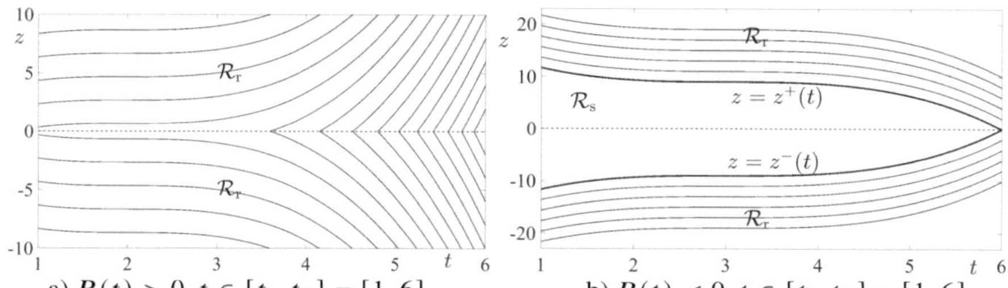

a) $R(t) > 0, t \in [t_0, t_f] = [1, 6]$ b) $R(t) < 0, t \in [t_0, t_f] = [1, 6]$

Fig. 1.3 The SGBC space decomposition for $N_R = 0$

where the time instant $\widetilde{t}$ is defined as follows. If the function $|z_f| + \int_{t_f}^{t} R(\sigma)d\sigma$ changes its sign in the interval (t_0, t_f), then $\widetilde{t}$ is the largest (the first from the right) point of such a change in this interval. Otherwise, $\widetilde{t} = t_0$. Varying $z_f \neq 0$, we obtain the family of all the trajectories $\widetilde{z}(t)$. This family is symmetric with respect to the t-axis, i.e., for any $z_f \neq 0$, the trajectories with $z(t_f) = z_f$ and $z(t_f) = -z_f$ are symmetric to each other with respect to the t-axis. All these trajectories cover some set of the SGBC space $\mathcal{S}_{[t_0,t_f]}$, which is the aforementioned regular region $\mathcal{R}_r$. If $\mathcal{R}_r$ coincides with the SGBC space $\mathcal{S}_{[t_0,t_f]}$, then the aforementioned singular region $\mathcal{R}_s$ is an empty set. If $\mathcal{R}_r \subset \mathcal{S}_{[t_0,t_f]}$, then $\mathcal{R}_s = \mathcal{S}_{[t_0,t_f]} \backslash \mathcal{R}_r \neq \emptyset$. In this case, the boundary of $\mathcal{R}_r$ is formed by two curves $z = z^+(t) \geq 0$ and $z = z^-(t) = -z^+(t)$, symmetrical with respect to the t-axis. These curves can be of two types. The curves of the first type are arcs of the trajectories $\widetilde{z}(t)$ tangent (at least once) to t-axis and, by definition, they belong to $\mathcal{R}_r$. The curves of the second type are obtained as limits of the trajectories (1.47) for $z_f \to +0$ and $z_f \to -0$, respectively, and they also belong to $\mathcal{R}_r$. The boundaries $z^+(t)$ and $z^-(t)$, $t \in \mathcal{T}$ are called the positive and negative boundaries of $\mathcal{R}_r$, where $\mathcal{T} \subseteq [t_0, t_f]$ is their existence set. The set $\mathcal{T}$ can be a single subinterval or a set of a number of disconnected subintervals in $[t_0, t_f]$. Note, that $z^+(t)$ and $z^-(t)$ also constitute the upper and lower boundaries, respectively, of $\mathcal{R}_s$. The structure of $\mathcal{R}_r$ and $\mathcal{R}_s$ (and, therefore, the SGBC space decomposition) depends on the number N_R, mentioned in the assumption AII, and on the sign of the function $R(t)$ in a proper left-hand vicinity of the point $t = t_f$. Below, several examples of such a decomposition are presented for different N_R and signs of $R(t)$ in the aforementioned set of t. In Fig. 1.3a, $\mathcal{R}_r = \mathcal{S}_{[t_0,t_f]}$, $\mathcal{R}_s = \emptyset$, while in Fig. 1.3b, $\mathcal{R}_r = \{(t, z) : t \in [t_0, t_f], z \geq z^+(t) \text{ or } z \leq z^-(t)\}$, $\mathcal{R}_s = \{(t, z) : t \in [t_0, t_f), z^-(t) < z < z^+(t)\}$.

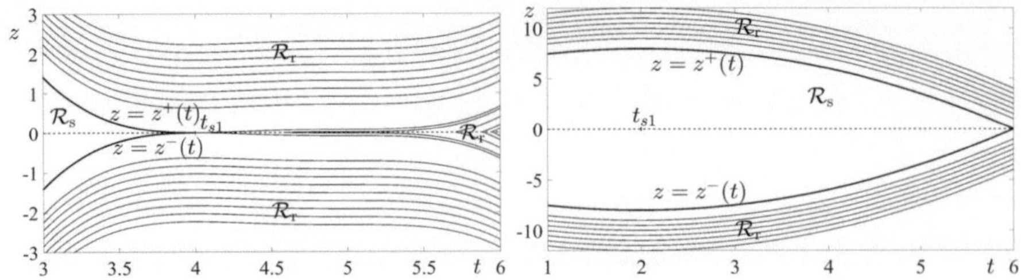

a) $R(t) < 0, t \in [t_0, t_{s1}) = [3, 4)$
$R(t) > 0, t \in (t_{s1}, t_f] = (4, 6]$

b) $R(t) > 0, t \in [t_0, t_{s1}) = [1, 2)$
$R(t) < 0, t \in (t_{s1}, t_f] = (2, 6]$

Fig. 1.4 The SGBC space decomposition for $N_R = 1$

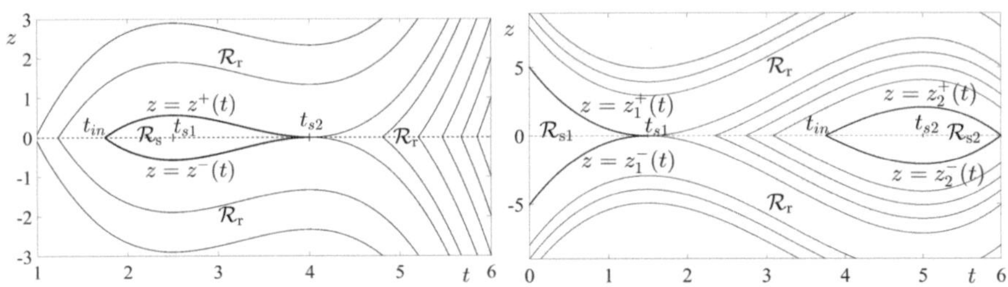

$R(t) > 0, t \in [t_0, t_{s1}) = [1, 2.5)$
a) $R(t) < 0, t \in (t_{s1}, t_{s2}] = (2.5, 4)$
$R(t) > 0, t \in (t_{s2}, t_f] = (4, 6]$

$R(t) < 0, t \in [t_0, t_{s1}) = [0, 1.5)$
b) $R(t) > 0, t \in (t_{s1}, t_{s2}] = (1.5, 5)$
$R(t) < 0, t \in (t_{s2}, t_f] = (5, 6]$

Fig. 1.5 The SGBC space decomposition for $N_R = 2$

In Fig. 1.4a, $\mathcal{R}_r = \big\{(t, z) : t \in [t_0, t_{s1}), z \geq z^+(t)$ or $z \leq z^-(t)\big\} \bigcup \big\{(t, z) : t \in [t_{s1}, t_f], z \in \mathbb{R}\big\}$, $\mathcal{R}_s = \big\{(t, z) : t \in [t_0, t_{s1}), z^-(t) < z < z^+(t)\big\}$, while in Fig. 1.4b, $\mathcal{R}_r = \big\{(t, z) : t \in [t_0, t_f], z \geq z^+(t)$ or $z \leq z^-(t)\big\}$, $\mathcal{R}_s = \big\{(t, z) : t \in [t_0, t_f), z^-(t) < z < z^+(t)\big\}$.

In Fig. 1.5a, $\mathcal{R}_r = \big\{(t, z) : t \in [t_0, t_{in}], z \in \mathbb{R}\big\} \bigcup \big\{(t, z) : t \in (t_{in}, t_{s2}), z \geq z^+(t)$ or $z \leq z^-(t)\big\} \bigcup \big\{(t, z) : t \in [t_{s2}, t_f], z \in \mathbb{R}\big\}$, $\mathcal{R}_s = \big\{(t, z) : t \in (t_{in}, t_{s2}), z^-(t) < z < z^+(t)\big\}$, while in Fig. 1.5b, $\mathcal{R}_r = \big\{(t, z) : t \in [t_0, t_{s1}), z \geq z^+(t)$ or $z \leq z^-(t)\big\} \bigcup \big\{(t, z) : t \in [t_{s1}, t_{in}], z \in \mathbb{R}\big\} \bigcup \big\{(t, z) : t \in (t_{in}, t_f], z \geq z^+(t)$ or $z \leq z^-(t)\big\}$, $\mathcal{R}_s = \mathcal{R}_{s1} \bigcup \mathcal{R}_{s2} = \big\{(t, z) : t \in [t_0, t_{s1}), z^-(t) < z < z^+(t)\big\} \bigcup \big\{(t, z) : t \in (t_{in}, t_f), z^-(t) < z < z^+(t)\big\}$. Note that $t_{in} \in (t_0, \hat{t})$ is the largest solution of the equation $\int_{\hat{t}}^{t_{in}} R(\sigma)d\sigma = 0$, where $\hat{t} = t_{s2}$ (case a)) and $\hat{t} = t_f$ (case b)). Now, based on the SGBC space decomposition, we have (by virtue of [6]) the following assertions.

Lemma 1.1 *The SGBC has the saddle point* $(u_{sg}^0(t, z), v_{sg}^0(t, z))$, $(t, z) \in \mathcal{S}_{[t_0, t_f]}$. *For* $(t, z) \in \mathcal{R}_r$, *the controls* $u_{sg}^0(\cdot) \in \mathcal{Q}^z$ *and* $v_{sg}^0(\cdot) \in \mathcal{Q}^z$ *are*

$$
\begin{aligned}
u_{\mathrm{sg}}^0(t,z) &= \begin{cases} -\mathrm{sign}(z)\mathrm{sign}\big(h_u(t)\big), & t\in[t_0,t_f],\ \ z\neq 0,\\ -\mathrm{sign}\big(h_u(t)\big), & t\in[t_0,t_f],\ \ z=0,\end{cases}\\
v_{\mathrm{sg}}^0(t,z) &= \begin{cases} \mathrm{sign}(z)\mathrm{sign}\big(h_v(t)\big), & t\in[t_0,t_f],\ \ z\neq 0,\\ \mathrm{sign}\big(h_v(t)\big), & t\in[t_0,t_f],\ \ z=0.\end{cases}
\end{aligned}
\tag{1.48}
$$

For $(t,z)\in\mathcal{R}_{\mathrm{s}}$, the pair of the controls $(u_{\mathrm{sg}}^0(t,z), v_{\mathrm{sg}}^0(t,z))$ is arbitrary admissible. The saddle-point trajectory of the SGBC, starting at any position $(\bar{t},\bar{z})\in\mathcal{R}_{\mathrm{r}}$, remains in $\mathcal{R}_{\mathrm{r}}$ till the end of the game. For any given starting position $(\bar{t},\bar{z})\in\mathcal{S}_{[t_0,t_f]}$ of the SGBC, its value is

$$
J_z^0(\bar{t},\bar{z}) = \begin{cases} |\bar{z}| + \int_{\bar{t}}^{t_f} R(\sigma)d\sigma, & \text{if } (\bar{t},\bar{z})\in\mathcal{R}_{\mathrm{r}},\\ z^+(\bar{t}) + \int_{\bar{t}}^{t_f} R(\sigma)d\sigma, & \text{if } (\bar{t},\bar{z})\in\mathcal{R}_{\mathrm{s}}.\end{cases}
\tag{1.49}
$$

For any connected set $\mathcal{S}_{\mathrm{s}}\subseteq\mathcal{R}_{\mathrm{s}}\neq\emptyset$, $J_z^0(t,z)=\mathrm{const}\ \forall(t,z)\in\mathcal{S}_{\mathrm{s}}$. Moreover, if $(t_f,0)\in\mathrm{clo}\big(\mathcal{S}_{\mathrm{s}}\big)$, then $J_z^0(t,z)=0\ \forall(t,z)\in\mathcal{S}_{\mathrm{s}}$.

Lemma 1.2 *The VGBC has the saddle point $\big(u^0(t,x), v^0(t,x)\big)$, $(t,x)\in[t_0,t_f]\times\mathbb{R}^n$. For all (t,x) such that $\big(t,\zeta(t,x)\big)\in\mathcal{R}_{\mathrm{r}}$, the controls $u^0(\cdot)\in\mathcal{G}^x$ and $v^0(\cdot)\in\mathcal{G}^x$ are $u^0(t,x)=u_{\mathrm{sg}}^0\big(t,\zeta(t,x)\big)$, $v^0(t,x)=v_{\mathrm{sg}}^0\big(t,\zeta(t,x)\big)$, where $\zeta(t,x)$ is given by (1.42), $u_{\mathrm{sg}}^0(t,z)$ and $v_{\mathrm{sg}}^0(t,z)$ are given in (1.48). For all (t,x) such that $(t,\zeta(t,x))\in\mathcal{R}_{\mathrm{s}}$, the pair of the controls $(u^0(\cdot), v^0(\cdot))$ is arbitrary admissible. For any given starting position $(\bar{t},\bar{x})$ of the VGBC such that $\big(\bar{t},\zeta(\bar{t},\bar{x})\big)\in\mathcal{R}_{\mathrm{r}}$, its saddle-point trajectory $x^0(t)=x^0(t;\bar{t},\bar{x})$ satisfies the inclusion $\big(t,\zeta(t,x^0(t))\big)\in\mathcal{R}_{\mathrm{r}}$, $t\in[\bar{t},t_f]$. For any given starting position $(\bar{t},\bar{x})\in[t_0,t_f]\times\mathbb{R}^n$ of the VGBC, its value is*

$$
J_x^0 = J_x^0(\bar{t},\bar{x}) = \begin{cases} |\zeta(\bar{t},\bar{x})| + \int_{\bar{t}}^{t_f} R(\sigma)d\sigma, & \text{if } \big(\bar{t},\zeta(\bar{t},\bar{x})\big)\in\mathcal{R}_{\mathrm{r}},\\ z^+(\bar{t}) + \int_{\bar{t}}^{t_f} R(\sigma)d\sigma, & \text{if } (\bar{t},\zeta(\bar{t},\bar{x}))\in\mathcal{R}_{\mathrm{s}}.\end{cases}
\tag{1.50}
$$

For any connected set $\mathcal{S}_{\mathrm{s}}\subseteq\mathcal{R}_{\mathrm{s}}\neq\emptyset$, $J_x^0(t,x)=\mathrm{const}\ \forall(t,x):(t,\zeta(t,x))\in\mathcal{S}_{\mathrm{s}}$. Moreover, if $(t_f,0)\in\mathrm{clo}\big(\mathcal{S}_{\mathrm{s}}\big)$, then $J_x^0(t,x)=0\ \forall(t,x):(t,\zeta(t,x))\in\mathcal{S}_{\mathrm{s}}$.

1.3.2 Scalar Zero-Sum Linear-Quadratic Differential Game with Cheap Controls

1.3.2.1 Game Formulation and Main Definitions

Consider the differential game with the following dynamics equation and the cost functional

$$
\frac{dz(t)}{dt} = h_u(t)u(t) + h_v(t)v(t),\quad t\in[t_0,t_f],\quad z(t_0)=z_0,
\tag{1.51}
$$

$$
J\big(u(\cdot),v(\cdot)\big) = \big(z(t_f)\big)^2 + \varepsilon\int_{t_0}^{t_f}\Big[\lambda\big(u(t)\big)^2 - \big(v(t)\big)^2\Big]dt,
\tag{1.52}
$$

where $z(t)$ is a scalar state variable; $u(t)$ and $v(t)$ are players' scalar controls; t_0 and t_f, $(t_0 < t_f)$ are given time instants; the given scalar functions $h_u(t)$, $h_v(t)$ are continuous for $t \in [t_0, t_f]$; z_0 is a given initial value of the state variable; $\lambda > 0$ is some given number; $\varepsilon > 0$ is a small cheap control parameter. The player with the control $u(t)$ (the minimizer) aims to minimize the cost functional (1.52), while the player with the control $v(t)$ (the maximizer) aims to maximize this cost functional. As the information pattern in this game, we assume that both players are aware of all the game's data, presenting in (1.51)–(1.52), and of the current game's position $\big(t, z(t)\big)$. In what follows, we call the differential game, consisting of the equation (1.51), the cost functional (1.52), the aforementioned objectives of the players and the information pattern, the Linear-Quadratic Game with Cheap Controls (LQGCC).

Consider the set Ω of all functions $\omega(t, z) : [0, t_f] \times \mathbb{R} \to \mathbb{R}$, which are measurable w.r.t. $t \in [0, t_f]$ for any fixed $z \in \mathbb{R}$ and satisfy the local Lipschitz condition w.r.t. $z \in \mathbb{R}$ uniformly in $t \in [0, t_f]$. Based on the results of the book [5], we introduce into the consideration the following definitions.

Definition 1.9 Let UV be the set of all pairs of functions $\big(u(t, z), v(t, z)\big)$, $(t, z) \in [0, t_f] \times \mathbb{R}$, such that the following conditions are valid: (i) $u(t, z) \in \Omega$, $v(t, z) \in \Omega$; (ii) the initial-value problem (1.51) for $u(t) = u(t, z)$, $v(t) = v(t, z)$ and any $z_0 \in \mathbb{R}$ has the unique absolutely continuous solution $z_{uv}(t; z_0)$ in the entire interval $[t_0, t_f]$; (iii) $u\big(t, z_{uv}(t; z_0)\big) \in L^2[t_0, t_f]$; (iv) $v\big(t, z_{uv}(t; z_0)\big) \in L^2[t_0, t_f]$. The pair $\big(u(t, z), v(t, z)\big)$ is called an admissible pair of the players' feedback controls in the LQGCC.

For a given $u(t, z) \in \Omega$, consider the set $\Upsilon_v\big(u(t, z)\big) \stackrel{\triangle}{=} \big\{v(t, z) \in \Omega : \big(u(t, z), v(t, z)\big) \in UV\big\}$. Let $\Sigma_u \stackrel{\triangle}{=} \big\{u(t, z) \in \Omega : \Upsilon_v\big(u(t, z)\big) \neq \emptyset\big\}$. Similarly, for a given $v(t, z) \in \Omega$, consider the set $\Upsilon_u\big(v(t, z)\big) \stackrel{\triangle}{=} \big\{u(t, z) \in \Omega : \big(u(t, z), v(t, z)\big) \in UV\big\}$. Let $\Sigma_v \stackrel{\triangle}{=} \big\{v(t, z) \in \Omega : \Upsilon_u\big(v(t, z)\big) \neq \emptyset\big\}$.

Definition 1.10 The minimizer's control $u^*(\cdot) \in \Sigma_u$ is called optimal in the LQGCC, if for any $z_0 \in \mathbb{R}$ and for any $u(\cdot) \in \Sigma_u$, it satisfies the inequality: $\sup_{v(\cdot) \in \Upsilon_v(u^*(\cdot))} J\big(u^*(\cdot), v(\cdot)\big) \le \sup_{v(\cdot) \in \Upsilon_v(u(\cdot))} J\big(u(\cdot), v(\cdot)\big)$. The value $J_{\rm u}(z_0) = \sup_{v(\cdot) \in \Upsilon_v(u^*(\cdot))} J\big(u^*(\cdot), v(\cdot)\big)$ is called the LQGCC upper value.

Definition 1.11 The maximizer's control $v^*(\cdot) \in \Sigma_v$ is called optimal in the LQGCC, if for any $z_0 \in \mathbb{R}$ and for any $v(\cdot) \in \Sigma_v$, it satisfies the inequality: $\inf_{u(\cdot) \in \Upsilon_u(v^*(\cdot))} J\big(u(\cdot), v^*(\cdot)\big) \ge \inf_{u(\cdot) \in \Upsilon_u(v(\cdot))} J\big(u(\cdot), v(\cdot)\big)$. The value $J_{\rm l}(z_0) = \inf_{u(\cdot) \in \Upsilon_u(v^*(\cdot))} J\big(u(\cdot), v^*(\cdot)\big)$ is called the LQGCC lower value.

Definition 1.12 If $\big(u^*(\cdot), v^*(\cdot)\big) \in UV$ and $J_{\rm u}(z_0) = J_{\rm l}(z_0) \stackrel{\triangle}{=} J^*(z_0) \quad \forall z_0 \in \mathbb{R}$, then $J^*(z_0)$ is called the LQGCC value and the pair $\big(u^*(\cdot), v^*(\cdot)\big)$ is called the LQGCC saddle point. Solution $z^*(t)$ of the problem (1.51) for $u(t) = u^*(t, z)$, $v(t) = v^*(t, z)$ is called the LQGCC saddle-point trajectory.

1.3.2.2 Saddle-Point Solution of the LQGCC Uniform in the Cheap Control Parameter

Consider the following terminal-value problem for the function $K(t)$:

$$\frac{dK(t)}{dt} = \frac{\big(h_u(t)\big)^2/\lambda - \big(h_v(t)\big)^2}{\varepsilon} K^2(t), \quad t \in [t_0, t_f], \quad K(t_f) = 1. \tag{1.53}$$

Based on the results of [1, 2, 5], we directly have the following assertion.

Proposition 1.1 *If, for some given $\lambda > 0$ and $\varepsilon > 0$, the terminal-value problem (1.53) has the solution $K(t, \lambda, \varepsilon)$ in the entire interval $[t_0, t_f]$, then the LQGCC has the saddle point $\big(u^*(t, z, \lambda, \varepsilon), v^*(t, z, \lambda, \varepsilon)\big)$, $(t, z) \in [t_0, t_f] \times \mathbb{R}$, and $u^*(t, z, \lambda, \varepsilon) = -h_u(t)K(t, \lambda, \varepsilon)z/(\varepsilon\lambda)$, $v^*(t, z, \lambda, \varepsilon) = h_v(t)K(t, \lambda, \varepsilon)z/\varepsilon$.*

Formal solution of the problem (1.53) yields the expression

$$K(t, \lambda, \varepsilon) = \left[1 + \frac{1}{\varepsilon}\int_t^{t_f} \left(\frac{\big(h_u(\sigma)\big)^2}{\lambda} - \big(h_v(\sigma)\big)^2\right) d\sigma\right]^{-1}, \quad t \in [t_0, t_f]. \tag{1.54}$$

Let us denote

$$H(t) \triangleq \int_t^{t_f} \big(h_u(\sigma)\big)^2 d\sigma \bigg/ \int_t^{t_f} \big(h_v(\sigma)\big)^2 d\sigma, \quad t \in [t_0, t_f). \tag{1.55}$$

Taking into account the positiveness of the integrals $\int_t^{t_f} \big(h_u(\sigma)\big)^2 d\sigma$ and $\int_t^{t_f} \big(h_v(\sigma)\big)^2 d\sigma$ for all $t \in [t_0, t_f)$, as well as Proposition 1.1 and the equation (1.54), we have (quite similarly to [18, 19]) the following assertion.

Proposition 1.2 *Let $\lambda_{\inf} \triangleq \inf_{t\in[t_0,t_f)} H(t) > 0$. Then, the solution $K(t, \lambda, \varepsilon)$ of the terminal-value problem (1.53) exists in the entire interval $[t_0, t_f]$ for all arbitrarily small values $\varepsilon > 0$ if and only if the positive number λ satisfies the inequality $\lambda \le \lambda_{\inf}$. Moreover, for all aforementioned ε and λ, the components of the LQGCC saddle point have the form*

$$u^*(t, z, \lambda, \varepsilon) = -\frac{h_u(t)}{\varepsilon\lambda + \int_t^{t_f} \Big[\big(h_u(\sigma)\big)^2 - \lambda\big(h_v(\sigma)\big)^2\Big] d\sigma} z, \quad t \in [t_0, t_f], \tag{1.56}$$

$$v^*(t, z, \lambda, \varepsilon) = \frac{\lambda h_v(t)}{\varepsilon\lambda + \int_t^{t_f} \Big[\big(h_u(\sigma)\big)^2 - \lambda\big(h_v(\sigma)\big)^2\Big] d\sigma} z, \quad t \in [t_0, t_f]. \tag{1.57}$$

1.4 Robust Capture Controls and Capture Zones

In this section, based on the results of Sect. 1.3, two types of the robust capture control are designed. The first type (bang-bang control) is designed using the results of Sect. 1.3.1, while the second type (continuous control) is designed using the results of Sect. 1.3.2.

1.4.1 Scalar Pursuit-Evasion Game

1.4.1.1 Derivation of Scalar Pursuit-Evasion Game

Note that the PEGBC (see Sect. 1.2.2.2) is a particular case of the VGBC (see Sect. 1.3.1.1). Hence, applying the results of Sect. 1.3.1.3, we can transform the PEGBC to a scalar differential game. Namely, using the data of the PEGBC in (1.33)–(1.35), the function $\zeta(t, x)$, given by (1.42), becomes as:

$$\zeta(t, x, \phi_{PC}, \phi_{E0}) = \sum_{i=1}^{6} \varphi_{1i}(t)x_i + \int_t^{t_f} \varphi_{12}(\sigma) f_2(\sigma, \phi_{PC}, \phi_{E0})d\sigma,$$
$$(t, x) \in [t_0, t_f] \times \mathbb{R}^6, \tag{1.58}$$

$$\varphi_{11}(t) = 1, \varphi_{12}(t) = t_f - t,$$
$$\varphi_{13}(t) = -\cos(\phi_{PC}) \int_{t_f}^{t} \left[\exp\left(\frac{t-\xi}{\tau_P}\right) \frac{1}{V_P(\xi)} \int_{t_f}^{\xi} V_P(\sigma)d\sigma \right] d\xi,$$
$$\varphi_{14}(t) = \cos(\phi_{E0}) \int_{t_f}^{t} \left[\exp\left(\frac{t-\xi}{\tau_E}\right) \frac{1}{V_E(\xi)} \int_{t_f}^{\xi} V_E(\sigma)d\sigma \right] d\xi,$$
$$\varphi_{15}(t) = \cos(\phi_{PC}) \left(V_P(t)(t_f - t) + \int_{t_f}^{t} V_P(\sigma)d\sigma \right),$$
$$\varphi_{16}(t) = -\cos(\phi_{E0}) \left(V_E(t)(t_f - t) + \int_{t_f}^{t} V_E(\sigma)d\sigma \right). \tag{1.59}$$

Using the results of Sect. 1.3.1.3 (see the Eqs. (1.43)–(1.44)) and the Eqs. (1.34), (1.35), (1.58), we transform the PEGBC to the Scalar Pursuit-Evasion Game with Bounded Controls (SPEGBC) consisting of the system

$$dz(t)/dt = h_u(t, \phi_{PC})u(t) + h_v(t, \phi_{E0})v(t), \quad t \in [t_0, t_f], \quad z(t_0) = z_0,$$
$$h_u(t, \phi_{PC}) = a_{P,c}^{\max}(t)\varphi_{13}(t)/\tau_P, \quad h_v(t, \phi_{E0}) = a_{E,c}^{\max}(t)\varphi_{14}(t)/\tau_E, \tag{1.60}$$

the control constraints (1.32) and the cost function $J_z\big(u(\cdot), v(\cdot)\big) = |z(t_f)|$. In (1.60), $z_0 = x_{20}(\phi_{PC}, \phi_{E0})\varphi_{12}(t_0) + x_{30}\varphi_{13}(t_0) + x_{40}\varphi_{14}(t_0) + \alpha_{P0}(\phi_{PC})\varphi_{15}(t_0)$. We also assume that $\cos(\phi_{PC}) \neq 0$ and $\cos(\phi_{E0}) \neq 0$ which yields that $h_u(t, \phi_{PC}) \neq 0$ and $h_v(t, \phi_{E0}) \neq 0$ for all $t \in [t_0, t_f)$.

1.4.1.2 Saddle-Point Solutions of the SPEGBC and the PEGBC

The saddle-point solutions of the SPEGBC and the PEGBC are obtained similarly to such solutions of the SGBC and the VGBC (see Sect. 1.3.1.4), i.e., based on the decomposition of the SPEGBC space. The latter depends on the behaviour of the determining function $R(t)$ (see the assumption AII). For the SPEGBC,

$$R(t) = R(t, \phi_{PC}, \phi_{E0}) = a_{E,c}^{\max}(t)|\varphi_{14}(t)|/\tau_E - a_{P,c}^{\max}(t)|\varphi_{13}(t)|/\tau_P. \tag{1.61}$$

Like in Sect. 1.3.1.4, we assume that this function has a finite number of distinct zeros in the interval (t_0, t_f). Similarly to Sect. 1.3.1.4, using the behaviour of the trajectories of the problem (1.46) with $R(t) = R(t, \phi_{PC}, \phi_{E0})$, we decompose the SPEGBC space into the regular $\mathcal{R}_r(\phi_{PC}, \phi_{E0})$ and singular $\mathcal{R}_s(\phi_{PC}, \phi_{E0})$ regions including the obtaining the positive $z^+(t, \phi_{PC}, \phi_{E0})$ and negative $z^-(t, \phi_{PC}, \phi_{E0})$ boundaries of $\mathcal{R}_r(\phi_{PC}, \phi_{E0})$ (the upper and lower boundaries of $\mathcal{R}_s(\phi_{PC}, \phi_{E0})$). Thus, for the SPEGBC and the PEGBC, Lemmas 1.1, 1.2 are valid with replacing $h_u(t)$, $h_v(t)$, $\mathcal{R}_r$, $\mathcal{R}_s$, $\zeta(t, z)$, $z^+(t)$, $\mathbb{R}^n$ by $h_u(t, \phi_{PC})$, $h_v(t, \phi_{E0})$, $\mathcal{R}_r(\phi_{PC}, \phi_{E0})$, $\mathcal{R}_s(\phi_{PC}, \phi_{E0})$, $\zeta(t, z, \phi_{PC}, \phi_{E0})$, $z^+(t, \phi_{PC}, \phi_{E0})$, $\mathbb{R}^6$, respectively.

1.4.1.3 Robust Capture Control and Capture Zone of the Scalar Engagement (1.60), (1.32)

It should be noted that the system (1.60) can be considered as a scalar particular case of the system (1.36) with zero coefficient for the scalar state variable, the scalar coefficients $h_u(t, \phi_{PC})$, $h_v(t, \phi_{E0})$ for the controls and zero non-homogeneity function. Therefore, the notions of the robust capture control and its capture zone for the engagement (1.60), (1.32) are quite similar to such notions for the engagement (1.36), (1.32). Definitions 1.1–1.4, valid for the engagement (1.36), (1.32), are directly reformulated for the engagement (1.60), (1.32). The existence of a robust capture control means the fulfilment of the conditions of Definition 1.3 for the engagement (1.36), (1.32) or the fulfilment of similar conditions in the case of the engagement (1.60), (1.32).

1.4.2 Main Assertions

Theorem 1.1 *Let the functions $a_{P,c}^{\max}(t)$ and $a_{E,c}^{\max}(t)$ be twice continuously differentiable in some left-hand vicinity of $t = t_f$ including this point. Let the following condition be satisfied:*

$$\gamma_f(\phi_{PC}, \phi_{E0}) \stackrel{\triangle}{=} |\cos(\phi_{E0})|a_{E,c}^{\max}(t_f)/\tau_E - |\cos(\phi_{PC})|a_{P,c}^{\max}(t_f)/\tau_P < 0. \quad (1.62)$$

Then, there exists a robust capture control for the engagement (1.60), (1.32).

Proof First, let us show that $\exists \hat{t} \in [t_0, t_f) : R(t) = R(t, \phi_{PC}, \phi_{E0}) < 0 \, \forall t \in (\hat{t}, t_f)$. Due to (1.59), (1.61), $R(t_f) = 0$. Let us show the existence of $\hat{t}$ by a proper sign of $d^k R(t_f)/dt^k \neq 0$, ($k > 0$ is the smallest integer). Using (1.59), (1.61), we have by direct calculation: $dR(t_f)/dt = 0$, $d^2 R(t_f)/dt^2 = \gamma_f(\phi_{PC}, \phi_{E0}) < 0$. Hence, taking into account that $R(t)$ is twice continuously differentiable in some left-hand vicinity of $t = t_f$ including $t = t_f$, we obtain the existence of the aforementioned point $\hat{t}$. Since $R(t, \phi_{PC}, \phi_{E0}) < 0$, $t \in (\hat{t}, t_f)$, then, due to the results of Sects. 1.3.1.4 and 1.4.1.2, there exist segments of the upper $z^+(t, \phi_{PC}, \phi_{E0}) = \int_{t_f}^{t} R(\sigma, \phi_{PC}, \phi_{E0})d\sigma$ and lower $z^-(t, \phi_{PC}, \phi_{E0}) = -z^+(t, \phi_{PC}, \phi_{E0})$, $t \in (\hat{t}, t_f)$ boundaries of the singular region $\mathcal{R}_s(\phi_{PC}, \phi_{E0})$ starting from the point $(t_f, 0)$ in the backward time. The latter means that $\mathcal{S}_s(\phi_{PC}, \phi_{E0}) \stackrel{\triangle}{=} \{(t, z) : t \in (\hat{t}, t_f), z^-(t, \phi_{PC}, \phi_{E0}) < z < z^+(t, \phi_{PC}, \phi_{E0})\} \subset \mathcal{R}_s(\phi_{PC}, \phi_{E0})$ and $(t_f, 0) \in \text{clo}(\mathcal{S}_s(\phi_{PC}, \phi_{E0}))$. Therefore, by virtue of Lemma 1.1 and the results of Sect. 1.4.1.2, the SPEGBC value equals zero for any starting position $(\bar{t}, \bar{z}) \in \mathcal{S}_s(\phi_{PC}, \phi_{E0})$ of this game. The latter means that the pursuer's optimal feedback control in the SPEGBC transfers any such a point $(\bar{t}, \bar{z})$ to the point $(t_f, 0)$ against any admissible (in the sense of Definition 1.1) evader's control. Hence, due to Definitions 1.2, 1.3 and the results of Sect. 1.4.1.3, we directly have the existence of a robust capture control in the SPEGBC from any $(\bar{t}, \bar{z}) \in \mathcal{S}_s(\phi_{PC}, \phi_{E0})$. □

Theorem 1.1 establishes the condition of the existence of a robust capture control for the engagement (1.60), (1.32). Now, we are going to design two types of such a control and to construct the corresponding capture zones. Due to the proof of Theorem 1.1, $\int_{t_f}^{t} R(\sigma, \phi_{PC}, \phi_{E0})d\sigma \geq 0$ for all $t \in [\hat{t}, t_f]$. However, $\hat{t} \in [t_0, t_f)$ is not, in general, the minimal possible such a time instant. By $t_{\min}$, we denote the minimal value of $t \in [t_0, t_f)$ keeping the following inequality:

$$z_c(t, \phi_{PC}, \phi_{E0}) \stackrel{\triangle}{=} \int_{t_f}^{t} R(\sigma, \phi_{PC}, \phi_{E0})d\sigma \geq 0 \quad \forall t \in [t_{\min}, t_f]. \qquad (1.63)$$

Remark 1.6 If $t_{\min} > t_0$, then the function $z_c(t, \phi_{PC}, \phi_{E0})$ changes its sign at $t = t_{\min}$ and $z_c(t, \phi_{PC}, \phi_{E0}) < 0$ in some left-hand vicinity of this point. Moreover, $R(t, \phi_{PC}, \phi_{E0}) > 0$ in some left- and right-hand neighbourhoods of $t = t_{\min}$. Since this function has a finite number of distinct zeros in the interval (t_0, t_f), the function

$z_c(t)$ can have a finite number of distinct zeros in the interval $(t_{\min}, t_f)$. However, not to overload the paper, we focus on the case $z_c(t, \phi_{PC}, \phi_{E0}) > 0$, $t \in (t_{\min}, t_f)$ which appears in many versions of the pursuit-evasion engagement between two flying vehicles (see e.g. [8, 10, 15, 17]).

Let us consider the set $C_z \triangleq \{(t, z) : t \in [t_{\min}, t_f],\ |z| \le z_c(t)\}$. Also, we consider the control $u^0_{\text{sg}}(t, z)$ having the form in (1.48) with $h_u(t) = h_u(t, \phi_{PC})$.

Theorem 1.2 *Let the condition (1.62) be valid. Then, $u^0_{\text{sg}}(t, z)$ is a robust capture control from any point $(\bar{t}, \bar{z}) \in C_z$ for the engagement (1.60), (1.32). Moreover, the set C_z is the capture zone of $u^0_{\text{sg}}(t, z)$, i.e., $C_z = C_z\big(u^0_{\text{sg}}(t, z)\big)$.*

Proof First of all, let us note that, due to Definition 1.2, Remark 1.2 and the results of Sect. 1.4.1.3, $u^0_{\text{sg}}(t, z)$ is an admissible control in the engagement (1.60), (1.32). Similarly to the proof of Theorem 1.1, we can show that the curves $z = z_c(t, \phi_{PC}, \phi_{E0})$ and $z = -z_c(t, \phi_{PC}, \phi_{E0})$, $t \in [t_{\min}, t_f]$ are segments of the upper and lower boundaries of $\mathcal{R}_{\text{s}}(\phi_{PC}, \phi_{E0})$. Therefore, $\text{int}\big(C_z\big) \subseteq \mathcal{R}_{\text{s}}(\phi_{PC}, \phi_{E0})$. Moreover, $\text{int}\big(C_z\big)$ is the maximal part of $\mathcal{R}_{\text{s}}(\phi_{PC}, \phi_{E0})$ which closure is a simply connected set and the point $(t_f, 0) \in \text{clo}\big(C_z\big)$. Using this observation, we obtain (similarly to the proof of Theorem 1.1) that the aforementioned control $u^0_{\text{sg}}(t, z)$ is the robust capture control in the SPEGBC from any $(\bar{t}, \bar{z}) \in C_z$.

To prove that C_z is the capture zone of $u^0_{\text{sg}}(t, z)$, we should show that $u^0_{\text{sg}}(t, z)$ is not a robust capture control from any point $(\bar{t}, \bar{z}) \in \mathcal{S}_{[t_0,t_f]}\backslash C_z$ for the engagement (1.60), (1.32). First, consider the case where $\bar{t} \in [t_{\min}, t_f)$. In this case, the aforementioned point $(\bar{t}, \bar{z}) \in \mathcal{R}_{\text{r}}(\phi_{PC}, \phi_{E0})$ and its second coordinate satisfies the inequality $|\bar{z}| > z_c(\bar{t}, \phi_{PC}, \phi_{E0})$. Hence, due to Lemma 1.1 (the Eq. (1.49)), the results of Sect. 1.4.1.2 and the Eq. (1.63), the SPEGBC value is positive for the starting position $(\bar{t}, \bar{z})$ of this game. The latter means that the saddle-point trajectory of the SPEGBC $z = z^0(t; \bar{t}, \bar{z})$, $t \in [\bar{t}, t_f]$ starting at $(\bar{t}, \bar{z})$ ends at (t_f, z_f) with $|z_f| > 0$. Note that $z = z^0(t; \bar{t}, \bar{z})$, $t \in [\bar{t}, t_f]$ is the unique solution of the differential equation in (1.60) with the initial condition $z(\bar{t}) = \bar{z}$ and the controls $u(t) = u^0_{\text{sg}}(t, z)$ and $v(t) = v^0_{\text{sg}}(t, z)$, where $v^0_{\text{sg}}(t, z)$ has the form in (1.48) with $h_v(t) = h_v(t, \phi_{E0})$. Let $\widetilde{v}^0_{\text{sg}}(t) \triangleq v^0_{\text{sg}}\big(t, z^0(t; \bar{t}, \bar{z})\big)$, $t \in [\bar{t}, t_f]$. Then, $|\widetilde{v}^0_{\text{sg}}(t)| = 1$, $t \in [\bar{t}, t_f]$. Thus, $\widetilde{v}^0_{\text{sg}}(t)$, $t \in [\bar{t}, t_f]$ is an admissible evader's control. Furthermore, the unique solution of the differential equation in (1.60) with the initial condition $z(\bar{t}) = \bar{z}$ and the controls $u(t) = u^0_{\text{sg}}(t, z)$ and $v(t) = \widetilde{v}^0_{\text{sg}}(t)$ coincides with $z^0(t; \bar{t}, \bar{z})$, $t \in [\bar{t}, t_f]$. This observation, along with the aforementioned inequality $|z^0(t_f; \bar{t}, \bar{z})| > 0$ and equality $|\widetilde{v}^0_{\text{sg}}(t)| = 1$, $t \in [\bar{t}, t_f]$, means that $u^0_{\text{sg}}(t, z)$ is not a robust capture control from any point $(\bar{t}, \bar{z}) \in \mathcal{S}_{[t_0,t_f]}\backslash C_z$ where $\bar{t} \in [t_{\min}, t_f)$. Proceed to the case where in the aforementioned point $(\bar{t}, \bar{z}) \in \mathcal{S}_{[t_0,t_f]}\backslash C_z$ the coordinate $\bar{t} \in [t_0, t_{\min})$. Since $R(t, \phi_{PC}, \phi_{E0}) > 0$ in some vicinity of $t = t_{\min}$ (see Remark 1.6), then $|z^0(t_{\min}; \bar{t}, \bar{z})| > z_c(t_{\min}; \bar{t}, \bar{z}) = 0$, where $z = z^0(t; \bar{t}, \bar{z})$, $t \in [\bar{t}, t_f]$ is the saddle-point trajectory of the SPEGBC starting at $(\bar{t}, \bar{z})$. Based on this observation, we prove that $u^0_{\text{sg}}(t, z)$ is not a robust capture control from any point $(\bar{t}, \bar{z}) \in \mathcal{S}_{[t_0,t_f]}\backslash C_z$ in the case $\bar{t} \in [t_0, t_{\min})$ quite similarly to the above considered case $\bar{t} \in [t_{\min}, t_f)$. This completes the proof of the theorem. □

Control $u^0_{\text{sg}}(t,z)$, considered in Theorem 1.2, is a discontinuous (bang-bang) function in z for all $t \in [t_0, t_f)$. Below, we propose a continuous with respect to z robust capture control $u_{\text{sg,c}}(t,z)$. Construction of $u_{\text{sg,c}}(t,z)$ is based on a function $L(t,z)$, $(t,z) \in [t_0,t_f] \times \mathbb{R}$ satisfying the following conditions:

AL1. For any $t \in [t_0, t_f]$, the function $L(t,z)$ is odd with respect to $z \in \mathbb{R}$.

AL2. For any $t \in [t_0, t_f]$, the function $L(t,z)$ and its partial derivative $\partial L(t,z)/\partial z$ are continuous with respect to $z \in \mathbb{R}$.

AL3. For any $t \in [t_0, t_f]$, $\partial L(t,z)/\partial z \geq 0$.

AL4. For any $t \in (t_{\min}, t_f)$, $L\big(t, z_c(t, \phi_{PC}, \phi_{E0})\big) \geq 1$.

Consider the pursuer's control

$$\begin{aligned} &u_{\text{sg,c}}(t,z) = -\text{sign}\big(h_u(t,\phi_{P0})\big)\text{sat}\big(L(t,z)\big), \quad (t,z) \in [t_0,t_f] \times \mathbb{R}, \\ &\text{sat}(L) = L, \ |L| \leq 1; \ \text{sat}(L) = \text{sign}(L), \ |L| > 1. \end{aligned} \tag{1.64}$$

Theorem 1.3 *Let the condition (1.62) and the conditions AL1-AL4 be valid. Then, the control (1.64) is a robust capture control from any point $(\bar{t}, \bar{z}) \in C_z$ and the set C_z is the capture zone of this control for the engagement (1.60), (1.32).*

Proof First of all, let us note that, due to Definition 1.2, Remark 1.2 and the results of Sect. 1.4.1.3, $u_{\text{sg,c}}(t,z)$ is an admissible control in the engagement (1.60), (1.32). Due to Lemma 1.1 and the results of Sect. 1.4.1.2, we have the following: (i) if $t_{\min} = t_0$ and $z_c(t_0, \phi_{PC}, \phi_{E0}) > 0$, then $u_{\text{sg,c}}(t,z)$ coincides with the pursuer's optimal feedback control of the SPEGBC for any $(t,z) \in C_z$; (ii) if either $t_{\min} > t_0$ or $t_{\min} = t_0$ and $z_c(t_0, \phi_{PC}, \phi_{E0}) = 0$, then $u_{\text{sg,c}}(t,z)$ coincides with the pursuer's optimal feedback control of the SPEGBC for any $(t,z) \in C_z \backslash (t_{\min}, 0)$. This observation and the fact that the value of the SPEGBC, starting at any $(\bar{t}, \bar{z}) \in C_z$, equals zero directly yield the following conclusions: in the case (i) $u_{\text{sg,c}}(t,z)$ is a robust capture control from any point $(\bar{t}, \bar{z}) \in C_z$; in the case (ii) $u_{\text{sg,c}}(t,z)$ is a robust capture control from any point $(\bar{t}, \bar{z}) \in C_z \backslash (t_{\min}, 0)$. Moreover, in the case (ii), due to Definitions 1.1, 1.2, Remark 1.2 and the results of Sect. 1.4.1.3, the trajectory of the differential equation in (1.60) starting from $(t_{\min}, 0)$ and governed by $u(t) = u_{\text{sg,c}}(t,z)$ and an admissible $v(t)$ remains in C_z till $t = t_f$. The latter means that in the case (ii) $u_{\text{sg,c}}(t,z)$ is a robust capture control from $(t_{\min}, 0)$ too. Thus, $u_{\text{sg,c}}(t,z)$ is a robust capture control from any point $(\bar{t}, \bar{z}) \in C_z$.

Proceed to the proof that C_z is the capture zone of $u_{\text{sg,c}}(t,z)$. This proof is similar to that of Theorem 1.2 based on the observations: (a) $u_{\text{sg,c}}(t,z)$ is a pursuer's optimal control of the SPEGBC in the set $\big\{(t,z) \in \mathcal{S}_{[t_0,t_f]} \backslash C_z, \, t \in [t_{\min}, t_f)\big\}$; (b) $u_{\text{sg,c}}(t,z)$ is not, in general, a pursuer's optimal control of the SPEGBC in the set $\big\{(t,z) \in \mathcal{S}_{[t_0,t_f]} \backslash C_z, \, t \in [t_0, t_{\min})\big\}$; (c) the pursuer's optimal control $u^0_{\text{sg}}(t,z)$ of the SPEGBC is not a robust capture control from any point $(t,z) \in \mathcal{S}_{[t_0,t_f]} \backslash C_z$. Thus, the theorem is proven. □

Using the controls $u^0_{\text{sg}}(t,z)$, $u_{\text{sg,c}}(t,z)$ and the set C_z, mentioned in Theorems 1.2, 1.3, we construct the following controls and set for the engagement (1.36), (1.32):

$$\begin{aligned} u^0(t,x) &= u^0_{\text{sg}}\big(t, \zeta(t,x,\phi_{PC},\phi_{E0})\big), \quad (t,x) \in [t_0,t_f] \times \mathbb{R}^6, \\ u_{\text{c}}(t,x) &= u_{\text{sg,c}}\big(t, \zeta(t,x,\phi_{PC},\phi_{E0})\big), \quad (t,x) \in [t_0,t_f] \times \mathbb{R}^6, \\ C_x &\stackrel{\triangle}{=} \big\{(t,x) \in [t_0,t_f] \times \mathbb{R}^6 : \big(t, \zeta(t,x,\phi_{PC},\phi_{E0})\big) \in C_z\big\}, \end{aligned} \tag{1.65}$$

where $\zeta(t,x,\phi_{PC},\phi_{E0})$ is given by the Eq. (1.58).

Remark 1.7 Due to the results of Sect. 1.2.2, 1.4.1.2, 1.4.1.3, as well as Lemmas 1.1, 1.2, Theorems 1.2, 1.3, and the connection (1.58) between the state variables of the engagements (1.60), (1.32) and (1.36), (1.32), we have that $u^0(t,x)$ and $u_{\text{c}}(t,x)$ are the robust capture controls for the engagement (1.36), (1.32), and C_x is their capture zone.

1.4.3 Construction of $L(t,z)$ Based on Linear-Quadratic Differential Game with Cheap Controls

First, we construct an auxiliary function. This construction, consisting of three stages, is based on the results of Sect. 1.3.2. At the first stage, we use the control $u^*(t,z,\lambda,\varepsilon)$ (see the Eq. (1.56)) and replace there $h_u(t)$ and $h_v(t)$ with $h_u(t,\phi_{PC})$ and $h_v(t,\phi_{E0})$ (see the Eq. (1.60)). Also, we replace t_0 with $t_{\min}$ defined in (1.63). At the second stage, multiplying the resulting expression by $-\text{sign}\big(h_u(t,\phi_{PC})\big)$, we obtain the following function for all $(t,z) \in [t_{\min},t_f] \times \mathbb{R}$:

$$\widetilde{L}(t,z,\lambda,\varepsilon) = \frac{|h_u(t,\phi_{PC})|}{\varepsilon\lambda + \int_t^{t_f} \big[\big(h_u(\sigma,\phi_{PC})\big)^2 - \lambda\big(h_v(\sigma,\phi_{E0})\big)^2\big] d\sigma} z. \tag{1.66}$$

To check the feasibility of the function $\widetilde{L}(t,z,\lambda,\varepsilon)$ for all arbitrarily small values $\varepsilon > 0$, we introduce the following function of $t \in [t_{\min},t_f)$ and the value:

$$\begin{aligned} H_{uv}(t,\phi_{PC},\phi_{E0}) &\stackrel{\triangle}{=} \int_t^{t_f} \big(h_u(\sigma,\phi_{PC})\big)^2 d\sigma \Big/ \int_t^{t_f} \big(h_v(\sigma,\phi_{E0})\big)^2 d\sigma, \\ \lambda_{\inf}(\phi_{PC},\phi_{E0}) &\stackrel{\triangle}{=} \inf_{t \in [t_{\min},t_f)} H_{uv}(t,\phi_{PC},\phi_{E0}). \end{aligned} \tag{1.67}$$

Proposition 1.3 *The value $\lambda_{\inf}(\phi_{PC},\phi_{E0})$ is positive. Moreover, if the positive number λ satisfies the inequality $\lambda \le \lambda_{\inf}(\phi_{PC},\phi_{E0})$, then $\widetilde{L}(t,z,\lambda,\varepsilon)$ is feasible for all arbitrarily small values $\varepsilon > 0$.*

Proof The first statement of the proposition is proven quite similarly to [18] (see Lemma 1), while the second statement is a direct consequence of Propositions 1.1 and 1.2. □

At the third stage in the construction of the auxiliary function, we calculate the limit for $\varepsilon \to +0$ of $\widetilde{L}(t,z,\lambda,\varepsilon)$, $(t,z) \in [t_{\min},t_f) \times \mathbb{R}$. This limit has the form

$$L_{\text{aux}}(t,z,\lambda) \triangleq \widetilde{L}(t,z,\lambda,0) = \frac{|h_u(t,\phi_{PC})|}{\int_t^{t_f}\left[\left(h_u(\sigma,\phi_{PC})\right)^2 - \lambda\left(h_v(\sigma,\phi_{E0})\right)^2\right]d\sigma}z. \tag{1.68}$$

Now, using the auxiliary function $L_{\text{aux}}(t,z,\lambda)$, we are going to construct the function $L(t,z)$ satisfying the conditions AL1–AL4. For this purpose, we distinguish two possible cases of the time instant $t_{\min}$: (a) $t_{\min} \geq t_0$ and $z_c(t_{\min},\phi_{PC},\phi_{E0}) = 0$, $R(t_{\min}) \neq 0$; (b) $t_{\min} = t_0$ and $z_c(t_{\min},\phi_{PC},\phi_{E0}) > 0$.

Case (a). Assume that $\lambda_{\inf}(\phi_{PC},\phi_{E0}) = H_{uv}(t_{\min},\phi_{PC},\phi_{E0}) \triangleq \lambda_{\min}$. Let $\widetilde{\gamma}_{\text{a}} \triangleq \inf_{t\in(t_{\min},t_f)} L_{\text{aux}}\big(t, z_c(t,\phi_{PC},\phi_{E0}),\lambda_{\min}\big) > 0$, $\gamma_{\text{a}} \triangleq \max\{1, 1/\widetilde{\gamma}_{\text{a}}\}$. Based on these assumption and values, we represent $L(t,z)$ as:

$$L(t,z) = \begin{cases} \gamma_{\text{a}} L_{\text{aux}}(t,z,\lambda_{\min}), & t \in (t_{\min},t_f), \quad z \in \mathbb{R}, \\ z, & t \in [t_0,t_{\min}]\bigcup\{t = t_f\}, \quad z \in \mathbb{R}. \end{cases} \tag{1.69}$$

Case (b). Consider $\widetilde{\gamma}_{\text{b}}(\lambda) \triangleq \inf_{t\in[t_{\min},t_f)} L_{\text{aux}}\big(t, z_c(t,\phi_{PC},\phi_{E0}),\lambda\big) > 0$, $\gamma_{\text{b}}(\lambda) \triangleq \max\{1, 1/\widetilde{\gamma}_{\text{b}}(\lambda)\}$, where $0 < \lambda < \lambda_{\inf}(\phi_{PC},\phi_{E0})$ is any prechosen number. Using these values, we construct $L(t,z)$ as:

$$L(t,z) = \begin{cases} \gamma_{\text{b}}(\lambda) L_{\text{aux}}(t,z,\lambda), & t \in [t_0,t_f), \quad z \in \mathbb{R}, \\ z, & t = t_f, \quad z \in \mathbb{R}. \end{cases} \tag{1.70}$$

It can be verified directly that the functions (1.69) and (1.70) satisfy the conditions AL1-AL4.

1.4.4 Example

Consider the pursuit-evasion engagement of two flying vehicles, modeled by (1.60), (1.32) and (1.36), (1.32), with the data: $V_P(t) = 2000\exp(t/16)$ m/sec, $V_E(t) \equiv 2600$ m/sec, $t_0 = 0$, $\phi_{E0} = 5\pi/6$, $R_0 = 30000$ m, $\tau_P = 0.1$ sec, $\tau_E = 0.15$ sec, $a_{P,c}^{\max}(t) = (25t + 100)$ m/sec^2, $a_{E,c}^{\max}(t) = \big(125\arctan(t) + 50\big)$ m/sec^2. For these data, the Eq. (1.2) has the unique solution $t_C = t_{C,min} = 6.845579$ sec which, along with (1.3), (1.22), yields $\cos(\phi_{PC}) = 0.853688$, $\sin(\phi_{PC}) = 0.520785$, $t_f = 6.845579$ sec. Using the aforementioned data and the Eqs. (1.59), (1.61), we obtain that the determining function in this example $R(t.\phi_{PC},\phi_{E0})$ changes its sign twice at $t_{s1} = 1.03811$ and $t_{s2} = 3.76126$, and this function is negative in the interval $(t_{s2},t_f) = (3.76126, 6.845579)$. In Fig. 1.6a, the regular $\mathcal{R}_{\text{r}}$ and singular $\mathcal{R}_{\text{s1}}\bigcup\mathcal{R}_{\text{s2}}$ regions of the SPEGBC space are presented for the aforementioned data, $t_{in} = 1.883973$. Due to (1.59), (1.60) and the data of the example, $h_u(t,\phi_{PC}) \leq 0, t \in [0,t_f]$. Therefore, the robust capture control $u_{\text{sg}}^0(t,z)$, mentioned in Theorem 1.2, is of the form: $u_{\text{sg}}^0(t,z) = \text{sign}(z)$ for $z \neq 0$ and $u_{\text{sg}}^0(t,z) = 1$ for $z = 0$. The robust capture control $u_{\text{sg,c}}(t,z)$, mentioned in Theorem 1.3, has the form $u_{\text{sg,c}}(t,z) = \text{sat}\big(L(t,z)\big)$

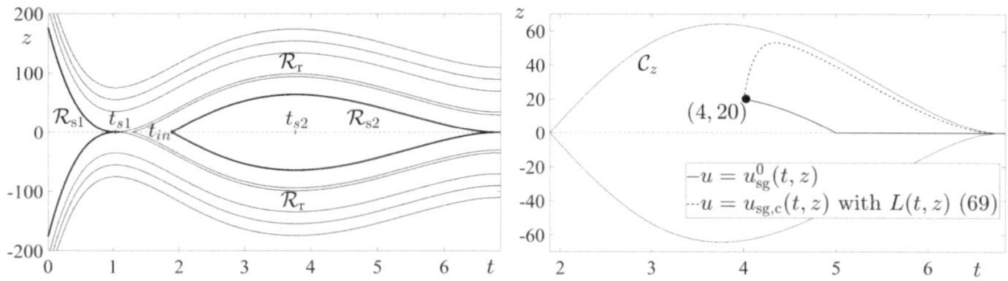

a) Regular and singular regions of the SPEGBC space

b) Capture zone and trajectories of the engagement (1.60), (1.32)

Fig. 1.6 The SPEGBC space decomposition

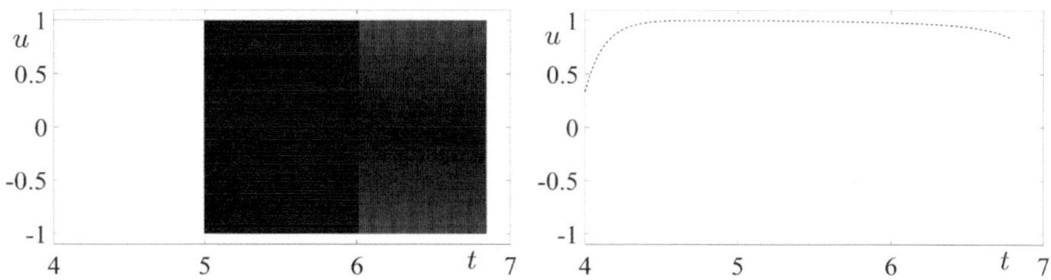

a) $u = u^0_{sg}(t, z)$

b) $u = u_{sg,c}(t, z)$

Fig. 1.7 Time realizations of the robust capture controls

where $L(t, z)$ is of the form (1.69) with $\lambda_{min} = 0.953762$, $\gamma_a = 1.4456$. In Fig. 1.6b, the capture zone C_z of $u^0_{sg}(t, z)$ and $u_{sg,c}(t, z)$ for the pursuit-evasion engagement (1.60), (1.32) is presented. Also, the trajectories of this engagement, starting from the point $(4, 20) \in C_z$ and generated by the controls $u^0_{sg}(t, z)$ and $u_{sg,c}(t, z)$, are depicted. It is seen that, starting in C_z, these trajectories remain in this set till the end of the engagement. In Fig. 1.7, the time realizations of the aforementioned robust capture controls along the corresponding trajectories are presented. From Figs. 1.6b and 1.7, it is seen that the control $u^0_{sg}(t, z)$ yields the sliding mode trajectory along the t-axis for $t \in [5, t_f]$, and this control produces the chattering of its time realization for $t \in [5, t_f]$. The time realization of the control $u_{sg,c}(t, z)$ is chattering free which is its considerable advantage in comparison with the control $u^0_{sg}(t, z)$. Using the above derived controls $u^0_{sg}(t, z)$, $u_{sg,c}(t, z)$, the Eq. (1.65) and Remark 1.7, we obtain the robust capture controls for the engagement (1.36), (1.32) as: $u^0(t, x) = \text{sign}\big(\zeta(t, x, \phi_{PC}, \phi_{E0})\big)$ for $\zeta(t, x, \phi_{PC}, \phi_{E0}) \neq 0$ and $u^0(t, x) = 1$ for $\zeta(t, x, \phi_{PC}, \phi_{E0}) = 0$; $u_c(t, x) = \text{sat}\Big(L\big(t, \zeta(t, x, \phi_{PC}, \phi_{E0})\big)\Big)$. The capture zone C_x of these controls is given in (1.65) with the set C_z given in Fig. 1.6b.

1.5 Conclusions

In this paper, the problem of a planar pursuit with the capture of the maneuvering evader was studied subject to the condition that the pursuer cannot predict the evader's behaviour. To solve this problem, the notions of the pursuer's robust capture control and its capture zone were introduced and the solutions of two auxiliary zero-sum differential games (linear dynamics game with bounded controls and linear-quadratic game with cheap controls) were presented. Using these notions and solutions, two pursuer's robust capture controls were designed. One of these controls (bang-bang one) yields a sliding mode trajectory of the pursuit-evasion engagement and the chattering of its time realization along this trajectory. The trajectory of the engagement, generated by the other (continuous in the state variable) control, is sliding-mode-free, and the control's time realization is chattering-free.

References

1. Basar, T., Olsder, G.J.: Dynamic Noncooparative Game Theory, 2nd edn. SIAM Books, Philadelphia, PA, USA (1999)
2. Ben-Asher, J., Yaesh, I.: Advances in Missile Guidance Theory, Progress in Astronautics and Aeronautics, vol. 180. AIAA Inc, Reston, VA, USA (1998)
3. Glizer, V.Y.: Optimal planar interception with fixed end conditions: closed-form solution. J. Optim. Theory Appl. **88**, 503–539 (1996)
4. Glizer, V.Y.: Homicidal chauffeur game with target set in the shape of a circular angular sector: conditions for existence of a closed barrier. J. Optim. Theory Appl. **101**, 581–598 (1999)
5. Glizer, V.Y., Kelis, O.: Singular Linear-Quadratic Zero-Sum Differential Games and H_∞ Control Problems: Regularization Approach. Birkhäuser, Switzerland (2022)
6. Glizer, V.Y., Turetsky, V.: Complete solution of a differential game with linear dynamics and bounded controls. Appl. Math. Res. Express. AMRX **2008** (2008)
7. Glizer, V.Y., Turetsky, V.: Robust Controllability of Linear Systems. Nova Science Publishers, New York, NY, USA (2012)
8. Glizer, V.Y., Turetsky, V., Fridman, L., Shinar, J.: History-dependent modified sliding mode interception strategies with maximal capture zone. J. Franklin Inst. **349**, 638–657 (2012)
9. Guelman, M., Shinar, J.: Optimal guidance law in the plane. J. Guid. Control Dyn. **7**, 471–476 (1984)
10. Gutman, S.: On optimal guidance for homing missiles. J. Guid. Control Dyn. **3**, 296–300 (1979)
11. Gutman, S.: Applied Min-Max Approach to Missile Guidance and Control, Progress in Astronautics and Aeronautics, vol. 209. AIAA Inc, Reston, VA, USA (2006)
12. Gutman, S., Leitmann, G.: Optimal strategies in the neighbourhood of a collision course. AIAA J. **14**, 1210–1212 (1976)
13. Isaacs, R.: Differential Games. John Wiley, New York, NY, USA (1965)
14. Krasovskii, N.N., Subbotin, A.I.: Game-Theoretical Control Problems. Springer, New York, NY, USA (2011)
15. Shima, T., Shinar, J.: Time-varying linear pursuit-evasion game models with bounded controls. J. Guid. Control Dyn. **25**, 425–432 (2002)
16. Shinar, J., Glizer, V.Y., Turetsky, V.: The effect of pursuer dynamics on the value of linear pursuit-evasion games with bounded controls. In: Křivan, V., Zaccour, G. (eds.) Advances in Dynamic Games, Annals of the International Society of Dynamic Games, vol. 13, pp. 313–350. Springer, Switzerland (2013)

17. Turetsky, V., Glizer, V.Y.: Continuous feedback control strategy with maximal capture zone in a class of pursuit games. Int. Game Theory Rev. **7**, 1–24 (2005)
18. Turetsky, V., Glizer, V.Y.: Robust solution of a time-variable interception problem: a cheap control approach. Int. Game Theory Rev. **9**, 637–655 (2007)
19. Turetsky, V., Shinar, J.: Missile guidance laws based on pursuit-evasion game formulations. Automatica J. IFAC **39**, 607–618 (2003)

Chapter 2
Simplicial Fermat–Weber Problems Are Dual to Minimum Cost Flow Problems

Andrei Comăneci

Abstract We show that Fermat–Weber problems under simplicial distances, having both sources and destinations as input, are equivalent to specific minimum cost flow problems. In particular, this enables the efficient computation of geometric medians involving Hilbert's projective metric for a simplex.

Keywords Fermat–Weber problem · Simplicial distance · Minimum cost flow · Hilbert simplex geometry · Tropical distance

2.1 Introduction

The Fermat–Weber problem is a classical location problem where one tries to find a point that minimizes the overall distance or travel time to a set of given points. The name is attributed to the mathematician Pierre de Fermat, as he was the first to propose the problem for three points in the Euclidean plane [19], and to the location economist Alfred Weber who popularized the general version in industrial situations [32].

To describe it formally, we firstly need to consider a *distance* d on an ambient space X, i.e. a function $\mathsf{d} : X \times X \to \mathbb{R}_{\geq 0}$ such that

- $\mathsf{d}(x, y) = 0$ if and only if $x = y$; and
- $\mathsf{d}(x, z) \leq \mathsf{d}(x, y) + \mathsf{d}(y, z)$ for all $x, y, z \in X$.

Notice that we do not assume the symmetry of our distances which "unpleasantly limits many applications" [15, p. xv]. Indeed, symmetry is not assumed in most location theory works; see, e.g., [13, 33]. This was especially used for models of movement on inclined planes [18, 28] or in the presence of winds [12].

Due to the lack of symmetry, it becomes crucial to differentiate between input points, as this distinction plays a significant role in whether we are measuring the distance *to* a point or *from* a point. Following Plastria [30], we will consider that we have a set S of *sources* and a set D of *destinations* (either possibly empty). For each

---

A. Comăneci (✉)
Technische Universität Berlin, Berlin, Germany
e-mail: andrei.comaneci@yandex.com

A. Ott et al. (eds.), *Applications of Mathematics in Sciences, Engineering, and Economics*, Springer Proceedings in Mathematics & Statistics 515,
https://doi.org/10.1007/978-3-032-01279-1_2

source $s \in S$ or destination $d \in D$ we assign a positive weight w_s or ω_d, respectively. With these notations, the Fermat–Weber problem is:

$$\text{minimize}_{x \in X} \sum_{s \in S} w_s \mathsf{d}(s, x) + \sum_{d \in D} \omega_d \mathsf{d}(x, d). \tag{2.1}$$

An optimal solution of (2.1) is called a *Fermat–Weber point* or *geometric median*.

In this note, we will focus on studying the Fermat–Weber problem when d is a distance whose unit ball is a simplex; this will be referred to as the *simplicial Fermat–Weber problem*. Such distances were mostly studied in the context of Voronoi diagrams [2, 3, 9, 22], while some appeared briefly in other location problems [29]. What is more, the simplicial Fermat–Weber problem gained recent interest in tropical geometry as it can be used for applications to evolutionary biology [6–8].

One of the main results of [6] was to show that for a specific simplicial distance, the simplicial Fermat–Weber problem with $D = \emptyset$ is dual to optimal transport, which gave a location-theoretic interpretation of the approach from [31]. Transportation problems are minimum cost flow problems in a bipartite graph and this note will show that a similar result holds when S and D are both non-empty, but on a tripartite graph. The proof will be similar, based on linear programming duality.

The second main result of this note will be to show that the Fermat–Weber problem based on the symmetric tropical distance can be transformed to a simplicial Fermat–Weber problem, as a round-trip Weber problem [12]. This will allow for the approach of Lin and Yoshida [21] to be used for large datasets, which attracted attention in computational biology [25]. In fact, the tropical distance is isometric to Hilbert's simplex distance [5, 11] which was also used in clustering compositional data [24] and their anomaly detection [20]. Note that (2.1) occurs in k-medians clustering.

After the introduction, the article proceeds with a section discussing simplicial distances and their transformation into a standard form. The third section focuses on the main result of the paper, utilizing linear programming duality to establish a connection between simplicial Fermat–Weber problems and flow problems. In the subsequent section, we examine the geometric median from Hilbert's simplex geometry and demonstrate its relation to simplicial Fermat–Weber problems. The article concludes with a summary of the findings and their practical relevance.

2.2 Simplicial Distances

In location theory, it is common to consider distances coming from gauges [13, 23]. Gauges are generalizations of norms and they can be introduced by specifying the shape of the unit ball.[1] Formally, given a subset K of a real vector space X, the *gauge* of K is the function $\gamma_K : X \to [0, \infty]$ given by

[1] See [17, Chapter C] for more details.

$$\gamma_K(x) := \inf\{\mu > 0 : x \in \mu K\}.$$

In this paper, our focus will be on gauges that are convex and take finite values. These types of gauges are commonly found when dealing with a convex body that includes the origin within its interior. They are also known as *asymmetric norms* [4] and induce a distance d_K on X defined as $\mathsf{d}_K(x, y) := \gamma_K(y - x)$ and is referred to as a *gauge distance*. We remark that d_K is symmetric if and only if $K = -K$.

By a *simplicial distance* we understand a distance induced by the gauge of a simplex. Concretely, let A be a simplex in $\mathbb{R}^{n-1}$ containing $\mathbf{0}$ in its interior and let $a_1, \ldots, a_n$ be its vertices. Then d_A is a simplicial distance.

Example 2.1 Let us write a formula for the simplicial gauge induced by the triangle A with vertices $(0, 1)$, $(1, 0)$, and $(-1, -1)$. This triangle is also the set of points satisfying the inequalities:

$$x_1 + x_2 \leq 1, \quad -2x_1 + x_2 \leq 1, \quad \text{and} \quad x_1 - 2x_2 \leq 1.$$

Therefore, we can get an algebraic description of the gauge as follows:

$$\begin{aligned}
\gamma_A(x) &= \inf\{\mu > 0 : x \in \mu A\} \\
&= \inf\{\mu > 0 : x_1 + x_2 \leq \mu,\ -2x_1 + x_2 \leq \mu,\ x_1 - 2x_2 \leq \mu\} \\
&= \max(x_1 + x_2,\ -2x_1 + x_2,\ x_1 - 2x_2) \\
&= x_1 + x_2 - 3 \cdot \min(x_1, x_2, 0).
\end{aligned}$$

Accordingly, the simplicial distance d_A has the expression:

$$\mathsf{d}_A(x, y) = (y_1 - x_1) + (y_2 - x_2) - 3 \cdot \min(y_1 - x_1, y_2 - x_2, 0).$$

Many of the simplicial distances are isometric, which can be quantified by the “skewness” of the simplex with respect to the origin. To better understand this, we will use a standard form that incorporates more symmetry and construct an isometry to a distance on $\mathbb{R}^n/\mathbb{R}\mathbf{1}$. To do this, we will consider positive values $\lambda_1, \ldots, \lambda_n$ such that $\sum_i \lambda_i = 1$ and $\sum_i \lambda_i a_i = \mathbf{0}$. This allows us to define a linear transformation $\psi : \mathbb{R}^n/\mathbb{R}\mathbf{1} \to \mathbb{R}^{n-1}$ such that $\psi(e_i) = \lambda_i a_i$. Thus, $\psi^{-1}(A) = \text{conv}\{e_i/\lambda_i : i \in [n]\} =: \Delta(\lambda)$. What is more,

$$\begin{aligned}
\gamma_A(\psi(x)) &= \inf\{\mu > 0 : \psi(x) \in \mu A\} \\
&= \inf\{\mu > 0 : x \in \mu\psi^{-1}(A) = \mu\Delta(\lambda)\} \\
&= \gamma_{\Delta(\lambda)}(x).
\end{aligned}$$

Specifically, ψ is an isometry between $(\mathbb{R}^n/\mathbb{R}\mathbf{1}, \mathsf{d}_{\Delta(\lambda)})$ and $(\mathbb{R}^{n-1}, \mathsf{d}_A)$. It can be inferred that the study of simplicial distances can be simplified to the case of $\mathsf{d}_{\Delta(\lambda)}$ for some λ in the interior of the standard simplex, Δ_n°. Therefore, we can refer to $\mathsf{d}_{\Delta(\lambda)}$ as a *standard simplicial distance*. It is worth noting that the standard form is preferred for its connections to tropical geometry, as demonstrated in [6, 7, 9] and the above observations are also present in [8].

For the standard formulation, we obtain the closed form

$$\gamma_{\triangle(\lambda)}(x) = \sum_i \lambda_i x_i - \left(\sum_i \lambda_i\right) \min_j x_j = \sum_i \lambda_i x_i - \min_j x_j \qquad (2.2)$$

where the last equality is due to the assumption $\sum_i \lambda_i = 1$. To verify this, we can note that $\gamma_{\triangle(\lambda)}$ must the maximum of n linear forms by following similar steps as in Example 2.1. Indeed, we have $\langle \lambda - e_i, e_j/\lambda_j \rangle = 1 - \delta_{ij}/\lambda_j = 1$ for distinct indices i, j, so $\triangle(\lambda)$ is defined by the inequalities $\langle \lambda - e_i, x \rangle \leq 1$. Combining the previous observations, equation (2.2) is immediate.

2.3 Simplicial Fermat–Weber Problems and Their Dual

In this section, our focus is on the problem (2.1) restricted to simplicial distances. We recall that we need two finite sets of points S and D in $\mathbb{R}^n/\mathbb{R}\mathbf{1}$ consisting of sources and destinations. Furthermore, each input point has an associated positive weight. With these factors in mind, the *simplicial Fermat–Weber problem* is the following optimization problem:

$$\text{minimize}_{x \in \mathbb{R}^n/\mathbb{R}\mathbf{1}} \quad \sum_{s \in S} w_s \mathsf{d}_{\triangle(\lambda)}(s, x) + \sum_{d \in D} \omega_d \mathsf{d}_{\triangle(\lambda)}(x, d), \qquad (2.3)$$

where $\lambda \in \mathbb{R}^n_{>0}$ with $\sum_i \lambda_i = 1$. In the following, we will use the notations: $w_S := \sum_{s \in S} w_s$ and $\omega_D := \sum_{d \in D} \omega_d$.

Since $\mathsf{d}_{\triangle(\lambda)}$ is piecewise linear, we can solve (2.3) through linear programming. But to facilitate the construction of a linear programming formulation, we identify $\mathbb{R}^n/\mathbb{R}\mathbf{1}$ with the hyperplane $\mathcal{H}_\lambda := \{x \in \mathbb{R}^n : \sum_i \lambda_i x_i = 0\}$. Accordingly, we have $\mathsf{d}_{\triangle(\lambda)}(x, y) = \max_j (x_j - y_j)$ for every $x, y \in \mathcal{H}_\lambda$.

To formulate a linear program for solving (2.3), we introduce new variables t_s for $s \in S$ and u_d for $d \in D$ which should represent, respectively, the distances to and from a point x. The relation $t_s = \mathsf{d}_{\triangle(\lambda)}(s, x) = \max_j (s_j - x_j)$ has to be relaxed to a set of constraints $s_j - x_j \leq t_s$ for each $j \in [n]$ to be able to integrate it in a linear program.[2] With these observations, the linear programming formulation of (2.3) is:

$$\begin{array}{lll} \text{minimize } \sum_{s \in S} w_s t_s + \sum_{d \in D} \omega_d u_d & & \\ \text{subject to } s_j - x_j \leq t_s\,, & \text{for } s \in S \text{ and } j \in [n] & \\ \qquad\qquad x_j - d_j \leq u_d\,, & \text{for } d \in D \text{ and } j \in [n] & \\ \qquad\qquad \lambda_1 x_1 + \cdots + \lambda_n x_n = 0. & & \end{array} \qquad (2.4)$$

[2] Recall that we identified $\mathbb{R}^n/\mathbb{R}\mathbf{1}$ with $\mathcal{H}_\lambda$, so we assume that S and D are subsets of $\mathcal{H}_\lambda$.

Note that for every optimal solution $(x^\star, t^\star, u^\star)$ of (2.4) we must have $t_s^\star = \max_j(s_j - x_j^\star) = \mathsf{d}_{\triangle(\lambda)}(s, x^\star)$ and $u_d^\star = \max_j(x_j^\star - d_j) = \mathsf{d}_{\triangle(\lambda)}(x^\star, d)$ for each $s \in S$ and $d \in D$. Indeed, if it were not the case, then a small reduction of some variables $t_s^\star$ or $u_d^\star$ would provide another feasible solution which decreases the objective function. Thus, $x^\star$ is a solution of (2.3).

To arrive to a flow problem, we will write its dual. So we introduce the dual variables y_{sj} for the source constraint $s_j - x_j \le t_s$, the variable z_{jd} for the destination constraint $x_j - d_j \le u_d$, and μ for the equality constraint. Hence, the dual formulation is:[3]

$$
\begin{aligned}
\text{maximize } & \sum_{s\in S}\sum_{j\in[n]} s_j y_{sj} + \sum_{d\in D}\sum_{j\in[n]}(-d_j)z_{jd} & & \\
\text{subject to } & \sum_{j\in[n]} y_{sj} = w_s\,, & & \text{for } s \in S \\
& \sum_{j\in[n]} z_{jd} = \omega_d\,, & & \text{for } d \in D \\
& \lambda_j\mu + \sum_{s\in S} y_{sj} - \sum_{d\in D} z_{jd} = 0\,, & & \text{for } j \in [n] \\
& y_{sj} \ge 0\,, & & \text{for } s \in S \text{ and } j \in [n] \\
& z_{jd} \ge 0\,, & & \text{for } d \in D \text{ and } j \in [n].
\end{aligned}
\tag{2.5}
$$

Summing the equality constraints in (2.5), we get $\left(\sum_j \lambda_j\right)\cdot \mu + w_S - \omega_D = 0$. Since we assumed $\sum_j \lambda_j = 1$, we infer $\mu = \omega_D - w_S$ for every feasible solution so we will make this change in (2.5). Additionally, we multiply the objective function with -1 to obtain an equivalent minimization problem:

$$
\begin{aligned}
\text{minimize } & \sum_{s\in S}\sum_{j\in[n]} (-s_j) y_{sj} + \sum_{d\in D}\sum_{j\in[n]} d_j z_{jd} & & \\
\text{subject to } & \sum_{j\in[n]} y_{sj} = w_s\,, & & \text{for } s \in S \\
& \sum_{j\in[n]} z_{jd} = \omega_d\,, & & \text{for } d \in D \\
& \sum_{s\in S} y_{sj} - \sum_{d\in D} z_{jd} = \lambda_j(w_S - \omega_D)\,, & & \text{for } j \in [n] \\
& y_{sj} \ge 0\,, & & \text{for } s \in S \text{ and } j \in [n] \\
& z_{jd} \ge 0\,, & & \text{for } d \in D \text{ and } j \in [n].
\end{aligned}
\tag{2.6}
$$

In this form, one can identify a minimum cost flow problem [1]. The nodes of the corresponding digraph are split in three classes: S, $[n]$, and D. There is an arc from every node of S to every node of $[n]$ and from every node of $[n]$ to every node of D. Each source $s \in S$ has supply w_s and each destination $d \in D$ has demand ω_d. The intermediary nodes $j \in [n]$ have an excess proportional to λ_j and they can be

---

[3] One should consult [10, §5.1.3] for details on obtaining dual linear programs.

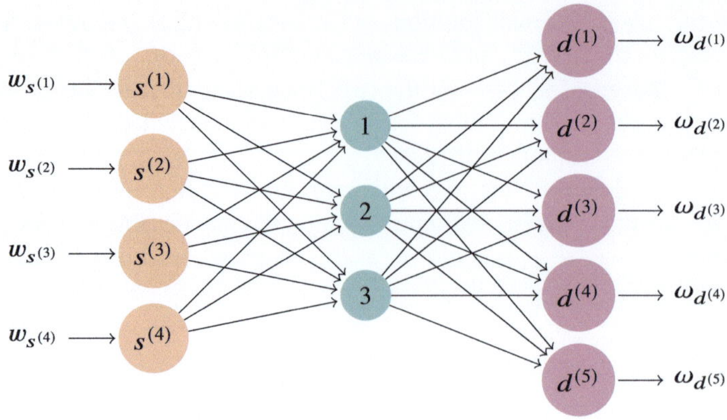

Fig. 2.1 Digraph corresponding to a simplicial Fermat–Weber problem in two dimensions with four sources and five destinations. The nodes from S are orange, the nodes from $[n]$ are green, while the nodes from D are purple. Supply and demand for nodes in S and D are depicted as entering and leaving, respectively, while the excess at green nodes is omitted

labeled as supply or demand nodes depending on the difference between the weights on sources and destinations. The costs of the arcs are given by the points in S and D. The corresponding digraph is depicted in Fig. 2.1.

Minimum cost flow problems have been extensively studied and there are several efficient algorithms that have been developed for them. One popular choice in practice is the network simplex method, as described in Chap. 11 of [1].

Remark 2.1 When $S = \emptyset$ or $D = \emptyset$, the optimization problem (2.6) is a transportation problem [10, Sect. 8.1]. This was already noted in [6] for a particular simplicial distance, while a more general situation is described in [8, Sect. 7.1]. This observation was exploited for efficient computations of the corresponding geometric median. A method to compute the set of all Fermat–Weber points is detailed in [8, Sect. 7.1.4] and extends straightforwardly to our more general setting.

The presentation of this section summarizes to the following statement.

> The simplicial Fermat–Weber problem admits a linear programming formulation which is dual to a minimum cost flow problem.

2.4 Geometric Medians with Hilbert's Simplex Distance

In 1895, Hilbert [16] generalized the distance of the Cayley–Klein model of hyperbolic geometry which gave rise to a new type of metric geometry [26]. Essentially, a metric is defined on a bounded convex (relatively) open domain Ω, called the

Hilbert's projective metric on Ω, by taking the logarithm of the cross-ratio of the two input points and the two points that form the intersection of $\partial\Omega$ with the line through the input points. The Cayley–Klein model is simply the Hilbert geometry of the unit circle. In this paper, we will be interested in the case when $\Omega = \Delta_n^\circ$—the interior of the standard simplex. The corresponding geometry will be called the *Hilbert simplex geometry*, following Nielsen and Sun [24]. Concretely, the distance function is $\mathsf{d}_{\mathrm{HS}} : \Delta_n^\circ \times \Delta_n^\circ \to \mathbb{R}$ given by

$$\mathsf{d}_{\mathrm{HS}}(x, y) = \max_i \log(x_i/y_i) - \min_i \log(x_i/y_i).$$

It turns out that this is isometric to a gauge distance [11]. To see this, one has to map Δ_n° to $\mathcal{H}_1$ through the centered logratio transformation $\mathrm{clr} : \Delta_n^\circ \to \mathcal{H}_1$ [27, §4.3]:

$$\mathrm{clr}(x) := (\log x_1, \ldots, \log x_n) - \log(x_1 \ldots x_n)^{1/n}\mathbf{1}.$$

Under this transformation, the Hilbert's simplex distance is isometric to

$$\mathsf{d}_{\mathrm{trop}}(x, y) := \max_i(x_i - y_i) - \min_i(x_i - y_i). \tag{2.7}$$

In tropical geometry, this was firstly introduced in [5] and it is dubbed the *tropical distance*. Note that the above formula is well-defined on $\mathbb{R}^n/\mathbb{R}\mathbf{1}$ and it is a distance induced by the *tropical norm* $\|x\|_{\mathrm{trop}} := \max_i x_i - \min_i x_i$.

With these considerations, a geometric median under the Hilbert's simplex distance corresponds (via clr) to the solution of the Fermat–Weber problem

$$\mathrm{minimize}_{x\in\mathbb{R}^n/\mathbb{R}\mathbf{1}} \quad \sum_{v\in V} w_v \mathsf{d}_{\mathrm{trop}}(v, x) \tag{2.8}$$

where $V \subset \mathbb{R}^n/\mathbb{R}\mathbf{1}$ is the input set. Since $\mathsf{d}_{\mathrm{trop}}$ is symmetric, we do not distinguish between sources and destinations.

Looking at (2.2) and (2.7), we notice that $\mathsf{d}_{\mathrm{trop}}(x, y) = \mathsf{d}_{\triangle(\lambda)}(x, y) + \mathsf{d}_{\triangle(\lambda)}(y, x)$ for every $\lambda \in \Delta_n^\circ$. Consequently, problem (2.8) is equivalent to

$$\mathrm{minimize}_{x\in\mathbb{R}^n/\mathbb{R}\mathbf{1}} \quad \sum_{v\in V} w_v \mathsf{d}_{\triangle(\lambda)}(v, x) + \sum_{v\in V} w_v \mathsf{d}_{\triangle(\lambda)}(x, v), \tag{2.9}$$

which is a simplicial Fermat–Weber problem where sources and destinations coincide. Such a situation is called a *round-trip Weber problem* in [12].

The Fermat–Weber problem (2.8) was firstly studied in [21] where a direct linear program with $m + n$ variables and $mn(n-1)$ constraints were used for computations, where $m = |V|$, while formulation (2.9) reduces the number of constraints to $2mn$ by adding m extra variables. The computational studies from [6, Sect. 5.3] show that the former approach is infeasible for large datasets in contrast to simplicial Fermat–Weber problems.

Remark 2.2 The main idea was to factorize (in the max-plus algebra) the piecewise linear (PL) function $\|x\|_{\text{trop}}$ as $\gamma_{\triangle(\lambda)}(x) + \gamma_{\triangle(\lambda)}(-x)$, going from one PL function with $n(n-1)$ regions of linearity to two PL functions with n regions of linearity. The split allowed to decrease considerably the number of constraints in the LP formulation while adding a few extra variables. A similar idea can be considered for more general PL functions as long as we can factorize them in the max-plus algebra. This is related to decompositions of polytopes [14].

All in all, the result of this section can be summarized as follows.

> Computing geometric medians in Hilbert's simplex geometry as solutions to round-trip simplicial Fermat–Weber problems is an efficient approach.

2.5 Conclusion

Our results indicate that the simplicial Fermat–Weber problem can be effectively formulated using linear programming, and this formulation has a dual correspondence to minimum cost flows. These generalize the results from [6, 8]. By expressing geometric medians under Hilbert's simplex distance as solutions to round-trip simplicial Fermat–Weber problems, efficient computational methods can be employed for their determination. These findings provide valuable insights into the mathematical structure of Hilbert's simplex geometry and offer practical computational approaches for solving basic problems in this setting.

With the above results, one can compute efficiently tropical Fermat–Weber points as defined in [21]. However, for applications to the consensus problem in phylogenetics, [7] mentions that a regularized form is necessary. If ϕ is the objective function from (2.8), the previously mentioned work suggests considering the objective $\phi + \mu \mathsf{d}_{\triangle(\mathbf{1})}(\cdot, u)$ instead, where u is a point depending on the sample V and μ is a positive number smaller than every weight. Note that the regularized objective function is also of the type (2.3), so it can be efficiently computed.

Acknowledgements I express my gratitude to the audience for their insightful questions regarding my presentation at this conference. Special thanks to Andreas Ott for his inquiries about the differences between symmetric and asymmetric tropical distances. Further reflections led me to the generalizations from this work. Additionally, I extend my appreciation to "Facets of Complexity" (GRK 2434, Project-ID 385256563) for their financial support.

References

1. Ahuja, R.K., Magnanti, T.L., Orlin, J.B.: Network Flows. Algorithms, and Applications. Prentice Hall, Englewood Cliffs, NJ, Theory (1993)
2. Amini, O., Manjunath, M.: Riemann-Roch for sub-lattices of the root lattice A_n. Electron. J. Combin. **17**(1), 124,50 (2010)
3. Boissonnat, J.-D., Sharir, M., Tagansky, B., Yvinec, M.: Voronoi diagrams in higher dimensions under certain polyhedral distance functions. Discrete Comput. Geom. **19**(4), 485–519 (1998). https://doi.org/10.1007/PL00009366
4. Cobzaş, Ş: Functional analysis in asymmetric normed spaces. Front. Math. Basel: Birkhäuser (2013). https://doi.org/10.1007/978-3-0348-0478-3
5. Cohen, G., Gaubert, S., Quadrat, J.-P.: Duality and separation theorems in idempotent semi-modules. Linear Algebra Appl. **379**, 395–422 (2004). ISSN 0024-3795. https://doi.org/10.1016/j.laa.2003.08.010
6. Comăneci, A., Joswig, M.: Tropical medians by transportation. Math. Program. **205**(1-2 (A)), 813–839 (2024). https://doi.org/10.1007/s10107-023-01996-8
7. Comăneci, A.: Tropical convexity in location problems, 2023. Math. Methods Oper. Res. **100**, 509–534 (2024). https://doi.org/10.1007/s00186-024-00869-w
8. Comăneci, A.: Tropical consensus trees. PhD thesis, TU Berlin (2024)
9. Comăneci, A., Joswig, M.: Asymmetric tropical distances and power diagrams. Algebraic Combinatorics **6**(5), 1211–1233 (2023). https://doi.org/10.5802/alco.306
10. Dantzig, G.B., Thapa, M.N.: Linear Programming. 1: Introduction. Incl. 1 CD-ROM. Springer, New York, NY (1997). https://doi.org/10.1007/b97672
11. de la Harpe, P.: On Hilbert's metric for simplices. In: Geometric Group Theory, vol. 1. Proceedings of the Symposium Held at the Sussex University, Brighton (UK), July 14–19, 1991, pp. 97–119. Cambridge University Press, Cambridge (1993)
12. Drezner, T., Drezner, Z.: Asymmetric distance location model. INFOR: Inf. Syst. Oper. Res. **59**(1), 102–110 (2021). ISSN 0315-5986. https://doi.org/10.1080/03155986.2020.1789386
13. Durier, R., Michelot, C.: Geometrical properties of the Fermat-Weber problem. Eur. J. Oper. Res. **20**, 332–343 (1985). ISSN 0377-2217. https://doi.org/10.1016/0377-2217(85)90006-2
14. Gao, S., Lauder, A.G.B.: Decomposition of polytopes and polynomials. Discrete Comput. Geom. **26**(1), 89–104 (2001). https://doi.org/10.1007/s00454-001-0024-0
15. Gromov, M.: Metric structures for Riemannian and non-Riemannian spaces. Transl. from the French by Sean Michael Bates, vol. 152 of Prog. Math. Birkhäuser, Boston, MA (1999)
16. Hilbert, D.: Ueber die gerade Linie als kürzestes Verbindung zweier Punkte. Math. Ann. **46**, 91–96 (1895). https://doi.org/10.1007/BF02096204
17. Hiriart-Urruty, J.-B., Lemaréchal, C.: Fundamentals of Convex Analysis. Springer, Berlin (2001)
18. Hodgson, M.J., Wong, R.T., Honsaker, J.: The P-centroid problem on an inclined plane. Oper. Res. **35**, 221–233 (1987). https://doi.org/10.1287/opre.35.2.221
19. Krarup, J., Vajda, S.: On Torricelli's geometrical solution to a problem of Fermat. IMA J. Math. Appl. Bus. Ind. **8**(3), 215–224 (1997)
20. Legaria, U., Mota, S., Martinez, S., Cobá, A., Chable, A., Neme, A.: Anomaly detection in the probability simplex under different geometries. Inf. Geom. **6**(2), 385–412 (2023). ISSN 2511-2481. https://doi.org/10.1007/s41884-023-00107-y
21. Lin, B., Yoshida, R.: Tropical Fermat–Weber points. SIAM J. Discrete Math. **32**(2), 1229–1245 (2018). ISSN 0895-4801. https://doi.org/10.1137/16M1071122
22. Manjunath, M.: The Laplacian lattice of a graph under a simplicial distance function. Eur. J. Combin. **34**(6), 1051–1070 (2013). https://doi.org/10.1016/j.ejc.2013.01.010
23. Nickel, S., Puerto, J.: Location Theory. A Unified Approach. Springer, Berlin (2005). https://doi.org/10.1007/3-540-27640-8
24. Nielsen, F., Sun, K.: Clustering in Hilbert's projective geometry: the case studies of the probability simplex and the elliptope of correlation matrices. In: Geometric Structures of Information. Proceedings of the Conference on Geometric Science of Information, GSI 2017, pp. 297–331. Springer, Cham (2019). https://doi.org/10.1007/978-3-030-02520-5

25. Page, R., Yoshida, R., Zhang, L.: Tropical principal component analysis on the space of phylogenetic trees. Bioinformatics **36**(17), 4590–4598 (2020). ISSN 1367-4803. https://doi.org/10.1093/bioinformatics/btaa564
26. Papadopoulos, A., Troyanov, M. (eds.): Handbook of Hilbert Geometry, vol. 22 of IRMA Lect. Math. Theor. Phys. European Mathematical Society (EMS), Zürich (2014). https://doi.org/10.4171/147
27. Pawlowsky-Glahn, V., Egozcue, J.J., Tolosana-Delgado, R.: Modeling and Analysis of Compositional Data. Statistics in Practice. John Wiley & Sons Ltd, Chichester (2015)
28. Plastria, F.: A destination optimality in asymmetric distance Fermat-Weber problems. Ann. Oper. Res. **40**, 355–369 (1992). https://doi.org/10.1007/BF02060487
29. Plastria, F.: Fully geometric solutions to some planar center location problems. Stud. Locat. Anal. **7**, 171–183 (1994)
30. Plastria, F.: Asymmetric distances, semidirected networks and majority in Fermat-Weber problems. Ann. Oper. Res. **167**, 121–155 (2009). https://doi.org/10.1007/s10479-008-0351-0
31. Tokuyama, T., Nakano, J.: Efficient algorithms for the Hitchcock transportation problem. SIAM J. Comput. **24**(3), 563–578 (1995). ISSN 0097-5397. https://doi.org/10.1137/S0097539792236729
32. Weber, A.: Über Den Standort Der Industrien. J. C. B. Mohr, Tübingen (1909)
33. Witzgall, C.: Optimal location of a central facility: mathematical models and concepts, 1964. NBS Technical Report 8388

Chapter 3
Forecasting Lightning Probabilities Derived from the Lightning Potential Index Using Neural Networks

Manuel Baumgartner, Guido Schröder, and Cristina Primo

Abstract Forecasting the occurrence of thunderstorms is a well-known challenge in weather prediction. Since a thunderstorm is by definition accompanied by at least one lightning, the goal is to forecast the occurrence of at least one lightning within a pre-defined area and a pre-defined time interval. This study reports on the development of a postprocessing method that is based on neural networks to translate the information from the lightning potential index (LPI) derived from the ICON-EU ensemble prediction system into a calibrated lightning probability. The ground truth from which the postprocessing method learns this translation function is given by lightning observations from the LINET network. The setup of the postprocessing method is presented together with results regarding the calibration and sharpness of the final product and the diurnal cycle of the predictions.

Keywords Machine learning · Neural networks · Lightning probability · Thunderstorm · Statistical postprocessing

3.1 Introduction

Forecasting the occurrence of thunderstorms is a well-known challenge in weather prediction. Since any thunderstorm is by definition accompanied by at least one lightning, any lightning forecast may also be seen as a forecast for thunderstorms.

Given the possible significant impact of thunderstorms on many areas of human activities it is not surprising that numerous groups have conducted research in its detection and forecasting. Moreover, the recent popularity of neural networks (NN)

M. Baumgartner (✉) · G. Schröder · C. Primo
Deutscher Wetterdienst, Offenbach, Germany
e-mail: Manuel.Baumgartner@dwd.de

G. Schröder
e-mail: Guido.Schroeder@dwd.de

C. Primo
e-mail: Cristina.Primo-Ramos@dwd.de

A. Ott et al. (eds.), *Applications of Mathematics in Sciences, Engineering, and Economics*,
Springer Proceedings in Mathematics & Statistics 515,
https://doi.org/10.1007/978-3-032-01279-1_3

(see, e.g., [1] for a general introduction to NN) has triggered research that tries to improve the lightning forecasts or detection by employing NN. As examples, the authors in [2] aim to detect and forecast lightning occurrence from satellite images for short leadtimes. While this application considers the nowcasting of lightning occurrence, the goal in this study is to derive lightning probabilities from numerical weather prediction (NWP) models. Similar approaches were reported in [3–5] where the former two also employ neural networks for their prediction and the latter a more classical machine-learning method. All these methods may be summarized under the term "(statistical) postprocessing models" which, loosely speaking, encompasses all systems that take the output of NWP (ensemble) models and try to correct systematic biases and errors as well as to construct new forecast variables that were not part of the NWP output such as lightning probabilities (see, e.g., [6, 7] for a general introduction to statistical postprocessing). The aforementioned studies follow a similar approach in that the authors try to learn their postprocessing method by using suitable variables from the NWP-model as predictors that are known (or suspected) to yield guidance on the occurrence of thunderstorms. This is a rather classical approach and may be summarized as an "ingredients-based method" [8] that is also commonly used by human forecasters in their thunderstorm forecasts.

The main difference of the approach in the present study to the approaches of the aforementioned studies is that the present lightning forecast is subdivided into one part that runs directly within the NWP-model and the second part runs as the postprocessing model. The first part consists of the computation of the so-called subgrid scale lightning potential index (LPI, see Sect. 3.2.1 for further details and references) that is implemented as a diagnostic variable directly within the NWP-model ICON that runs operationally at the German Meteorological Service (Deutscher Wetterdienst; DWD). The second part is the main subject of the present study, i.e. the postprocessing using NN to translate the LPI forecast from the NWP model into a lightning probability. In order to learn this translation function lightning observations from the LINET-network are used as ground-truth. Using observed lightnings as ground-truth enables the postprocessing-model to additionally correct systematic errors that are inherent in the NWP model forecasts.

It would be desirable to construct a (statistical) postprocessing model that predicts the probability of occurrence of at least one lightning given a point on the Earth's surface and a time instant, i.e. a point probability. However, since it may be considered as random that a lightning will occur at precisely the given point and time, it is expected that such a model can only be slightly better compared to forecasting the climatological probability, but that probability is not considered useful. In order to define a more predictable target quantity, the precision requirements of the desired forecast should be relaxed. Instead of making predictions for a single point on the surface, a circular area around that point can be considered. Similarly, instead of making predictions for a single time instant, a larger time interval can be considered. As a rule of thumb, the larger the spatial radius and the longer the time interval, the more predictable the occurrence of lightning. However, taking the radius too large (e.g. 1000 km) and/or the time interval too long (e.g. one week) results in a well predictable event but a less useful product. Therefore, the challenge is to define a radius and a time interval that yields both a predictable and useful event.

In the present study, a radius of 40 km is considered together with a time interval of 3 h. As a result of these choices, the target quantity is the probability that at least one lightning occurs within 40 km around the given location within 3 h.

The target definition defined above follows the philosophy of neighborhood methods where the radius can be interpreted as a spatial searching length for variables whose exact spatial location is difficult to predict. This way, the predictability of the variable is increased, see [9] for a discussion.

In the present study, a description of the data is provided in Sect. 3.2 while Sect. 3.3 is devoted to the postprocessing method. Results of applying the postprocessing method are given in Sects. 3.4 and 3.5 concludes the study with a short summary and an envisioned application of the lightning probabilities.

3.2 Numerical Model and Observation Data

The methodology to predict lightning probabilities is based on two sources of data: the numerical weather prediction model ICON and the observation data LINET.

3.2.1 Numerical Model and Lightning Potential Index

DWD operationally runs a global ensemble prediction system based on the numerical weather prediction model ICON [10, 11] with a refined grid over Europe, called ICON-EU-EPS. The distinguishing feature of an ensemble prediction system is that not only a single deterministic forecast is produced but a set of deterministic forecasts ("ensemble members") obtained by introducing perturbations in the initial state and some physical processes within the model. This technique has been used for decades in the meteorological field to estimate the forecast uncertainty from different sources of uncertainties; see, e.g. [12–15] for more details. The ICON-EU-EPS consists of 40 members with a horizontal resolution of about 13 km in the European domain (23.5° W–62.5° E and 29.5° N–70.5° N). Instead of a regular latitude-longitude grid the ICON model employs a triangular grid that allows for a nearly constant grid spacing. For this work, the daily four main runs of the ICON-EU-EPS are considered for the postprocessing. These runs are initialized each day at 00, 06, 12, 18 UTC.

For training the NN, the two static variables land-sea fraction and geographical latitude are used. The land-sea-fraction is a variable with values between zero (only sea) and one (only land) that indicates the subgrid-distribution of land and sea within the model grid cells. The only non-static variable included in the training is the subgrid scale lightning potential index (LPI). The LPI was developed by [16] for convection permitting models. It is a vertical integral over the charge separation zone within a convective cloud where the updraft as well as the proportion of liquid water droplets, snow, ice and graupel are combined. However, in ICON-EU-EPS a subgrid scale version of the LPI is implemented [17] since (deep) convection is not resolved. Similarly to [18], it is based on the subgrid scale vertical updraft profile in regions where deep convection is triggered. This profile is taken from the convection parameterisation (see [19] for details on this parameterisation). In essence, the LPI takes into account all informations from the model such as the amount of cloud droplets and ice particles within a simulated convective cloud that are involved in the cloud charge separation process, eventually leading to a lightning. It is shown in [17] that this index is indeed correlated with the observed occurrence of lightning and therefore is qualified as a model guidance for lightning. The drawback of the LPI from the model is that its values are not easily interpretable since they are not lightning probabilities and additionally show biases, see the discussion in Sect. 3.4.2.

From the current operational configuration of ICON-EU-EPS, the LPI is available hourly up to leadtime +48 h and 3-hourly afterwards up to leadtime +72 h. It is preprocessed before being fed into the neural network, see Sect. 3.2.3 for details.

3.2.2 Observation Data

Observations of lightnings over most of Europe are provided by the lightning observation network LINET [20]. Although the LINET network provides additional information about the observed lightnings, only the horizontal position of the lightning and its time of observation are taken into account for further data processing. The observed lightnings are considered as the ground-truth for the final postprocessing model. Note, the spatial observation domain of the LINET network (see the blue box in Fig. 3.3b) is given by 12° W–32° E and 33° N–66° N, hence it is smaller than the domain of the ICON-EU-EPS model.

3.2.3 Data Preparation

Data preparation refers to the computations applied to the input data that are required for the postprocessing method. There are four main aspects of the data preparation that are outlined below:

1. gridding of the observational data,
2. temporal aggregation of the data,
3. spatial aggregation of the data,
4. computation of further predictors.

The objective of the temporal and spatial aggregation is to bring the data into a shape that is compatible with the target quantity described in Sect. 3.1. For the LPI, temporal and spatial aggregation is applied to each member. It is emphasized that, after gridding of the observed lightnings, the same temporal and spatial aggregation is employed to the observations as for the LPI. The static variables (land-sea-fraction and latitude) are not aggregated. To fix the notation, let $(q_{p,t}^{(m)})_{p\in P,\, t\in T_i,\, m\in M}$ denote the LPI output from the NWP model where P denotes the set of all gridpoint indices, $T_i = \{0,\ 1,\ \ldots,\ 48,\ 51,\ 54,\ldots,\ 72\}$ the set of all temporal indices of the model output (leadtime in hours) and $M = \{1,\ 2,\ldots,\ 40\}$ the set of all members. Similarly, let $(o_{p,t})_{p\in P,\, t\in T_o}$ denote the gridded lightning data (see the following Sect. 3.2.3.1) and $T_o = \{1,\ 2,\ldots,\ 72\}$ the set of all hours from the model initialization time up to +72 h. For further reference, let $T = \{3,\ 4,\ \ldots,\ 48,\ 51,\ 54,\ldots,\ 72\}$ denote the final set of temporal indices.

3.2.3.1 Gridding of the Observed Lightning

The raw lightning data consists of the horizontal positions and the times of the observed lightnings. The first step is to assign each observed lightning to the nearest gridpoint of the ICON grid. Therefore, the result is a gridded quantity where the value at each gridpoint is the number of lightnings assigned to that gridpoint. In the second step all lightnings per hour are summed up and the value is assigned to the end of the respective hour (e.g. summing up all lightnings between 5 UTC and 6 UTC and assigning the result to 6 UTC). The result of these two steps is the gridded quantity $(o_{p,t})_{p\in P,\, t\in T_o}$ that has the number of lightnings that were observed within one hour in the vicinity of each gridpoint from the model initialization time up to +72 h.

3.2.3.2 Temporal Aggregation

Temporal aggregation is the computation of the maximum value of a gridded dataset within a 3 h time interval for each gridpoint and each member. As an example, consider the target time 15 UTC. For each gridpoint the maximum of the values from 13, 14, 15 UTC is computed and the result is assigned as the new value at 15 UTC. Given that the first three hours do not provide a self-consistent 3 h time interval, these are not taken into account. Consequently, the temporal aggregation yields an output with the leadtimes +03 h, +04 h, …, +48 h, +51 h, +54 h, …, +72 h. More formally, the gridded lightning observations $(o_{p,t})_{p\in P,\, t\in T_o}$ are transformed according to $\hat{o}_{p,t} = \max\left\{o_{p,t-2},\ o_{p,t-1},\ o_{p,t}\right\}$ for $3 \le t \le 72$ and all $p \in P$ and then removing the temporal indices that are not in the set T. The LPI forecast $(q_{p,t}^{(m)})_{p\in P,\, t\in T_i,\, m\in M}$ is transformed according to $\hat{q}_{p,t}^{(m)} = \max\left\{q_{p,t-2}^{(m)},\ q_{p,t-1}^{(m)},\ q_{p,t}^{(m)}\right\}$ for $3 \le t \le 48$ and $\hat{q}_{p,t}^{(m)} = q_{p,t}^{(m)}$ for $48 < t \le 72$ and all $p \in P,\ m \in M$. The results are the gridded datasets $(\hat{o}_{p,t})_{p\in P,\, t\in T}$ and $(\hat{q}_{p,t}^{(m)})_{p\in P,\, t\in T,\, m\in M}$ on the same spatial and temporal grid given by the index sets P and T.

3.2.3.3 Spatial Aggregation

Spatial aggregation is done individually for each gridpoint and member on the temporally aggregated data $(\hat{o}_{p,t})_{p\in P,\, t\in T}$ and $(\hat{q}^{(m)}_{p,t})_{p\in P,\, t\in T,\, m\in M}$. For a given gridpoint $p \in P$ let $N(p) \subset P$ denote the set of all gridpoints that lie within a circle with radius 40 km around p. The spatially aggregated lightning observations are computed by $O_{p,t} = \max\left\{\hat{o}_{n,t} \mid n \in N(p)\right\}$ for all $t \in T$ and the spatially aggregated LPI is computed individually for each member $m \in M$ by $Q^{(m)}_{p,t} = \max\left\{\hat{q}_{n,t} \mid n \in N(p)\right\}$ for all $t \in T$. As a result, the temporally and spatially gridded datasets $(O_{p,t})_{p\in P,\, t\in T}$ and $(Q^{(m)}_{p,t})_{p\in P,\, t\in T,\, m\in M}$ are obtained. Note, the spatial and temporal aggregations as described in Sect. 3.2.3.2 are done through the computation of a maximum, so the aggregation order (first time or space) is not relevant. Attention should be paid if a different function is used (e.g. quantile).

Note further, the choice of computing the maximum of the values in the temporal and spatial aggregation is the reason why the target variable as described in Sect. 3.1 is the probability of the occurrence of *at least* one lightning *anywhere within* the given temporal and spatial domain.

3.2.3.4 Computation of Further Predictors

After the temporal and spatial aggregation of the LPI data for each ensemble member, the final LPI-related predictors are computed from $(Q^{(m)}_{p,t})_{p\in P,\, t\in T,\, m\in M}$. Instead of using each ensemble member separately, summary-statistics such as the mean and standard deviation are computed per gridpoint as it is common in other ensemble postprocessing techniques (such as EMOS and related techniques, e.g. [21]) together with quantiles. The complete list of predictors for all $p \in P$ and $t \in T$ is:

- Ensemble mean of the aggregated LPI

$$\overline{Q}_{p,t} = \frac{1}{|M|} \sum_{m\in M} Q^{(m)}_{p,t},$$

- Ensemble standard deviation of the aggregated LPI

$$Q^{\mathrm{sd}}_{p,t} = \sqrt{\frac{1}{|M|-1} \sum_{m\in M} \left(Q^{(m)}_{p,t} - \overline{Q}_{p,t}\right)^2},$$

- Ensemble quantiles (1 %, 10 %, 25 %, 50 %, 75 %, 90 %, 99 %) of the aggregated LPI data $(Q^{(m)}_{p,t})_{p\in P,\, t\in T,\, m\in M}$ over m,
- Land-sea-fraction,
- Latitude.

It should be noted that, apart from the static predictors land-sea-fraction and latitude, only the LPI-data is used by the postprocessing method. Consequently, a lightning probability with values that are significantly larger than zero can only occur if at least one of the LPI summary statistics is larger than zero, i.e. if deep convection has been triggered in the numerical model for at least one ensemble member.

Finally, the values of the target $(O_{p,t})_{p \in P,\, t \in T}$ in each gridpoint are binarized, i.e. set to one if at least one lightning occurred and zero otherwise. This step is an allowed simplification of the data since the goal is to forecast whether *at least one* lightning occurred and not the number of lightnings.

3.3 Postprocessing Method

The postprocessing method that is proposed in this study belongs to the class of "global methods", i.e. the same method is applied to all gridpoints. This is in contrast to various classical postprocessing methods where individual predictive equations are produced for each forecast location (ususally a location where synoptical observations are available), see, e.g., [22–24]. Moreover, the goal is to directly predict the desired lightning probabilities which is different from other postprocessing techniques where the parameters of an a-priori selected probability distribution are estimated and the forecast probabilities are subsequently derived from that distribution (e.g., [6, 7, 21, 22]).

The general steps to produce a forecast are described in Sect. 3.3.1 whereas the architecture and the training of the neural networks are described in Sects. 3.3.2 and 3.3.3. The technical implementation is based on Python 3 and the neural network infrastructure is taken from the library "scikit-learn" (see https://scikit-learn.org), i.e. the class "MLPClassifier" that allows to learn a NN for classification.

3.3.1 Forecasting Process

A schematic of the forecast process is shown in Fig. 3.1. Leadtimes are trained separately but all gridpoints are trained together. For a given leadtime and a given gridpoint, the processed data are collected and afterwards the individual lightning probability forecasts of five different NN are computed. Each of these NN (blue boxes in Fig. 3.1) have the same architecture (as described in the next Sect. 3.3.2) and predict the same target, i.e. the desired lightning probability as described in Sect. 3.1. This mini-ensemble of five slightly different NN is employed to stabilize the forecasts, mainly in case the values of the predictors are within an undersampled region of the training data. All NN of the mini-ensemble predict the same target quantity and the final forecast is constructed by computing the mean of the five individual forecasts. This is a common strategy in machine learning (known as a comitee-forecast, see [1, Chap. 14]) and also applied in other postprocessing studies that use NN such as [22, 30].

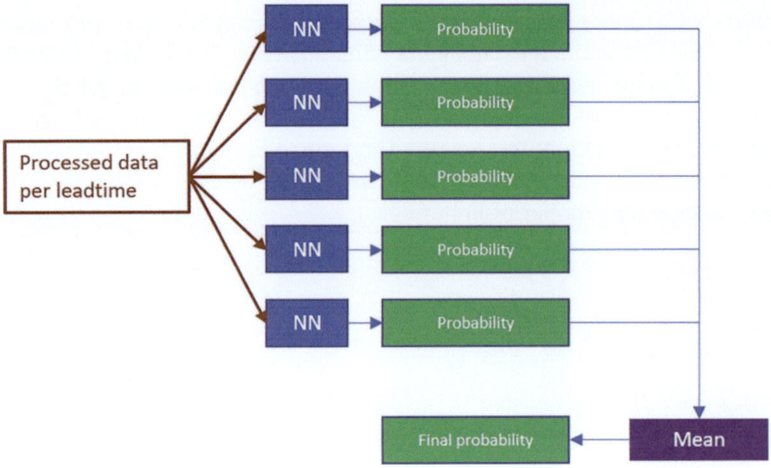

Fig. 3.1 Schematic of the forecasting process. Taking the processed data per leadtime, the mini-ensemble of neural networks (blue boxes named "NN") is applied to produce five individual forecast probabilities. The final forecast probability is then computed as the mean of the individual forecasts of the mini-ensemble

It is emphasized that a mini-ensemble is trained for each of the 54 leadtimes separately, so in total $54 \cdot 5 = 270$ networks are trained.

Currently the training of the NN is exclusively done with the data from the ICON-EU-EPS run at 00 UTC. In order to make forecasts for the model runs starting at 06, 12, 18 UTC the NN that were trained on the 00 UTC data are reused. Specifically, the time of the day is used to select the corresponding NN. As an example, consider the model started at 12 UTC. To make a forecast for 18 UTC, i.e. 12 UTC+6 h, the available NN for leadtime +18 h is used. This strategy is also applied for forecasts for the second and third day.

The reason for this strategy is that the raw numerical model has a bias in the diurnal cycle that is apparently stronger than the leadtime dependent bias (see Fig. 3.8 in Sect. 3.4.4).

3.3.2 *Architecture of the Neural Networks*

The architecture of a single NN (depicted as a blue box in Fig. 3.1) is shown in Fig. 3.2. The fully connected NN consists of an input layer taking the 11 predictors (see Sect. 3.2.3.4), two hidden layers with 30 and 50 neurons and an output layer with two neurons. The activation function for the hidden layers is the rectified linear unit (ReLu) and the output neurons are equipped with the softmax function [1, Chap. 4] to ensure a probabilistic output. In essence, this configuration of a NN is typical for a classification problem and in the current application the network tries to separate

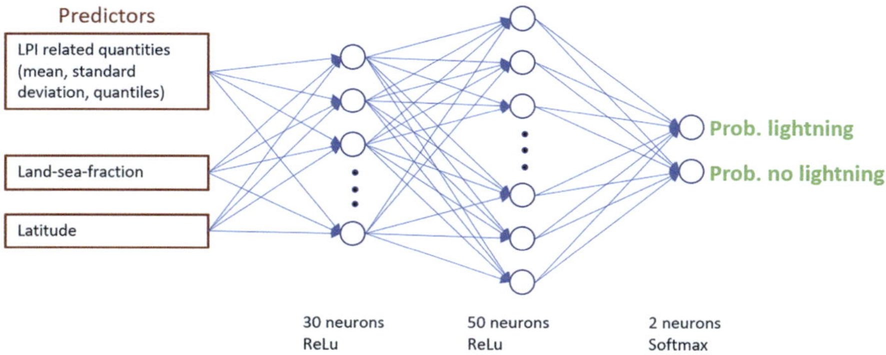

Fig. 3.2 Schematic of the neural network architecture that is the building block of the postprocessing method. The NN corresponds to a classification network trying to separate the classes "no lightning occurred" and "at least one lightning occurred". The two output neurons provide the probabilities for each class

the classes "no lightning occurred" and "at least one lightning occurred". More precisely, for each $p \in P$ and $t \in T$ the binarized targets $(O_{p,t})_{p \in P,\, t \in T}$ are first one-hot encoded into a vector $y_{p,t} \in \{(0, 1),\ (1, 0)\} \subset \mathbb{R}^2$ where a one in the first entry encodes the case of no lightning and a one in the second entry the case of an observed lightning. By virtue of the softmax function, the output of the NN is a vector $(1 - z, z) \in \mathbb{R}^2$ with $z \in [0, 1]$ and is interpreted as the probability for each class. Note, the value z corresponds to the desired lightning probability.

An obvious question is whether it is essential to use neural networks at all for the present forecasting problem. Initial experiments using only the ensemble mean and standard deviation as predictors and a classical logistic regression (e.g., [1, Chap. 4]) instead of neural networks revealed, that the logistic regression was not able to provide a meaningful discrimination between the two classes. A visual data inspection of the ensemble mean and standard deviation revealed, that the predictor data was indeed not linearly separable which is a requirement for the successful usage of a linear method as the logistic regression is. Therefore, it was decided to proceed with methods that are able to create nonlinear decision boundaries such as neural networks.

It appears as plausible that the explicit spatial aggregation of the LPI data described in Sect. 3.2.3.3 could be omitted by equipping the NN with some convolutional and max-pooling layers. However, as the data is given on a triangular grid and the number of neighbors is not the same for each gridpoint, this approach requires a highly non-trivial data handling and was therefore not taken.

3.3.3 Training of the Neural Networks

Training of a single NN is based on a standard training strategy for classification networks, i.e. the network parameters are optimized based on the cross-entropy error function [1, Chap. 4] using the Adam stochastic gradient optimizer [27] with a constant learning rate of 0.001. For a fixed leadtime $t \in T$, the whole training dataset consists of data pairs $\{(x_i, y_i)\}_{i \in I}$ where $x_i \in \mathbb{R}^{11}$ is a vector collecting the standardized predictors (see Sect. 3.2.3.4) and $y_i \in \{(0, 1), (1, 0)\} \subset \mathbb{R}^2$ are the one-hot encoded targets. The index set I encompasses all gridpoints from all model runs of ICON-EU-EPS for the selected leadtime. Standardization of the predictors is achieved by using a minimum-maximum-scaling, i.e. each predictor s is scaled as $\frac{s - s_{\min}}{s_{\max} - s_{\min}}$ where $s_{\min}$, $s_{\max}$ are the minimum and maximum values of the predictor on the training set. Standardizing of the predictors is good practice in machine learning since then numerical artifacts due to strongly differing values/magnitudes of the predictors are remedied [25, Chap. 11]. Moreover, it is ensured that the training data is shuffled on the basis of model runs to avoid possible too strong correlations between subsequent model runs.

There are at most 500 epochs during training but training is stopped if there is no improvement in the probability of detection (POD) score after 80 epochs and the model with the highest POD-score is used. During training the POD-score is computed on a validation dataset that consists of 20 % of the data of the first batch (each batch contains data from at most 400 model runs). This early-stopping criterion is used to avoid overfitting of the NN (see, e.g., [1, Chap. 5]).

For the training, the weights of the NN are initialized with random numbers. The mini-ensemble of NN is created using five different seeds of the random number generator. Since the error functions of NN are known to have many global minima (e.g. [1, Chap. 5]), it is expected that the final weights of the five members of the mini-ensemble represent five realizations of the "optimal" distribution of the weights.

The training and testing dataset in this study is based on the 00 UTC model run of ICON-EU-EPS for the timeranges given in Table 3.1. In total, there are about 9 months of training data and about 3 months of testdata and it is ensured that each dataset contains data from the summer and winter season. Note, the horizontal resolution of the model changed at 2022-11-06 from approximately 20 km (ICON-grid 28) to approximately 13 km (ICON-grid 37). Prior experiments showed that this model change was significant in the sense that the calibration of the neural networks trained with the old data was worse if one tried to make forecasts using the data of the new model version. However, adding the training data of the new model version restored the verification results.

As noted in Sect. 3.2.2, the LINET domain is smaller than the domain of ICON-EU-EPS, hence the data used for training and evaluation is limited to that smaller domain. As a result, the number of datapoints that were used for the training are $\sim 9.3 \cdot 10^6$ (including $\sim 1.8 \cdot 10^6$ datapoints for validation) and $\sim 3.7 \cdot 10^6$ datapoints for testing.

Table 3.1 Timeranges of the training and testing dataset with an indicator of the numerical grid and the approximate horizontal grid resolution

Dataset	Timerange	ICON-grid	Approx. hor. resolution (km)
Testing	2020-06-01 to 2020-06-30	28	20
Training	2020-07-01 to 2020-08-29	28	20
Training	2021-06-01 to 2021-08-29	28	20
Training	2022-07-06 to 2022-07-31	37	13
Testing	2022-08-01 to 2022-08-31	37	13
Training	2022-09-01 to 2022-10-31	37	13
Testing	2022-11-01 to 2022-11-30	37	13
Training	2022-12-01 to 2022-12-29	37	13

3.4 Results

This section is dedicated to the presentation of an example forecast together with evaluation results.

3.4.1 Example Forecast

A visualization of a (randomly chosen) forecast is presented in Fig. 3.3. Panel 3.3a shows the forecast for 2023-08-17 at 18 UTC based on the model run from 2023-08-17 at 00 UTC with leadtime +18 h. Panel 3.3b shows the same forecast but now overlaid with the observed lightnings between 15 UTC and 18 UTC, indicated as magenta dots. Moreover the LINET domain is indicated by the blue box. As can be seen from this example, the observed lightnings align with the regions of enhanced lightning probability, hence the model provided a good guidance for the occurrence of most of the lightnings. However, not each individual occurrence of lightning was captured by the model, e.g. there were some lightnings over north-western France where no enhanced lightning probability was forecasted. Given that the LPI is the only dynamical predictor (see Sect. 3.2.3.4), most likely the ICON-EU-EPS had no convection in that region or the convection was spatially too spreaded among the individual ensemble members such that the LPI statistics of the ensemble showed no or only a very small signal, resulting in a postprocessed probability below 5 %.

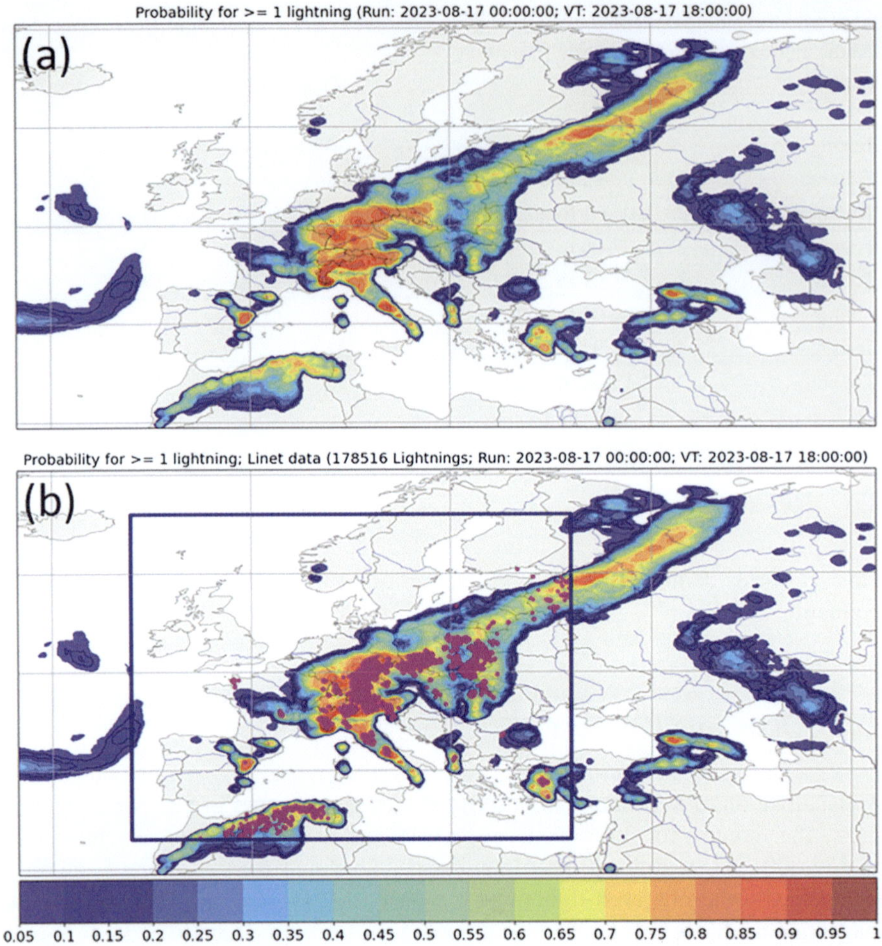

Fig. 3.3 Example forecast on the domain of the ICON-EU-EPS for 2023-08-17 at 18 UTC based on the model run from 2023-08-17 at 00 UTC with leadtime +18 h. Panel (**a**) forecasted lightning probabilities. Panel (**b**) same as in panel (**a**) overlaid with the observed lightnings (magenta dots) and the LINET domain (blue box)

3.4.2 Calibration and Sharpness

Calibration of postprocessing-methods, i.e. the statistical compatibility of its probability forecasts with the frequency of the observations, is usually assessed with the reliability diagram [28, Chap. 9]. Figure 3.4 shows reliability diagrams for the lead-times +18 h (panel 3.4a) and +30 h (panel 3.4b). The magenta curve corresponds to the predictions of the final model while the other curves indicate the reliabilities of the members of the mini-ensemble. In addition to the NN based probability forecasts, the thick black line shows the reliability of a raw model probability forecast based

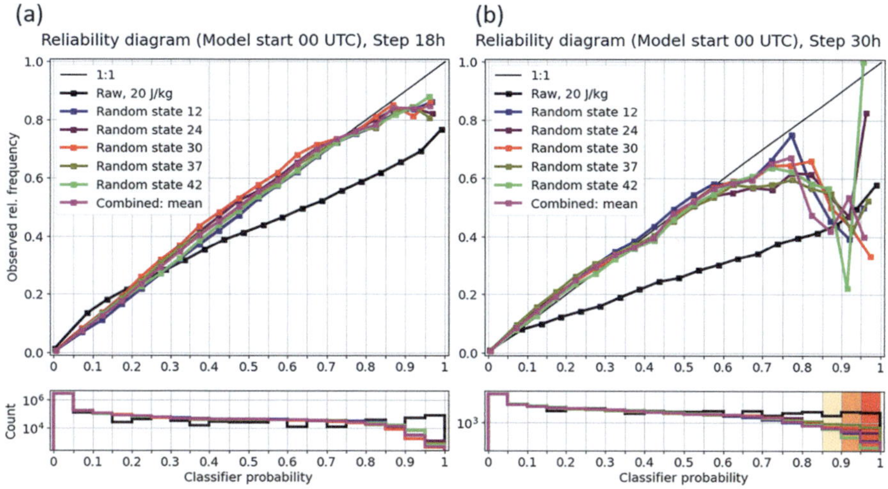

Fig. 3.4 Reliability diagrams for leadtimes +18 h (panel **a**) and +30 h (panel **b**). The final model prediction corresponds to the magenta curve while the other curves indicate the reliabilities of the members of the mini-ensemble. In addition, the thick black line indicates the reliability of the raw model probabilities based in the threshold 20 J/kg. The panels at the bottom indicate the count of the datapoints per bin with a reddish background-color indicating only a small number of datapoints

on the threshold 20 J/kg for the LPI. In this context, the raw forecast is produced by counting the number of ensemble members that have a LPI value[1] exceeding the pre-defined threshold and dividing this number by the total number of ensemble members. The threshold of 20 J/kg is a value that was previously employed for other products and is adopted here.

Several conclusions can be drawn from these reliability plots. Firstly, the reliability curves of the individual members of the mini-ensemble differ, indicating that indeed the training of the individual models resulted in different realizations of the NN. Secondly, the overall postprocessing-model is well-calibrated, in particular for the leadtimes that correspond to the hours in the afternoon where most thunderstorms occur due to the diurnal cycle (panel 3.4a). Calibration is worse during the night and morning hours (panel 3.4b); the reliability curves tend to level off at lower probabilities. Moreover, the number of forecasts with high probabilities is reduced during the night compared to the afternoon (bottom panels in Fig. 3.4). Finally, calibration of the postprocessing-model is significantly better in comparison to the raw model probabilities (thick black curve).

The patterns seen in Fig. 3.4 are representative for the other leadtimes, i.e. reliability for a leadtime close to the afternoon looks similar to Fig. 3.4a and for a leadtime closer to the night similar to Fig. 3.4b. Therefore, the diagrams for these intermediate leadtimes are omitted. Currently it remains unclear whether the worse calibration

[1] The same spatially and temporally aggregated LPI data is used as for the computation of the NN predictors.

during the night is due to a general inability of the NN and the predictors provided to produce a better calibrated forecast, due to an inherent low predictability, or due to a low lightning frequency during the night. In the latter case, the deficiency will be remedied in the future by using a longer timeseries for training.

Apart from the calibration of a postprocessing-method, also the sharpness of the forecasts are of relevance [28, Chap. 9]. For the present classifier, this translates into the ability to produce probabilities that are close to zero in case no lightning was observed and close to one in case at least one lightning was observed. Preferably, the intermediate probabilities should be issued only occasionally.

Figure 3.5 presents discrimination diagrams for leadtimes +18 h (panel 3.5a) and +30 h (panel 3.5b) together with discrimination diagrams for the corresponding raw probability forecast (panels 3.5c and d). The probability range is divided up into bins and the bars count the number of occurrences of forecasts with a probability within that bin. Moreover, blue bars are forecasts where no lightning was observed and red bars are forecasts where at least one lightning was observed. Ideally, only a blue bar at the left and a red bar at the right should be visible, indicating a perfect and sharp forecast. Comparing the forecasts of the postprocessing model (panels 3.5a, b) and the raw probability forecasts (panels 3.5c, d), the postprocessing model tends to forecast lower probabilities, i.e. the bars close to 100 % probability are significantly smaller in comparison to the raw probabilistic forecast (note the logarithmic and differing scaling of the vertical axes). This indicates a lower sharpness in the forecasts of the postprocessing model. However, the number of forecasts of high probabilities where no lightning was observed is much smaller for the postprocessing model, in particular during the afternoon (compare panels 3.5a, c).

It should be emphasized that the forecasts of a given model are only beneficial if the model yields sharp forecast subject to calibration [26]. The main difference between the current postprocessing model and the raw probability forecasts is the calibration: the calibration of the raw probabilities is much worse (see the thick black lines in Fig. 3.4). In that respect, the postprocessing outlined in this study diminishes the sharpness in comparison to the raw probabilities but yields much better forecasts since the forecasts are calibrated and still reasonably sharp.

3.4.3 Reuse of Trained Neural Networks

As already explained in Sect. 3.3.1 the current forecasting process involves a reusage of the NN trained on the modeldata from the run at 00 UTC to postprocess the other model runs that were initialized at 06, 12, 18 UTC. The main reason for this strategy is the fact that there was not a comparably large set of modeldata available for the model runs started at 06, 12, 18 UTC. In the future, data from these runs will be collected as well and then allow for a separate training of the NN for each model run. However, the obvious question is: Is there a degradation in the forecast performance

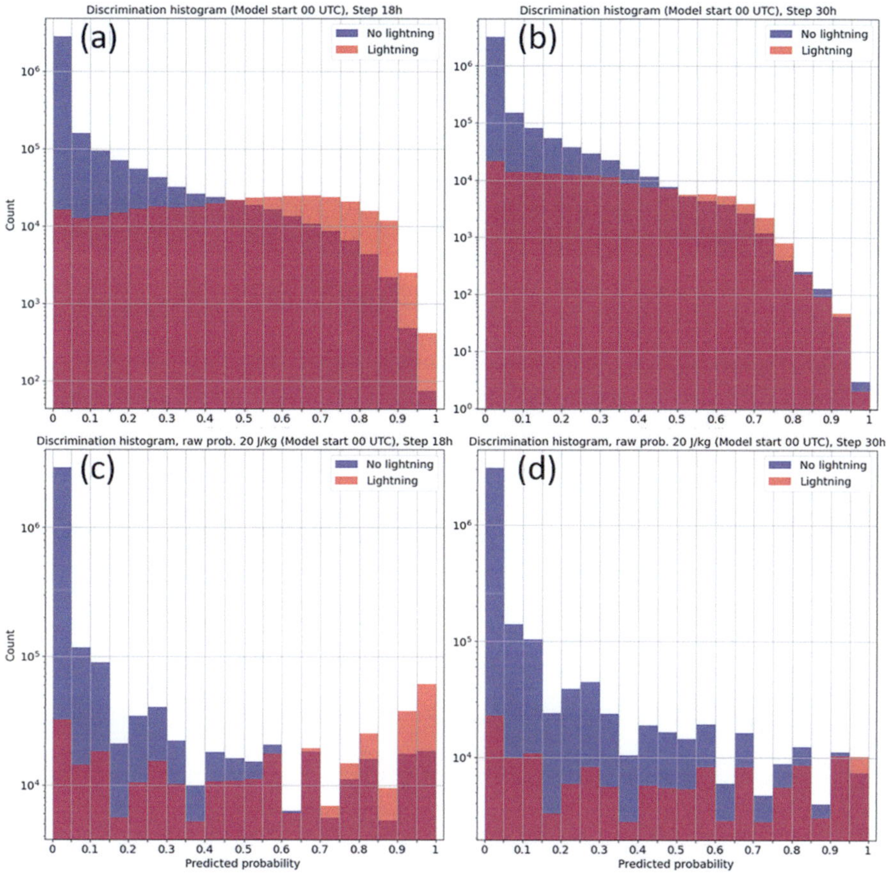

Fig. 3.5 Discrimination diagrams for the postprocessing model (panels **a**, **b**) and the raw probability forecast based on a threshold of 20 J/kg (panels **c**, **d**) for leadtimes +18 h (panels **a**, **c**) and +30 h (panels **b**, **d**). The bars indicate the number of times a forecast was issued with the forecasted probability falling into the respective bin. Blue bars indicate forecasts where actually no lightning was observed and red bars where at least one lightning was observed. Note the log-scaled and differing vertical axes

if the available NN are reused? To answer this question, a separate[2] test dataset covering the timerange from 2023-07-01 to 2023-08-27, i.e. about two months, is considered. Note that this dataset obviously has a bias towards the summer season. Figure 3.6 shows the reliability diagrams of all the eight forecasts from the available model runs that are valid at 06 UTC (panel a), 12 UTC (panel b), 18 UTC (panel c) and 24 UTC (panel d). The first aspect that should be noted is that the reliability curves for leadtimes +18 h and +30 h for the model run initialized at 00 UTC are not

[2] Separate from the testset indicated in Table 3.1 since LPI-data from ICON-EU-EPS runs with initialization times other than 00 UTC are not available for all dates listed in Table 3.1.

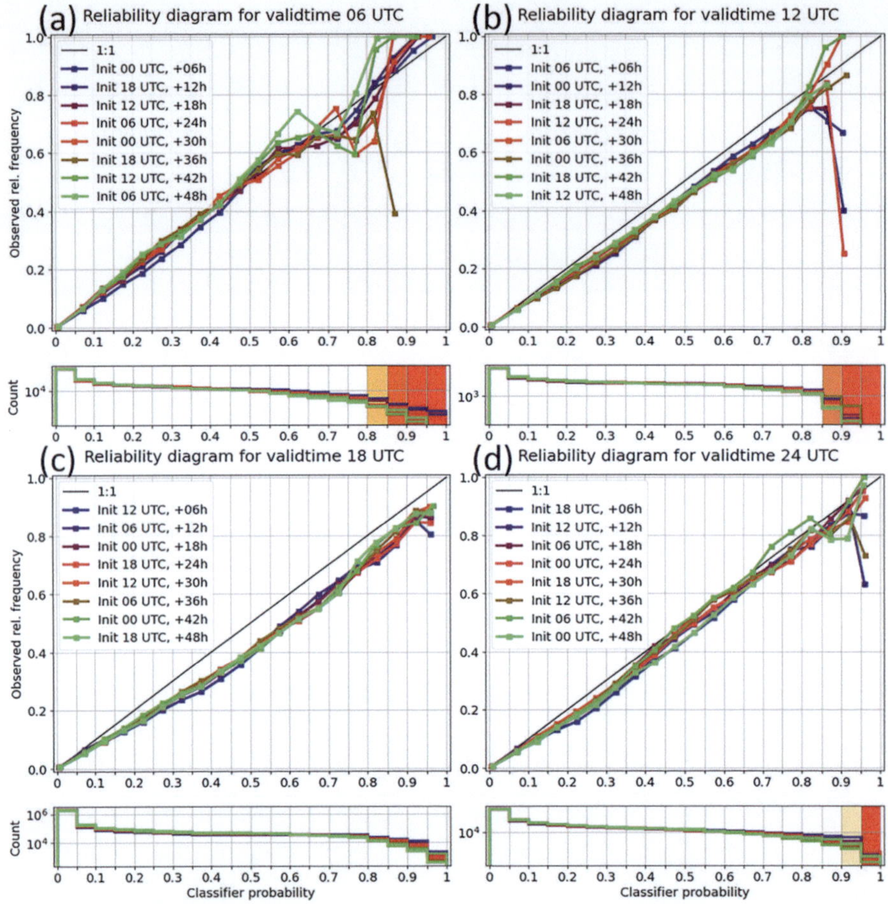

Fig. 3.6 Reliability diagrams for validtimes 06 UTC (panel **a**), 12 UTC (panel **b**), 18 UTC (panel **c**), 24 UTC (panel **d**) for different leadtimes. The panels at the bottom indicate the count of the datapoints per bin with a reddish background-color indicating only a small number of datapoints

identical to the curves from Fig. 3.4. The reason is the different underlying dataset and its length (about three months in Fig. 3.4 and about two months in Fig. 3.6). The more important conclusion that is drawn from this figure is the remarkable similarity between all the curves within each of the four panels. On the one hand, this implies that the reusage of the NN for other model runs does not degrade the forecast performance and, therefore, provides a justification in hindsight for their reusage. On the other hand, this additionally gives a hint that the NN mostly learned the expected physical behavior for a given time of the day rather than only a pure bias of the model run for the respective leadtime.

3.4.4 Brier-Score

The Brier-Score $\frac{1}{N}\sum_{i=1}^{N}(z_i - O_i)^2$ for a set $\{(z_i, O_i)\}_{i=1}^{N}$ of pairs of forecast probabilities $z_i \in [0, 1]$ and binarized targets $O_i \in \{0, 1\}$ together with its decomposition into the reliability component, resolution component and uncertainty component (see [28, Chap. 9]) gives a more condensed insight into the model performance. Here, N is the number of all gridpoints and forecasts in the testset for a given leadtime. The reliability component of the Brier-Score gives a summary of the calibration of the forecasts whereas the resolution component indicates the ability of the postprocessing method to create forecasts that are different from the climatological mean. The uncertainty component only depends on the observations and is independent from the forecasted probabilities; hence this component is omitted.

Figure 3.7a shows the Brier-score as a function of the leadtime for the first 48 h together with its decomposition into the reliability component (Fig. 3.7b, lower is better) and the resolution component (Fig. 3.7c, higher is better). Apart from the NN based postprocessing method (red curves), curves for the raw probabilities for the different thresholds 10 J/kg (blue curve), 20 J/kg (green curve), 30 J/kg (magenta curve) are included. For all leadtimes, the Brier-score is lowest for the NN based post-processing model (red curve in panel 3.7a). The reliability component is also lowest for the postprocessing model (red curve in panel 3.7b) reflecting the improvement in the calibration of the postprocessed forecast in comparison to the raw probability forecasts that was already discussed in Sect. 3.4.2. The resolution components are comparable for all the models, however for short leadtimes the postprocessed forecast shows an improved resolution over all the others (higher is better). During the first 34 h the postprocessed forecast has among others the highest resolution before the values get lower for larger leadtimes.

3.4.5 Diurnal Cycle

A well-known bias in numerical weather prediction models with parameterised convection is their tendency to initiate convection too early per day (e.g. [29]). As a consequence, it is interesting to investigate whether the postprocessed forecast is able to correct this bias. Figure 3.8 compares the mean predicted probability over all gridpoints and forecasts within the testset for the postprocessed model (red curve) and the raw probability forecasts based on several LPI-thresholds (blue, green, magenta curve) with the observed mean frequency as derived from the LINET observations (yellow curve). Apparently, the ICON-EU-EPS model also shows the aforementioned bias in the initiation of the convection since the blue, green and magenta curve all show increasing values a few hours earlier compared to the observation (yellow curve). In contrast, the postprocessing corrects this bias and the corresponding red

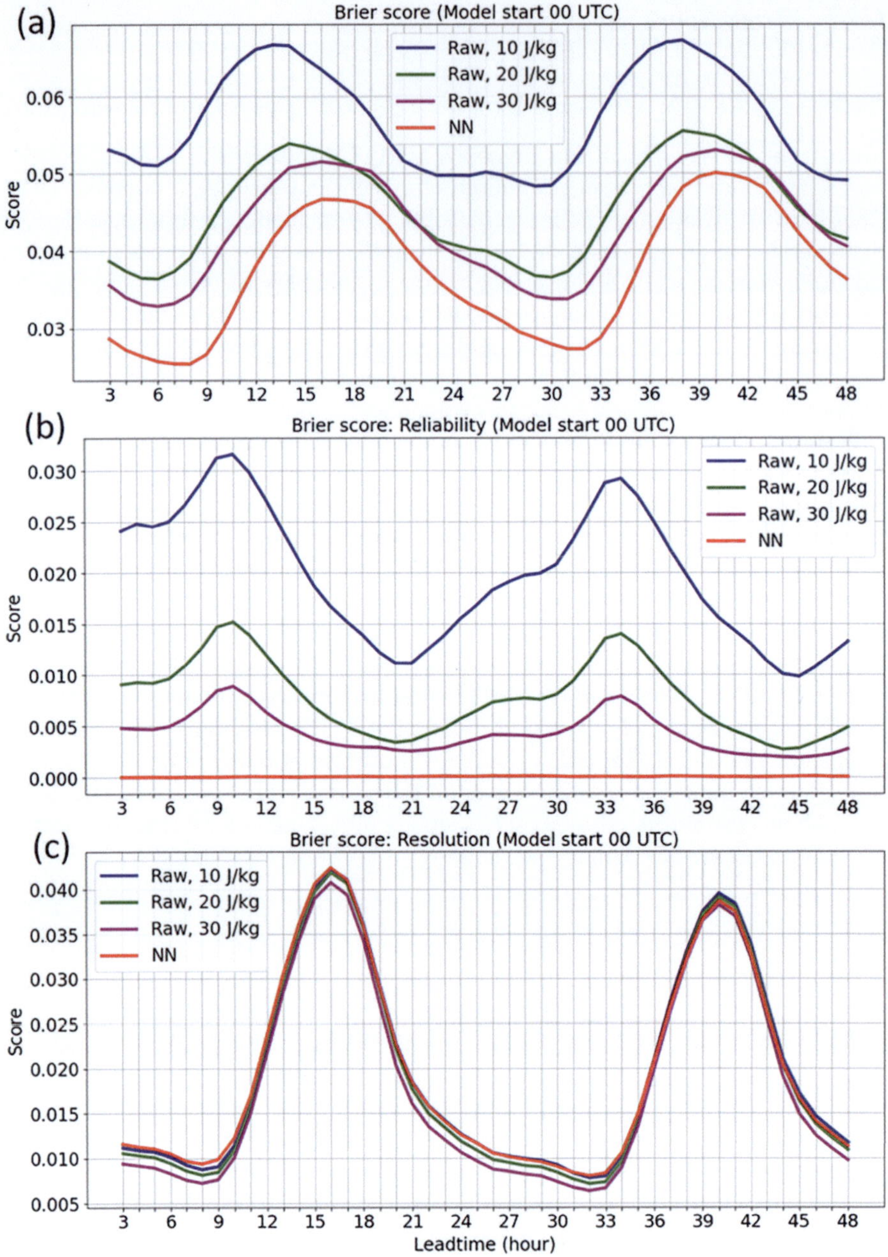

Fig. 3.7 Brier-score (panel **a**; lower is better) and its reliability-(panel **b**; lower is better) and resolution component (panel **c**; higher is better) from its decomposition as a function of leadtime up to +48 h. The red curve corresponds to the postprocessed forecast and the other curves correspond to the raw probability forecasts based on the LPI-thresholds 10 J/kg (blue curve), 20 J/kg (green curve) and 30 J/kg (magenta curve)

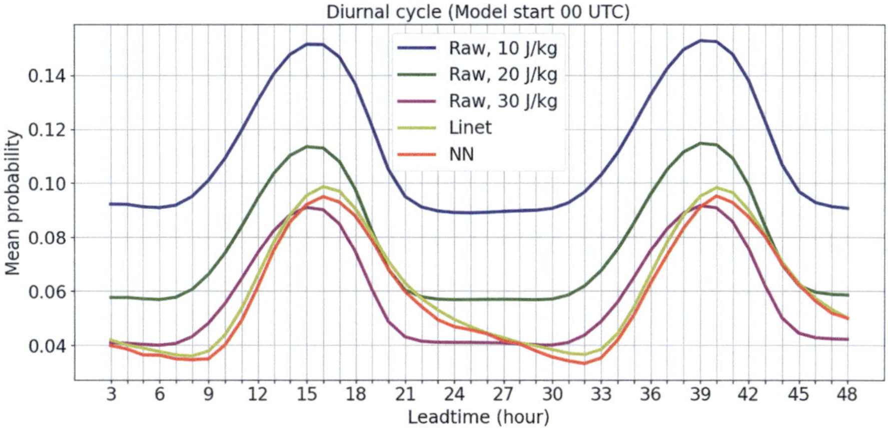

Fig. 3.8 Mean predicted probability as a function of the leadtime up to +48 h for the postprocessed forecast (red curve) and curves corresponding to the raw probability forecasts based on the LPI-thresholds 10 J/kg (blue curve), 20 J/kg (green curve) and 30 J/kg (magenta curve). The yellow curve indicates the diurnal cycle based on the observed lightnings from the LINET-network

curve lies much closer to the observations (yellow curve). Note that the yellow curve in Fig. 3.8 is based on the gridded and aggregated LINET data, see Sect. 3.2.3.

3.4.6 Uncertainty Propagation

There are various possible sources of errors that might enter the postprocessing outlined in this study and ultimately influence the final lightning probability forecast. Assuming the lightning observations as free of significant errors, the obvious first source of errors come from the ICON-EU-EPS model forecasts. Actually, the uncertainty inherent in the model forecasts is twofold: (1) model errors and biases, (2) sampling errors for training. There are no doubts that the model forecasts contain systematic errors and biases, e.g. the biases of the raw model probabilities visible in Figs. 3.4 and 3.8. Such errors are systematic enough and can be successfully corrected through a postprocessing as it is demonstrated in this study.

Sampling errors for the training of the postprocessing are more difficult to identify and correct. Obviously, longer timeseries of data will relax the sampling error in the training data because they will contain many different weather conditions[3] and therefore (modelled) configurations of the predictors. It was already hypothesized above that the degradation in reliability during the night (see Fig. 3.4) is due to sampling issues since thunderstorms are prevalent during the afternoon.

[3] Successive weather forecasts are correlated since the succession of real weather states is correlated in time. In effect, the number of model forecasts does not equal the number of different weather states.

Uncertainties that are specific to the NN approach are related to its architecture and training. During the development of the postprocessing some experimentation was done with different network architectures, i.e. using an increased number of neurons for the hidden layers or an increased number of hidden layers. However, no significant change in the resulting performance was found, hence it was decided to continue with the comparatively small networks described in Sect. 3.3.2.

Training of NN is commonly done by an inherently stochastic gradient descent strategy (e.g. [1, Chap. 5] or [27]) with two main sources of uncertainty: (1) computation of the gradients; (2) initialization of the weights. The computation of the gradients of the neural-network loss-function with respect to the network weights that should be optimized and the update of the weights is done batch-wise, i.e. not the full gradient is computed but a noisy approximation thereof. This has empirically proven as beneficial to allow the optimization process to not get stuck too early in an "unwanted local optimum". The Adam optimizer performs such a stochastic gradient descent [27].

The randomness in the initialization of the network weights to start the optimization process is respected in the current approach by employing the mini-ensemble of NN, i.e. using random values to initialize the weights. Ideally, this strategy leads to trained NN with different weights therefore allowing to estimate the uncertainty within the selected class of models defined by the fixed architecture of the NN. Arguably, a mini-ensemble size of 5 members seems rather small, but this sample size is in the range of other studies (e.g. [22, 30]). A rigorous assessment how the mini-ensemble size might be optimized is beyond the scope and the authors are not aware of such a study. The even more important usage of the mini-ensemble in practice is to identify forecast situations where the predictions of the individual NN differ significantly, which is expected for rare predictor configurations that were not or only occasionally seen during training.

3.5 Summary and Conclusions

This study reports on the development of a postprocessing system that is able to translate the information of the lightning potential index (LPI) from the ICON-EU ensemble system into a calibrated lightning probability. The postprocessing method mainly consists of a mini-ensemble of five fully-connected neural networks to predict the probability of occurrence of at least one lightning within a time interval of 3 h and a circular neighborhood with 40 km radius. Since point probabilities of lightning are rather unpredictable, probabilities that refer to a sufficiently large spatial and temporal domain are considered instead. These probabilities show sufficient predictability.

The postprocessed forecast shows a good overall calibration, although the calibration is better in the afternoon where most of the thunderstorms occur compared to the night. In addition, the postprocessed forecasts tend to be less sharp in comparison to the raw model probabilities, but their sharpness is still reasonable and, if measured by the resolution component of the Brier-Score, overall comparable for a

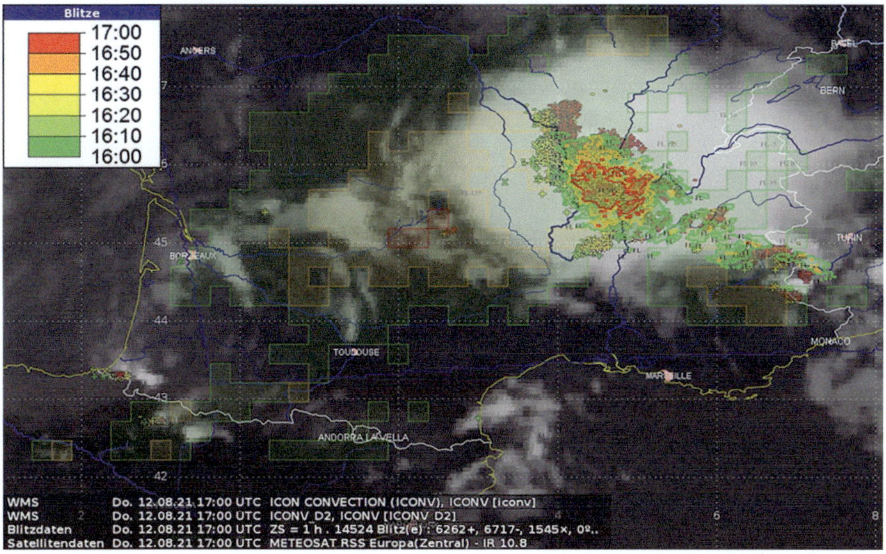

Fig. 3.9 Forecast from the iCONv product for 2021-08-12 at 17 UTC (light polygons with colors green, orange, red) overlaid on a satellite image. Small crosses depict the observed lightnings; their colorcode indicate the time the lightning was observed. A large convective system is shown over south-eastern France

wide range of leadtimes. Moreover, the postprocessing is able to correct the bias of the numerical weather prediction model in the diurnal cycle and agrees well with the observation.

At the time of writing, the implemented postprocessing model runs continuously at DWD and it will be soon put into operations. After that, it is possible to use the calibrated lightning forecast in other products such as the iCONv-product (see Fig. 3.9 for an example forecast). This product is designed to provide convection warning products for the aviation industry by disseminating polygons with the associated risk for convection (less likely, likely, very likely) by considering several exceeding probabilities of variables from ICON-EU-EPS. One of these variables is the raw lightning probability based on the LPI that can be substituted by the lightning probabilities from the postprocessing. In addition to the longer forecasts, iCONv is a seamless product that is able to take the nowcasting (see [31]) into account by a careful tuning of the parameters and thresholds.

In future work on the postprocessing method from this study it is envisioned to explore the possible advantage of a separate training of the required sets of neural networks for each of the model runs that are initialized at 06, 12, 18 UTC. It is expected that the separately trained neural networks are able to better correct the biases that are associated with an individual model run.

Acknowledgements We thank Susanne Theis, Axel Barleben and Ulrich Blahak for fruitful discussions during the development of the product.

References

1. Bishop, C.M.: Pattern Recognition and Machine Learning. Springer (2009). ISBN: 978-1-4939-3843-8. https://doi.org/10.1007/978-0-387-45528-0
2. Brodehl, S., Müller, R., Schömer, E., Spichtinger, P., Wand, M.: End-to-end prediction of lightning events from geostationary satellite images. Remote Sens. **14**(15), 3760 (2022). https://doi.org/10.3390/rs14153760
3. Vahid Yousefnia, K., Bölle, T., Zöbisch, I., Gerz, T.: A machine-learning approach to thunderstorm forecasting through post-processing of simulation data. Q. J. R. Meteorol. Soc. 1–16 (2024). https://doi.org/10.1002/qj.4777
4. Kamangir, H., Collins, W., Tissot, P., King, S.A.: A deep-learning model to predict thunderstorms within 400 km2 South Texas domains. Meteorol Appl. **27**, e1905 (2020). https://doi.org/10.1002/met.1905
5. Battaglioli, F., Groenemeijer, P., Tsonevsky, I., Púčik, T.: Forecasting large hail and lightning using additive logistic regression models and the ECMWF reforecasts. Nat. Hazards Earth Syst. Sci. **23**(12), 3651–3669 (2023)
6. Vannitsem, S., Bremnes, J.B., Demaeyer, J., Evans, G.R., Flowerdew, J., Hemri, S., Lerch, S., Roberts, N., Theis, S., Atencia, A., Ben-Bouallègue, Z., Bhend, J., Dabernig, M., De Cruz, L., Hieta, L., Mestre, O., Moret, L., PlenkoviÄ, I.O., Schmeits, M., Taillardat, M., Van den Bergh, J., Van Schaeybroeck, B., Whan, K., Ylhaisi, J.: Statistical postprocessing for weather forecasts: review, challenges, and avenues in a big data world. Bull. Am. Meteorol. Soc. **10**, E681–E699 (2021). https://doi.org/10.1175/BAMS-D-19-0308.1
7. Vannitsem, S., Wilks, D.S., Messner, J.: Statistical Postprocessing of Ensemble Forecasts. Elsevier (2018). ISBN: 978-0-12-812372-0
8. Doswell, C.A., Brooks, H.E., Maddox, R.A.: Flash flood forecasting: an ingredients-based methodology. Weather Forecast. **11**(4), 560–581 (1996). https://doi.org/10.1175/1520-0434(1996)011<0560:FFFAIB>2.0.CO;2
9. Schwartz, C.S., Sobash, R.A.: Generating probabilistic forecasts from convection-allowing ensembles using neighborhood approaches: a review and recommendations. Mon. Wea. Rev. **145**, 3397–3418 (2017). https://doi.org/10.1175/MWR-D-16-0400.1
10. Zängl, G., Reinert, D., RíÂpodas, P., Baldauf, M.: The ICON (ICOsahedral Non-hydrostatic) modelling framework of DWD and MPI-M: description of the non-hydrostatic dynamical core. Q. J. R. Meteorol. Soc. **141**, 563–579 (2015). https://doi.org/10.1002/qj.2378
11. Reinert, D., Prill, F., Frank, H., Denhard, M., Baldauf, M., Schraff, C., Gebhardt, C., Marsigli, C., Zängl, G.: DWD database reference for the global and regional ICON and ICON-EPS forecasting system. Technical report Version 2.2.2, Deutscher Wetterdienst. https://www.dwd.de/DWD/forschung/nwv/fepub/icon_database_main.pdf
12. Leutbecher, M., Palmer, T.N.: Ensemble forecasting. J. Comput. Phys. **227**(7), 3515–3539 (2008). https://doi.org/10.1016/j.jcp.2007.02.014
13. Buizza, R.: Introduction to the special issue on "25 years of ensemble forecasting". Q. J. R. Meteorol. Soc. **145**(S1) (2019). https://doi.org/10.1002/qj.3370
14. Molteni, E., Buizza, R., Palmer, T.N., Petroliagis, T.: The ECMWF ensemble prediction system: methodology and validation. Q. J. R. Meteorol. Soc. **122**, 73–119 (1996). https://doi.org/10.1002/qj.49712252905
15. Murphy, J.M.: The impact of ensemble forecasts on predictability. Q. J. R. Meteorol. Soc. **114**, 463–493 (1988). https://doi.org/10.1002/qj.49711448010
16. Lynn, B., Yair, Y.: Prediction of lightning flash density with the WRF model. Adv. Geosci. **23**, 11–16 (2010). https://doi.org/10.5194/adgeo-23-11-2010
17. Schröder, G., Göcke, T., Köhler, M.: Subgrid scale lightning potential index for ICON with parameterized convection. Reports ICON **10**, 3–28 (2022). https://doi.org/10.5676/DWD_pub/nwv/icon_010
18. Lopez, P.: 2016: A lightning parameterization for the ECMWF integrated forecasting system. Mon. Weather Rev. **144**, 3057–3075 (2016). https://doi.org/10.1175/MWR-D-16-0026.1

19. Bechtold, P., Semane, N., Lopez, P., Chaboureau, J.-P., Beljaars, A., Bormann, N.: Representing equilibrium and nonequilibrium convection in large-scale models. J. Atmos. Sci. **71**, 734–753 (2005). https://doi.org/10.1175/JAS-D-13-0163.1
20. Betz, H.D., Schmidt, K., Laroche, P., Blanchet, P., Oettinger, W.P., Defer, E., Dziewit, Z., Konarski, J.: LINET-an international lightning detection network in Europe. Atmos. Res. **91**(2), 564–573 (2009). https://doi.org/10.1016/j.atmosres.2008.06.012
21. Gneiting, T., Raftery, A.E., Westveld, A.H., Goldman, T.: Calibrated probabilistic forecasting using ensemble model output statistics and minimum CRPS estimation. Mon. Wea. Rev. **133**, 1098–1118 (2005). https://doi.org/10.1175/MWR2904.1
22. Schulz, B., Lerch, S.: Machine learning methods for postprocessing ensemble forecasts of wind gusts: a systematic comparison. Mon. Wea. Rev. **150**, 235–257 (2022). https://doi.org/10.1175/MWR-D-21-0150.1
23. Hess, R.: Statistical postprocessing of ensemble forecasts for severe weather at Deutscher Wetterdienst. Nonlin. Processes Geophys. **27**, 473–487 (2020). https://doi.org/10.5194/npg-27-473-2020
24. Glahn, H.R., Lowry, D.A.: The use of model output statistics (MOS) in objective weather forecasting. J. Appl. Meteorol. (1962–1982) **11**(8), 1203–1211 (1972). http://www.jstor.org/stable/26176961
25. Hastie, T., Tibshirani, R., Friedman, J.: The Elements of Statistical Learning. Springer (2009). ISBN: 978-0-387-84857-0. https://doi.org/10.1007/978-0-387-84858-7
26. Gneiting, T., Balabdaoui, F., Raftery, A.E.: Probabilistic forecasts, calibration and sharpness. J. R. Stat. Soc. B **69**(2), 243–268 (2007). https://doi.org/10.1111/j.1467-9868.2007.00587.x
27. Kingma, D.P., Ba, J.: Adam: A Method for Stochastic Optimization. arXiv preprint arXiv:1412.6980
28. Wilks, D.S..: Statistical Methods in the Atmospheric Sciences. Fourth Edition. Elsevier (2019). ISBN: 978-0-12-815823-4
29. Bechtold, P., Chaboureau, J.-P., Beljaars, A., Betts, A.K., Köhler, M., Miller, M., Redelsperger, J.-L.: The simulation of the diurnal cycle of convective precipitation over land in a global model. Q. J. R. Meteorol. Soc. **130**, 3119–3137 (2004). https://doi.org/10.1256/qj.03.103
30. Rasp, S., Lerch, S.: Neural networks for postprocessing ensemble weather forecasts. Mon. Wea. Rev. **146**, 3885–3900 (2018). https://doi.org/10.1175/MWR-D-18-0187.1
31. Müller, R., Barleben, A., Haussler, S., Jerg, M.: A novel approach for the global detection and nowcasting of deep convection and thunderstorms. Remote Sens. **14**(14), 3372 (2022). https://doi.org/10.3390/rs14143372

Chapter 4
Numerical Approaches to Modelling of Lava Dome Growth

Natalya Zeinalova, Alik Ismail-Zadeh, Igor Tsepelev, Oleg Melnik, and Frank Schilling

Abstract Non-explosive volcanic eruptions can result in the formation of lava domes when highly viscous magmas extrude on the surface. To distinguish between different mechanisms during the evolution of lava domes, isothermal and non-isothermal numerical models were used to explore the processes behind the evolution of the internal structure and morphological shape. To model the lava dome growth, we assume that the lava behaves as a viscous fluid, and use the Navier-Stokes, continuity, and the heat equations, along with relevant initial and boundary conditions. Additionally, an advection equation is employed to describe the movement of the interface between air and lava. The lava viscosity depends on the melt viscosity and the volume fraction of crystals. The latter is determined from an evolutionary equation that describes simplified crystal growth kinetics of degassing-induced crystallization. Numerical modelling employs finite volume and volume of fluid methods. The models demonstrate how lava dome dynamics are influenced by crystallization, the radiative-convective heat transfer, and the rheological stiffening driven by cooling. If the lava viscosity mainly depends on the volume fraction of crystals (as in the isothermal case study), the lava dome grows upward initially and later advances horizontally. The non-isothermal model shows that a highly viscous layer (carapace) forms within the uppermost part of the lava dome due to cooling, which restricts rapid dome broadening by creating steep slopes on the sides during long periods of dome

N. Zeinalova (✉) · A. Ismail-Zadeh · F. Schilling
Karlsruhe Institute of Technology, Institute of Applied Geosciences, Karlsruhe, Germany
e-mail: natalya.zeinalova@kit.edu

A. Ismail-Zadeh
e-mail: alik.ismail-zadeh@kit.edu

F. Schilling
e-mail: frank.schilling@kit.edu

I. Tsepelev
Institute of Mathematics and Mechanics, Ural Branch of the Russian Academy of Sciences, Yekaterinburg, Russia
e-mail: tsepelev@imm.uran.ru

O. Melnik
ISTerre, University Grenoble Alpes, Grenoble, France

A. Ott et al. (eds.), *Applications of Mathematics in Sciences, Engineering, and Economics*, Springer Proceedings in Mathematics & Statistics 515,
https://doi.org/10.1007/978-3-032-01279-1_4

building. Both models provide insights into the underlying processes and mechanisms of the evolution of lava domes, including their growth, carapace formation, and morphological modifications.

Keywords Fluid dynamics · Numerical modelling · Heat transfer · Crystallization · Lava dome

4.1 Introduction

Extrusion of a viscous magma on the surface can result in the growth of a lava dome depending on the magma's extrusion rate and its viscosity (see Table 4.1 for the volcanological terminology used in this paper). The lavas from dome-building volcanoes typically exhibit low average eruption rates and high viscosities, commonly attributed to their high crystallinity (e.g., [17]). Numerical modelling assists in understanding the dynamics of lava domes and, particularly, in explaining the impact of the rheological properties of magma (its crystal content, temperature, and the discharge rate) on the morphological shape of lava domes (e.g., [3, 4, 6, 19, 23, 24]).

Zeinalova et al. [23] developed a numerical model focusing on the gradual lava dome-building process at Volcán de Colima, examining how magma viscosity and discharge rates influenced the dome growth during years 2007–2009. Their viscosity model accounted for the crystal-dependent viscosity and the constant crystal-free melt viscosity. They showed that a highly viscous carapace is required to keep the model dome growing up and not spreading horizontally. To generate the carapace in a self-consistent way, Zeinalova et al. [24] developed thermomechanical models of a long-term lava dome evolution and analysed the lava dome growth depending on various heat sources and boundary conditions. Their model showed that cooling and carapace formation play a significant role during long duration of dome building.

Here, we present two case studies: (i) an isothermal case, where lava viscosity solely depends on the crystallization process, and (ii) a non-isothermal case, where viscosity is temperature- and crystallinity-dependent. Compared to the previous numerical models, which exhibit rapid (up to a few months) dome building with potential explosive destruction of the lava dome (e.g., [21, 22]), these case studies focus on a long (up to a year) dome-building, which was observed, for example, in Volcán de Colima, México [23, 25]. Following [24], we demonstrate that the radiative-convective heat flux leads to cooling of the lava at the interface with the surrounding air, resulting in the formation of a thick carapace. We compare the results of the two case studies to analyse the influence of temperature and heat transfer on the growth of the modelled lava domes.

Table 4.1 Volcanological terminology

Term	Definition
Carapace	The outer solid layer of a lava dome
Conduit	A channel that connects the magma chamber to the volcanic vent
Crater	A depression that usually lies directly above the vent from which volcanic material is ejected (source: online U.S. National Park Service)
Crystallization	A process at which crystals start to grow in magmas mainly due to cooling and/or degassing
Degassing	A process by which magma loses its volatiles
Discharge rate	A rate at which magma erupts
Lava	Magma emerging as a fluid onto Earth's surface. The term 'lava' is also used for the solidified rock formed by the cooling of a molten lava flow (source: online Encyclopedia Britannica)
Lava dome	Any steep-sided mound that is formed when lava reaching the Earth's surface is so viscous that it cannot flow away readily and accumulates around the vent (source: online Encyclopedia Britannica)
Magma	Molten or partially molten rocks
Magma chamber	A reservoir of magma in the Earth's crust where the magma may reside temporarily on its way from the uppermost mantle to the Earth's surface (source: online Collins Dictionary)
Rock	A naturally occurring and coherent aggregate of one or more minerals (source: online Encyclopedia Britannica)
Volcano	An opening where lava, tephra (small rocks), ash, and gases erupt onto the Earth's surface. Volcanic eruptions can last days, months, or even years (sources: online U.S. Geological Survey and the National Geographic Education Program)

4.2 Mathematical Statement

4.2.1 Geometry

The geometry of model domain $\Omega = \Omega_1 \cup \Omega_2$ (Fig. 4.1) corresponds to a configuration of the crater of Volcán de Colima [25], which was used in numerical modelling by [24]. The width of the crater is 257 m, while the conduit has 14.5 m in width and 30 m in depth. In the domain, we analyse a two-dimensional, viscous, incompressible, two-phase, immiscible fluid flow simulating an extrusion of lava (one fluid phase) into the surrounding air (another fluid phase) and its subsequent advancement. The sub-domain Ω_1 (the model conduit) is filled with the solidified rocks from previous eruption, and the sub-domain Ω_2 is filled with air. Fresh magma enters through the conduit, pushes the older rocks up to the surface, until it eventually reaches the surface itself, and contributes to the development of a lava dome. We note that the influence of the air phase on the lava dome growth is negligible due to a large ratio between densities/viscosities of the air and the lava.

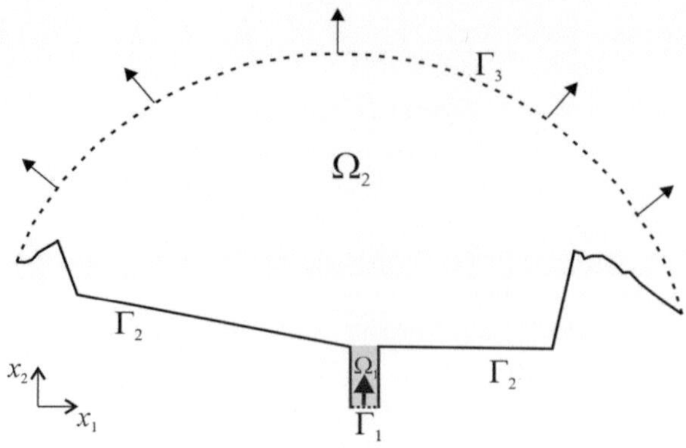

Fig. 4.1 Sketch of the model domain. The model conduit (sub-domain Ω_1) is occupied by older magma (grey-shaded), the thick arrow represents the upward ascent of new magma through Γ_1 (the dotted line). The bold line Γ_2 outlines the lava dome base. The upper boundary of the domain is represented by a dashed curve (Γ_3), facilitating the escape of the air from the model, with the direction of escape illustrated by arrows perpendicular to Γ_3

4.2.2 Governing Equations

The following governing equations describe the fluid flow approximating lava dynamics (e.g., [7, 19, 23, 24]):
the Navier-Stokes equations

$$\frac{\partial(\rho\mathbf{u})}{\partial t} + \langle\mathbf{u}, \nabla\rangle(\rho\mathbf{u}) - \nabla\cdot(\eta(\nabla\mathbf{u} + \nabla\mathbf{u}^T)) = -\nabla p - \rho\mathbf{g}\,, \tag{4.1}$$

the continuity equation

$$\nabla\cdot\mathbf{u} = 0\,, \tag{4.2}$$

and the equation for advection of the interface between the air and the lava

$$\frac{\partial\alpha}{\partial t} + \nabla\cdot(\alpha\mathbf{u}) = 0\,, \tag{4.3}$$

where t is the time; $\mathbf{u} = (u_1(t,\mathbf{x}), u_2(t,\mathbf{x}))$ is the velocity; $\mathbf{x} = (x_1, x_2) \in \Omega$ is the position vector in the Cartesian coordinates; ρ is the density; η is the viscosity; $p = p(\mathbf{x})$ is the pressure; $\mathbf{g} = (0, g)$, and g ($= 9.81$ m s^{-2}) is the acceleration due to gravity; $\alpha(t,\mathbf{x})$ takes the value of one for the lava and zero for the air at each point $\mathbf{x}$ and time t; ∇ , $\nabla\cdot$, ^{T} , and $\langle\cdot,\cdot\rangle$ denote the gradient operator, the divergence operator, the transposed matrix, and the scalar product of vectors, respectively. We consider that the model density and viscosity are presented as:

$$\rho = \rho_L \alpha(t, \mathbf{x})) + \rho_A (1 - \alpha(t, \mathbf{x}))\ , \quad \eta = \eta_L \alpha(t, \mathbf{x}) + \eta_A (1 - \alpha(t, \mathbf{x}))\ , \tag{4.4}$$

where ρ_A and η_A are the density and viscosity of the air, ρ_L and η_L are the lava density and its viscosity.

To study the influence of temperature on the development of lava domes, the heat equation is added to Eqs. (4.1)–(4.4) [24]:

$$\frac{\partial (c\rho T)}{\partial t} + \nabla \cdot (\mathbf{u}(c\rho T)) = \nabla \cdot (k \nabla T) - HF\ . \tag{4.5}$$

Here T is the temperature; the term HF accounts for the radiative and nonlinear convective heat transfer at the air-lava interface $HF = [\varepsilon\sigma(T^4 - T_A^4) + \tilde{\lambda}(T - T_A)^{4/3}]\delta_S$ (e.g., [19, 24], ε(=0.9; [15]) is the effective emissivity of the lava dome solid surface; σ is the Stefan-Boltzmann constant; $\tilde{\lambda}$ is the nonlinear convective heat transfer coefficient measured in (W m^{-2} K$^{-4/3}$) [12]; T_A is the air temperature; δ_S is the special function (approximating the Dirac delta function; measured in (m^{-1}) determined at interface (S) between the lava and the air. The model heat conductivity k and the specific heat capacity c are represented as:

$$k = k_L \alpha(t, \mathbf{x}) + k_A (1 - \alpha(t, \mathbf{x}))\ , \quad c = c_L \alpha(t, \mathbf{x}) + c_A (1 - \alpha(t, \mathbf{x}))\ , \tag{4.6}$$

where k_A and c_A are the heat conductivity and the specific heat capacity of the air, and k_L, and c_L are the thermal conductivity and specific heat capacity of the lava, respectively.

4.2.3 Lava Viscosity

The viscosity of lava η_L (measured in (Pa s)) is described as the product of the melt viscosity $\eta_{mt}(T)$ depending on temperature and the relative viscosity depending on the volume fraction of crystals $\eta_{cr}(\phi)$:

$$\eta_L(T, \phi) = \eta_{mt}(T) \times \eta_{cr}(\phi)\ . \tag{4.7}$$

In the isothermal case study, we assume that the melt viscosity η_{mt} is constant. In the non-isothermal case study, a temperature-dependent melt viscosity is described by the following experimental relationship [9]:

$$log_{10}\, \eta_{mt}(T_C) = a + \frac{b}{T_C - T_g}\ , \tag{4.8}$$

where T_C is the temperature in °C, $T_g = 710\,°\text{C}$ is the glass transition temperature, $a = 0.29$, and $b = 606.9\,°\text{C}$. Temperature T_C should be greater than the glass transition temperature T_g, if not, $T_C = T_g + 1$.

To define the dimensionless viscosity, we employ the model proposed by [2]:

$$\eta_{cr}(\phi) = (1+\varphi^{\delta})[1-(1-\xi)\mathrm{erf}(\frac{\sqrt{\pi}}{2(1-\xi)}\varphi(1+\varphi^{\gamma}))]^{-B\phi_*}, \quad (4.9)$$

where $\varphi = \phi/\phi_*$; ϕ is the volume fraction of crystals; ϕ_* is the specific volume fraction of crystals representing the critical solid fraction at the onset of the exponential increase of $\eta_{cr}(\phi)$ assumed to be $\phi_* = 0.591$ [2]; the empirical parameters $\delta = 7.24$, $\xi = 4.63 \times 10^{-4}$, and $\gamma = 5.76$ are sourced from [2, 10]; and erf$(\cdot)$ is the error function; and the theoretical value of B is the Einstein coefficient [11]. It was experimentally determined that this coefficient varies from 1.5 to 5 [8]; and we set B = 2.5 [2].

The volume fraction of crystals ϕ is determined from the evolutionary equation describing the simplified crystal growth kinetics of degassing-induced crystallization (e.g., [20]):

$$\frac{\partial \phi}{\partial t} + \nabla \cdot (\alpha \phi \mathbf{u}) = -\alpha \frac{\phi - \phi_{eq}}{\tau}, \quad (4.10)$$

with the initial condition for the volume fraction of crystals $\phi(t=0,\mathbf{x}) = 0$ for $\mathbf{x} \in \Omega_2$ and $\phi(t=0,\mathbf{x}) = 0.8$ for $\mathbf{x} \in \Omega_1$. Here, ϕ_{eq} is the volume fraction of crystals at the equilibrium state; τ is the characteristic time of crystal content growth needed by crystals to reach ϕ_{eq}. We assume that $\phi_{eq} = 0.83$, which is in the range of experimental constraints [1, 16].

Thus, in the numerical modelling, we employ Eqs. (4.1)–(4.4), (4.7), (4.9) and (4.10) in the isothermal case study, and Eqs. (4.1)–(4.10) in the non-isothermal case study, together with relevant initial and boundary conditions described below.

4.2.4 Initial and Boundary Conditions

At the initial time, the flow velocity is assumed to be $\mathbf{u}(t=0,\mathbf{x}) = \mathbf{u}_0$. The subdomain Ω_1 is filled by the older magma ($\alpha(t=0,\mathbf{x}) = 1$) with the volume fraction of crystals $\phi_{ct} = 0.8$ and temperature T_{ct} (in the non-isothermal case study). The subdomain Ω_2 is filled by the air ($\alpha(t=0,\mathbf{x}) = 0$) with the volume fraction of crystals $\phi = 0$ and temperature of the air T_A (in the non-isothermal case study).

The following conditions are applied on the model boundary $\Gamma = \Gamma_1 \cup \Gamma_2 \cup \Gamma_3$ (Fig. 4.1). A new magma with the volume fraction of crystals ϕ_{in} (assumed to be 0.4 in the modelling), and the temperature T_{in} (assumed to be 1250 K in the non-isothermal case) enters the model domain at a specified extrusion rate $\mathbf{u}_{ext} = (0, 5 \times 10^{-6})$ (m s^{-1}) through the part Γ_1 of the model boundary. No-slip condition $\mathbf{u} = 0$ and zero heat flux (in the non-isothermal case) are prescribed at Γ_2. To guarantee the incompressibility condition (Eq. 4.2) due to lava extrusion and thereby maintain the volume of the model domain, the air should be removed from the domain at the lava discharge rate. Hence, the outflow condition is determined at Γ_3 (the upper dashed curve) as $\mathbf{u}_{out} = u_{out}\mathbf{n}$, $\phi = 0$, and $\alpha = 0$, where $u_{out} = -|\Gamma_3|^{-1} \int_{\Gamma_1} \langle \mathbf{u}_{ext}, \mathbf{n} \rangle d\Gamma$, and $\mathbf{n}$ is the outward unit normal vector at a point on the model boundary.

4.3 Solution Method

The finite volume (e.g., [7]) and the volume of fluid (VOF; [5]) methods are used in the modelling. The VOF method allows for computationally inexpensive treatment of a moving interface between two fluid phases, e.g., the lava and the air. The cells containing the interface between lava and air are assigned α values ranging between zero and one, indicating the respective proportions of lava within them. The numerical solver operates within a discretized model domain consisting of approximately 60,000 hexahedral cells.

The pressure is discretized by the second-order staggered scheme PRESTO! (e.g., [14]). To approximate the Laplacian, we use the numerical scheme of the second-order accuracy, and the monotonic schemes are used to discretize convective terms in the equations (e.g., [7]). The pressure-velocity coupling is handled by the SIMPLE method [13], where the relaxation parameters are chosen to be 0.01 and 0.3 for the velocity and pressure, respectively. We assign 0.5 to the relaxation parameter for the function α, and the volume fraction of crystals. Considering a discontinuity between the lava and air viscosities, the choice of the relaxation parameters is critical and sometimes it was lowered to ensure the solution's stability. A time step is chosen in the range of 0.1–40 s depending on the stability and optimization of the velocity to assure a convergence of a set of linear algebraic equations (SLAE), which is obtained after the discretization of the Navier-Stokes equations. The implicit scheme maintains stability in computations even with relatively large time steps. The use of explicit schemes within the model results in numerical instability, despite employing small Courant numbers. In the modelling, we employ the conjugate-gradient method to solve SLAE. The numerical accuracy attains 10^{-3} for the velocity and pressure, and 10^{-6} for the function α and the volume fraction of crystals.

We note that the Reynolds number in the model of the lava dome dynamics remains small within the conduit and lava dome due to the high lava viscosity and hence slow flow. This results in insignificant influence of the inertial terms in the Navier-Stokes equations on the lava dynamics. Conversely, the air dynamics in the model exhibits a large Reynolds number, where inertial terms significantly affect the system. To ensure a laminar flow within the air subdomain and the stability of numerical solutions, we set $\mathbf{u}(t = 0, \mathbf{x}) = 0$ initially and use small time steps (ranging from 10^{-1} s to 10^{-3} s and spanning from 100 to 1,000 time steps) to determine the model velocity, which is then used to run subsequent computations with larger time steps. The presence of the air phase has a negligible effect on the growth of the lava dome, primarily due to the significant differences in the density and viscosity between the magma and the air.

A second-order upwind scheme has been implemented to solve the heat equation, where a relaxation parameter of 0.5 is assigned for T. The numerical accuracy achieved for the solution of SLAE to find T is 10^{-6}. The term HF in Eq. (4.5) is calculated in a thin layer of the cells adjacent to the interface between the lava and the air, and the special function δ_S is approximated by the ratio of the length of a relevant cell to the cell's area.

For numerical computations, we employ Ansys Fluent software.[1] Numerical experiments were conducted using the multiprocessor computer (bwHPC) at the Karlsruhe Institute of Technology. The processor type is Intel Xeon Gold 6230 and the processor frequency is 2.1 Ghz. The computational time for the numerical models varies based on several factors, such as the complexity of the mathematical problem, the quantity of computational cells, the number of cores utilized, and the time-step size. As an example: one day of the lava dome growth in the isothermal case study requires approximately four minutes using eight cores with a model time step of 40 s. In the non-isothermal case using the same parameters, the calculation time for one day of lava dome growth extends to around 6 min.

4.4 Numerical Results

4.4.1 Isothermal Case Study

Fresh magma enters the model domain via the boundary Γ_1 at the prescribed extrusion rate 5×10^{-6} m s^{-1}. The melt viscosity η_{mt} is set at 1.6×10^5 Pa s for the initial 150 days, and subsequently increased by a factor of 5 (reaching 8×10^5 Pa s) from day 150 to day 300. All other parameters are listed in Table 4.2. Figure 4.2 presents the results of the numerical isothermal modelling showing the snapshots of the extruded lava, which develops a dome structure for 300 days. During the initial 100 days the lava advances both vertically and horizontally, but later a more profound horizontal spreading of the lava can be observed. With the chosen characteristic time for crystal content growth of 5 days, the crystals achieve their equilibrium state already within the conduit. Following this, the volume fraction of crystals stabilizes, maintaining a constant value of 0.83 during the further lava dome growth. The subsequent increase of the melt viscosity leads to the increase of the lava viscosity and hence constrains the lateral boundary of the lava dome promoting the growth of the dome upwards. Over time, the dome exhibits a horizontal advancement with a vertical growth. The initiation of lava dome growth indicates a notable westward movement towards the right side of the model domain because of the slope of the crater's topography on the left side of the model domain.

The long-term evolution of the lava dome, where the lava rheology is solely dependent on the volume fraction of crystals, results in a nearly constant viscosity (see Fig. 4.2, day 300). The lava advances then as a fluid with a constant (even

[1] https://www.ansys.com/products/fluids/ansys-fluent.

Table 4.2 Model parameters and their values

Symbol	Parameter, unit	Value
B	The Einstein coefficient	2.5
c_A	Specific heat capacity of the air, J kg^{-1} K^{-1}	1006
c_L	Specific heat capacity of the lava, J kg^{-1} K^{-1}	1200
g	Acceleration due to gravity, m s^{-2}	9.81
k_A	Thermal conductivity of the air, J s^{-1} m^{-1} K^{-1}	0.025
k_L	Thermal conductivity of the lava, J s^{-1} m^{-1} K^{-1}	3.0
T_A	Air temperature, K	300
T_{in}	Temperature of the new magma entering the conduit, K	1250
T_{ct}	Initial temperature of the magma in the conduit, K	1150
$\mathbf{u}_{ext}$	Extrusion rate, m s^{-1}	5×10^{-6}
ε	Effective emissivity	0.9
ϕ_{ct}	Initial volume fraction of crystals in the magma in the model conduit	0.8
ϕ_{in}	Volume fraction of crystals of the new magma entering the model conduit	0.4
ϕ_{eq}	Volume fraction of crystals at the equilibrium state	0.83
ϕ_*	Specific volume fraction of crystals	0.591
η_A	Air viscosity, Pa s	0.001
$\tilde{\lambda}$	Nonlinear convective heat transfer coefficient, J s^{-1} m^{-2} K$^{-4/3}$	5.5
ρ_A	Air density, kg m^{-3}	1.225
ρ_L	Lava density, kg m^{-3}	2500
σ	Stefan-Boltzmann constant, J s^{-1} m^{-2} K^{-4}	5.67×10^{-8}
τ	Characteristic time of the crystal content growth, days	5
ξ	Rheological parameter	4.63×10^{-4}
δ	Rheological parameter	7.24
γ	Rheological parameter	5.76

increased after 150 days) viscosity and spreads horizontally due to gravity. This lava spreading occurs due to the absence of a carapace, capable of keeping the lava from horizontal advancement on the surface of the crater.

4.4.2 *Non-isothermal Case Study*

To generate a carapace on the top of the lava dome, we use a non-isothermal numerical model. The magma enters the model conduit through the boundary Γ_1 at the same (as in the isothermal case study) extrusion rate 5×10^{-6} m s^{-1} and at temperature 1250 K, and extrudes on the surface forming the lava dome. Other parameters of the

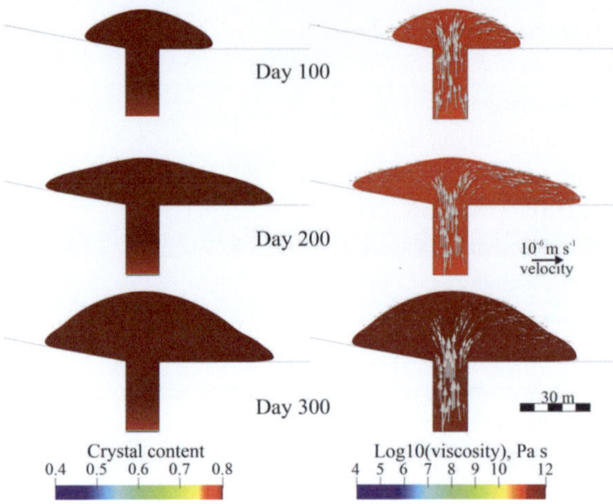

Fig. 4.2 The modelled crystal content (left column) and the viscosity overlaid by velocity vectors (white arrows) (right column) at days: 100 (upper panel), 200 (middle panel) and 300 (lower panel)

model are listed in Table 4.2. The melt viscosity depends on the temperature, and we do not increase it as in the isothermal case study. Figure 4.3 presents the results of the numerical modelling.

As in the isothermal case, the crystals achieve their equilibrium state already within the conduit (Fig. 4.3, the left column). This rapid crystallization leads to an increase in the lava viscosity. As the crystal-dependent viscosity attains the value $\eta_{cr}(\phi = 0.83) \sim 10^6$, the lava viscosity and hence the dynamics of the lava dome become dependent only on the temperature.

The radiative convective heat transfer at the interface between the lava and the air results in a decrease of the temperature (Fig. 4.3, the right column) and the viscosity increase (Fig. 4.3, the middle column) in the uppermost part of the dome, and, therefore, in the development of a carapace. This carapace retards the horizontal advancement of the lava dome and promotes steep lateral lava fronts. The lava remains sufficiently hot within the dome interior maintaining the lower viscosity (Fig. 4.3, the middle column) and exhibiting ductile behaviour during the entire duration of the dome growth, as shown in [24].

4.5 Discussion

The results of the numerical models related to the two case studies provide a valuable insight into the complex dynamics of lava dome growth under varying rheological properties, and clarify the role of the temperature, melt viscosity, crystal content, and heat transfer at the air-lava interface in the lava dome dynamics. In the isothermal case study, initially the modelled dome grows predominantly upward, and later it exhibits

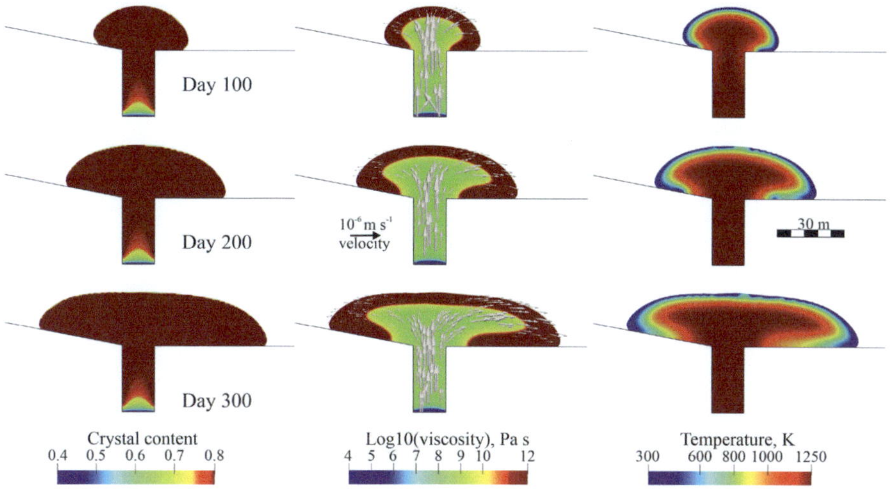

Fig. 4.3 The modelled crystal content (left column), the viscosity overlaid by velocity vectors (white arrows) (middle column) and the temperature (right column) at days 100 (upper panel), 200 (middle panel), and 300 (lower panel)

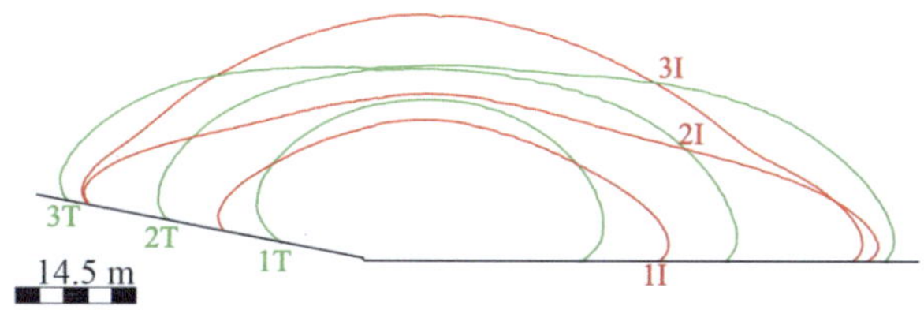

Fig. 4.4 Comparison of the morphological shapes of the modelled domes in exp.1 (isothermal case, red curves with index I) and exp. 2 (non-isothermal case, green curves with index T) at days 100 (index 1), 200 (index 2), and 300 (index 3).

more horizontal advancement resulting in a flattening of the dome (Fig. 4.4). If the melt viscosity (and hence the lava viscosity) increases, the dome resumes its vertical growth while still advancing horizontally. On the contrary, in the non-isothermal case, the carapace develops self-consistently due to the radiative-convective heat loss at the air/lava interface and it initially restricts lateral advancement of the modelled dome, resulting in a more compact shape within the first 200 days (Fig. 4.4). Subsequently, the lava dome starts to advance horizontally due to a stabilization of the extrusion and gravity forces, generating a flat top of the dome.

The model results highlight the substantial influence of the heat transfer and the carapace development on the lava dome morphology and its internal structure. While in the isothermal case, the viscosity increases almost uniformly within the dome, the radiative-convective heat exchange with the air in the non-isothermal case lowers the temperature and elevates the lava viscosity at the air-lava interface. Meanwhile, the lava within the dome interior remains rather hot and less viscous.

To quantitatively compare the morphological shapes of the two modelled domes (Fig. 4.4), we evaluate the closeness of the shape of the dome obtained in the isothermal case ($\tilde{F}$) with that obtained in the non-isothermal case (F). For this comparison, we employ the quality functional J based on the symmetric difference [18]:

$$J(\tilde{F}, F) = \frac{S((\tilde{L} \cup L)\backslash(\tilde{L} \cap L))}{S(\tilde{L})}, \tag{4.11}$$

where S is the area measured in (m^2), $\tilde{L}$ is the domain of modelled dome in the isothermal case, and L is the domain of the modelled dome in the non-isothermal case. The functional J takes the values 0.34, 0.39, and 0.37 for days 100, 200, and 300, respectively. The closeness of the function J to zero means the closeness of the shapes $\tilde{F}$ and F. The obtained values of the functional J show that the morphological shapes of the lava domes in the two case studies differs by about 37%.

The numerical results highlight two key points in dynamics of lava domes. In the isothermal model, where the lava viscosity is solely influenced by the crystallization process, a rapid crystallization results in the viscosity approaching its maximum value, and the lava behaves after that as a fluid of a constant viscosity spreading horizontally on the topography. In the non-isothermal case, the model accounts for the radiative-convective heat transfer at the air/lava interface, and this leads to an alteration in the morphological shape and the internal structure of the lava dome cause by the cold and more viscous carapace. The developed isothermal and non-isothermal models of the lava dome evolution can be used in numerical modelling of lava dome building and the dome carapace formation.

Author contribution AIZ and NZ (a PhD student) contributed to the conceptualization of the paper. NZ processed the data, developed a numerical model and performed all model experiments. AIZ supervised the study and assisted NZ in numerical modelling and data processing. IT contributed to the development of numerical codes and modelling. AIZ, FS, NZ and OM contributed to interpretation of numerical results. AIZ and NZ performed writing-original draft preparation and figures. All co-authors performed writing-review and editing.

Acknowledgements This contribution to the research project was funded by the Deutsche Forschungsgemeinschaft (project DFG IS203/14_1). OM acknowledges support from College de France, PAUSE program (C7H-PUB23A59).

References

1. Cashman, K.V.: Crystal size distribution (CSD) analysis of volcanic samples: advances and challenges. Front. Earth Sci. **8**, 291 (2020)
2. Costa, A., Caricchi, L., Bagdassarov, N.: A model for the rheology of particle-bearing suspensions and partially molten rocks. Geochem., Geophys., Geosystems **10**(3) (2009). https://doi.org/10.1029/2008GC002138
3. Hale, A.J., Wadge, G.: Numerical modelling of the growth dynamics of a simple silicic lava dome. Geophys. Res. Lett. **30**(19) (2003). https://doi.org/10.1029/2003GL018182
4. Harnett, C.E., Thomas, M.E., Purvance, M.D., Neuberg, J.: Using a discrete element approach to model lava dome emplacement and collapse. J. Volcanol. Geotherm. Res. **359**, 68–77 (2018). https://doi.org/10.1016/j.jvolgeores.2018.06.017
5. Hirt, C., Nichols, B.: Volume of fluid (VOF) method for the dynamics of free boundaries. J. Comput. Phys. **39**(1), 201–225 (1981). https://doi.org/10.1016/0021-9991(81)90145-5
6. Husain, T., Elsworth, D., Voight, B., Mattioli, G., Jansma, P.: Influence of extrusion rate and magma rheology on the growth of lava domes: insights from particle-dynamics modeling. J. Volcanol. Geotherm. Res. **285**, 100–117 (2014). https://doi.org/10.1016/j.jvolgeores.2014.08.013
7. Ismail-Zadeh, A., Tackley, P.: Computational Methods for Geodynamics. Cambridge University Press (2010). https://doi.org/10.1017/CBO9780511780820
8. Jeffrey, D.J., Acrivos, A.: The rheological properties of suspensions of rigid particles. AIChE J. **22**(3), 417–432 (1976). https://doi.org/10.1002/aic.690220303
9. Lavallée, Y., Varley, N.R., Alatorre-Ibargüengoitia, M.A., Hess, K.U., Kueppers, U., Mueller, S., Richard, D., Scheu, B., Spieler, O., Dingwell, D.B.: Magmatic architecture of dome-building eruptions at Volcán de Colima, Mexico. Bull. Volcanol. **74**(1), 249–260 (2012). https://doi.org/10.1007/s00445-011-0518-4
10. Lejeune, A.M., Richet, P.: Rheology of crystal-bearing silicate melts: an experimental study at high viscosities. J. Geophys. Res.: Solid Earth **100**(B3), 4215–4229 (1995). https://doi.org/10.1029/94JB02985
11. Mardles, E.W.J.: The viscosity of suspensions in non-aqueous liquids. Trans. Faraday Soc. **36**, 1007 (1940). https://doi.org/10.1039/TF9403601007
12. Neri, A.: A local heat transfer analysis of lava cooling in the atmosphere: application to thermal diffusion-dominated lava flows. J. Volcanol. Geotherm. Res. **81**(3–4), 215–243 (1998). https://doi.org/10.1016/S0377-0273(98)00010-9
13. Patankar, S., Spalding, D.: A calculation procedure for heat, mass and momentum transfer in three-dimensional parabolic flows. Int. J. Heat Mass Transf. **15**(10), 1787–1806 (1972). https://doi.org/10.1016/0017-9310(72)90054-3
14. Peyret, R.: Handbook of Computational Fluid Mechanics. Academic Press Limited, USA (1996)
15. Ramsey, M., Chevrel, M., Coppola, D., Harris, A.: The influence of emissivity on the thermo-rheological modeling of the channelized lava flows at Tolbachik volcano. Ann. Geophys. **61**(2018) (2019). https://doi.org/10.4401/ag-8077
16. Riker, J.M., Blundy, J.D., Rust, A.C., Botcharnikov, R.E., Humphreys, M.C.S.: Experimental phase equilibria of a Mount St. Helens rhyodacite: a framework for interpreting crystallization paths in degassing silicic magmas. Contrib. Mineral. Petrol. **170**(6) (2015). https://doi.org/10.1007/s00410-015-1160-5
17. Sheldrake, T.E., Sparks, R., Cashman, K.V., Wadge, G., Aspinall, W.P.: Similarities and differences in the historical records of lava dome-building volcanoes: implications for understanding magmatic processes and eruption forecasting. Earth-Sci. Rev. **160**, 240–263 (2016). https://doi.org/10.1016/j.earscirev.2016.07.013
18. Starodubtseva, Y.V., Starodubtsev, I.S., Ismail-Zadeh, A.T., Tsepelev, I.A., Melnik, O.E., Korotkii, A.I.: A method for magma viscosity assessment by lava dome morphology. J. Volcanol. Seism. **15**(3), 159–168 (2021). https://doi.org/10.1134/S0742046321030064

19. Tsepelev, I., Ismail-Zadeh, A., Melnik, O.: Lava dome morphology inferred from numerical modelling. Geophys. J. Int. **223**(3), 1597–1609 (2020). https://doi.org/10.1093/gji/ggaa395
20. Tsepelev, I., Ismail-Zadeh, A., Starodubtseva, Y., Korotkii, A., Melnik, O.: Crust development inferred from numerical models of lava flow and its surface thermal measurements. Ann. Geophys. **61**(2) (2019). https://doi.org/10.4401/ag-7745
21. Tsepelev, I.A., Ismail-Zadeh, A.T., Melnik, O.E.: 3D numerical modeling of the Summit Lake lava flow, Yellowstone, USA. Izvestiya, Phys. Solid Earth **57**(2), 257–265 (2021). https://doi.org/10.1134/S1069351321020129
22. Walter, T.R., Harnett, C.E., Varley, N., Bracamontes, D.V., Salzer, J., Zorn, E.U., Bretón, M., Arámbula, R., Thomas, M.E.: Imaging the 2013 explosive crater excavation and new dome formation at Volcán de Colima with TerraSAR-X, time-lapse cameras and modelling. J. Volcanol. Geotherm. Res. **369**, 224–237 (2019). https://doi.org/10.1016/j.jvolgeores.2018.11.016
23. Zeinalova, N., Ismail-Zadeh, A., Melnik, O., Tsepelev, I., Zobin, V.: Lava dome morphology and viscosity inferred from data-driven numerical modeling of dome growth at Volcán de Colima, Mexico during 2007-2009. Front. Earth Sci. **9** (2021). https://doi.org/10.3389/feart.2021.735914
24. Zeinalova, N., Ismail-Zadeh, A., Tsepelev, I., Melnik, O., Schilling, F.: Numerical thermomechanical modelling of lava dome growth during the 2007–2009 dome-building eruption at Volcán de Colima. Geophys. J. Int. **236**(1), 290–304 (2023). https://doi.org/10.1093/gji/ggad415
25. Zobin, V.M., Arámbula, R., Bretón, M., Reyes, G., Plascencia, I., Navarro, C., Téllez, A., Campos, A., González, M., León, Z., Martínez, A., Ramírez, C.: Dynamics of the january 2013–june 2014 explosive-effusive episode in the eruption of Volcán de Colima, México: insights from seismic and video monitoring. Bull. Volcanol. **77**(31) (2015). https://doi.org/10.1007/s00445-015-0917-z

Chapter 5
Enabling Microscale Simulations for Long-Term Aging of Li-Ion Batteries

Cyclic Extrapolation of the Solid Electrolyte Interphase Growth

Falco Schneider and Jochen Zausch

Abstract In this work we consider a microscale electrochemical model for Li-ion batteries which allows for detailed insights into local transport and degradation processes on the porous electrode microstructures. The model captures the growth of the solid electrolyte interphase (SEI), one of the major aging effects in Li-ion batteries. Due to the computational complexity of such a model, continuous simulation of long-term degradation studies is not applicable. Therefore, we extend on a previously introduced extrapolation scheme that allows to reduce the simulation time for cyclic operation protocols, which are commonly used to determine cell aging characteristics. We investigate convergence of the scheme with decreasing uniform extrapolation step sizes and compare different non-uniform step size controls for a fixed number of steps. Furthermore, a parameter study for the SEI model is performed to illustrate the impact on the aging results and analyse the robustness of the deployed numerical solver.

Keywords Li-ion batteries · Degradation · Solid electrolyte interphase · 3D-microstructure simulation · Cyclic extrapolation

5.1 Introduction

Li-ion batteries are currently a key technology to provide energy storage for mobile applications, such as battery electric vehicles or consumer electronics, but also act as stationary storage in power grids using renewable energies to buffer discrepancies between energy production and demand.

F. Schneider (✉) · J. Zausch
Fraunhofer Institute for Industrial Mathematics ITWM, Kaiserslautern, Germany
e-mail: falco.schneider@itwm.fraunhofer.de

J. Zausch
e-mail: jochen.zausch@itwm.fraunhofer.de

A. Ott et al. (eds.), *Applications of Mathematics in Sciences, Engineering, and Economics*, Springer Proceedings in Mathematics & Statistics 515,
https://doi.org/10.1007/978-3-032-01279-1_5

During operation and storage, the battery is exposed to a variety of degradation effects resulting in diminishing performance. One major contributor to capacity fade is the solid electrolyte interphase (SEI) [1, 9, 11]. The SEI is a thin, porous layer which forms at the interface between anode active material and liquid electrolyte as a product of irreversible reduction reactions of the electrolyte components. The capacity fade is the result of lithium being consumed by these reactions.

Performing aging studies by experiments poses certain challenges, since they need to be executed for extensive time periods [11, 21] to observe the gradual aging. Experimental insights by in-situ methods can be limited, while more detailed and elaborate, but also destructive, post-mortem analysis [26] prevents further use of the cell.

This issue can be addressed by using numerical simulations of physical models to supplement experimental studies and allow for additional insights into the internal transport and reaction processes of Li-ion batteries. Here, we consider a three-dimensional electrochemical model which characterizes the cell dynamics in terms of the Li-ion concentration and electric potentials across the spatially resolved porous electrodes. The model accounts for cell aging by incorporating the long-term growth of the SEI at the interface between anode active material and liquid electrolyte.

While these microscale simulations allow for detailed insights into the transport and degradation dynamics on the electrode microstructure and facilitate studying the impact of different structural designs, they come at the cost of substantial computational effort. Thus, fully resolved degradation studies of hundreds to thousands of cycles of battery operation are currently not applicable to obtain efficient aging predictions by these type of simulations. Examples for long-term SEI growth studies with a microscale model very similar to the present model can be found in [2, 22]. The two studies consider a rather small number of computational cells and mention a limited length of the cycling experiments due to numerical challenges, but simulation runtimes are not specified explicitly. Due to the performance, aging studies from the literature often deploy less detailed models in form of homogenized continuum scale models like the pseudo-two-dimensional Newman-type models [2] and single-particle models [20] or even (semi-)empirical models [25] to allow efficient continuous simulation of long-term cell aging.

Considering a broader context, cyclic aging studies are not unique to batteries, but structurally similar scenarios, where a sample being under cyclic load exhibits gradual aging over time, can also be found in the field of mechanical engineering, where high cycle fatigue studies are performed to analyze durability characteristics of components. There exist multiple publications in this field describing the acceleration of complex mechanical fatigue simulations by alternating between simulating load cycles and performing extrapolation of the damage and internal variables to accelerate fatigue assessment [13, 14, 27], which are also referred to as cycle jump schemes.

In this work we extend on a similar cyclic extrapolation technique which we previously utilized to accelerate cell aging simulations with the considered Li-ion battery model [23, 24]. Using the repetitive nature of cycling experiments and taking into account the local degradation information gathered from the continuous simulation of a single cycle, the spatially inhomogeneous aging of the next batch of cycles is predicted by extrapolation. In the following, we will analyze how the extrapolation behaves for decreasing uniform step sizes and compare the accuracy of different non-uniform stepping strategies for a fixed number of simulated cycles. The robustness of the numerical scheme is tested for different aging behaviours by performing a parameter study of the SEI model.

The article is structured as follows: Sect. 5.2 provides a summary of the considered microscale cell aging model for a Li-ion battery. The utilized numerical schemes are briefly described in Sect. 5.3. In Sect. 5.4, we discuss the obtained numerical results and solver performance for different variations of the extrapolation scheme and SEI model parameters while considering a realistic full cell configuration with strong nonlinearities, before ending with a conclusion in Sect. 5.5.

5.2 Microscale Model for SEI Growth in Li-Ion Batteries

The basis of the considered model [23, 24] is given by a thermodynamically consistent microscale model to capture the electrochemical behaviour of a Li-ion battery [12]. This is realized by solving for the Li-ion concentration c, the Maxwell potential ϕ and the electrochemical potential φ, which are characterized by nonlinear conservation equations. The microscale descriptor refers to the fact, that the porous electrode microstructures formed by the collection of active particles are explicitly resolved within the considered domain. Hence, the model offers a more detailed insight compared to the commonly used Newman-type models [3, 8], at the cost of a higher computational complexity. The considered model accounts for aging of the Li-ion battery by including the gradual growth of the SEI during operation (cyclic aging) and storage (calendar aging) at the interface of anode active material and liquid electrolyte [22]. The formation or onset of the SEI is not covered by the model and the existence of an initial layer is assumed.

In Sect. 5.2.1 we introduce the different material domains under consideration, while the required model equations and parameters are summarized in Sect. 5.2.2.

5.2.1 Material Composition and Domains

The fundamental unit required to capture the dynamics of a Li-ion battery consists of a single anode-cathode pair, see Fig. 5.1. It is crucial to include the full depth of the electrode pair from one current collector to the other, since this is the main direction of transport and the current collectors provide a direct connection to the external battery contacts. The cross-sectional area of the three-dimensional computational domain Ω can then be adjusted to achieve a representative volume element and

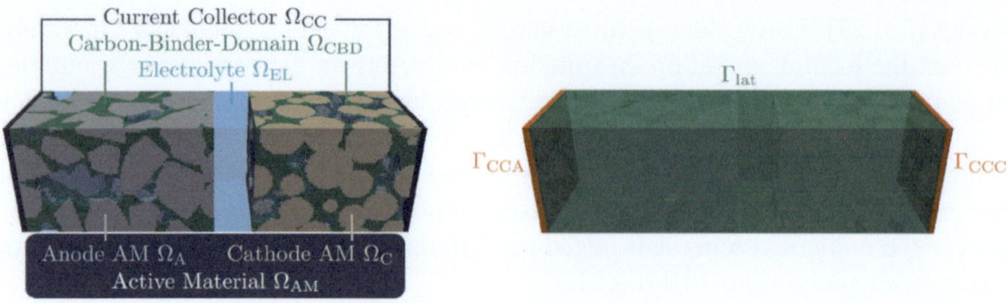

Fig. 5.1 Left: visualization of the considered battery microstructure and its material domains. The geometry shows the exact three-dimensional Cartesian grid used for our simulations consisting of $206 \times 60 \times 60$ voxels of uniform length 0.5μm. Right: illustration of the boundary of the computational domain, consisting of the lateral faces Γ_{lat} and the orthogonal outer current collector faces Γ_{CCA}, Γ_{CCC} for anode and cathode, respectively

match computational resources. This allows to effectively capture the dynamics of the full cell, while only considering a fraction of its geometry in the simulation, which is crucial for detailed microscale models.

In total, the model considers four different types of materials: The current collectors (CC), the active materials (AM) of anode and cathode, allowing storage of Li-ions by the chemical process of intercalation, the carbon-binder-domain (CBD) describing a mixture of binder for structural stability of the electrodes and a carbon powder additive for improved conductivity and the liquid electrolyte (EL) enabling the exchange of Li-ions between anode and cathode by ionic conduction.

The CBD exhibits a nanoporosity which allows for electrolyte transport through its pore space. This porosity is not spatially resolved in the geometry, instead it is modeled as a two phase material which captures the effective electric conductivity in its solid phase as well as the effective ionic conductivity and diffusivity in its pore space. For simplicity, we model the separator between the electrodes as bulk liquid electrolyte phase, but the model can also be modified to include porosity and tortuosity of the separator by modeling it as simplified case of the CBD material.

5.2.2 *Model Formulation*

The Li-ion concentration c and the potentials ϕ, φ within the different material domains are governed by two type of conservation laws

$$\begin{aligned} \partial_t c &= -\nabla \cdot \mathbf{N}, \\ 0 &= -\nabla \cdot \mathbf{J}, \end{aligned} \tag{5.1}$$

which describe conservation of Li-ions and conservation of charges in the system, respectively. The Li-ion flux density is denoted by $\mathbf{N}$, while $\mathbf{J}$ describes the cur-

Table 5.1 Summary of the considered variables and respective conservation laws for each material phase

Material phase	Variables	Conservation laws	Domain
CC	ϕ_s	$0 = -\nabla \cdot \mathbf{J}_s$	$\Omega_{CC} \subset \Omega$
AM	c_s ϕ_s	$\partial_t c_s = -\nabla \cdot \mathbf{N}_s$ $0 = -\nabla \cdot \mathbf{J}_s$	$\Omega_{AM} \subset \Omega$
CBD	c_e φ_e ϕ_s	$\partial_t (\varepsilon_e c_e) = -\nabla \cdot \mathbf{N}_e$ $0 = -\nabla \cdot \mathbf{J}_e$ $0 = -\nabla \cdot \mathbf{J}_s$	$\Omega_{CBD} \subset \Omega$
EL	c_e φ_e	$\partial_t c_e = -\nabla \cdot \mathbf{N}_e$ $0 = -\nabla \cdot \mathbf{J}_e$	$\Omega_{EL} \subset \Omega$

The variables and their corresponding governing equation are listed in matching order

rent density. The exact system of equations and variables depends on the material domain that is present at a given point in space. Table 5.1 lists the respective transport equations for the different materials under consideration. We use the index $s \in \{a, c\}$ to refer to variables or parameters of either anode or cathode solid materials.

Similarly, the fluxes are material dependent as well. The flux densities in the solid materials of CC, AM and CBD are given by Fickian diffusion and Ohm's law

$$\begin{aligned} \mathbf{N}_s &= -D_s \nabla c_s, \\ \mathbf{J}_s &= -\kappa_s \nabla \phi_s, \end{aligned} \tag{5.2}$$

while the electrolyte flux densities in the bulk phase and pore space of the CBD are based on moderately concentrated solution theory [12] and exhibit a direct coupling

$$\begin{aligned} \mathbf{N}_e &= -D_e \nabla c_e + \frac{t_+}{F} \mathbf{J}_e, \\ \mathbf{J}_e &= -RT \frac{t_+ - 1}{F c_e} \kappa_e \nabla c_e - \kappa_e \nabla \varphi_e, \end{aligned} \tag{5.3}$$

where R denotes the molar gas constant, F the Faraday constant and T the local temperature, which we assume to be constant in time and uniform across the domain Ω. The remaining parameters are listed with a short description in Table 5.6.

In order to capture the interaction between different materials, suitable interface conditions need to be described, see Table 5.2. These can be grouped into three type of conditions: zero-flux, perfect contact, i.e. continuity of fluxes and variables at the interface, and reactions. Reactive surfaces are only present at common interfaces between active material and electrolyte, i.e. bulk electrolyte or electrolyte inside the CBD. At the cathode interface $\Gamma_{CE} = \partial\Omega_C \cap (\partial\Omega_{EL} \cup \partial\Omega_{CBD})$ only intercalation, i.e. the transfer of Li-ions between electrolyte and active material, is considered, which is modeled by Butler-Volmer reaction kinetics [16]

Table 5.2 Overview of the considered interface conditions for all possible material interfaces

l \ r	CC	AM	CBD	EL
CC	–			
AM	$\mathbf{N}_s^l \cdot \mathbf{n} = 0$ * $\mathbf{J}_s^l \cdot \mathbf{n} = \mathbf{J}_s^r \cdot \mathbf{n}$ †	–		
CBD	$\mathbf{N}_e^l \cdot \mathbf{n} = 0$ * $\mathbf{J}_e^l \cdot \mathbf{n} = 0$ * $\mathbf{J}_s^l \cdot \mathbf{n} = \mathbf{J}_s^r \cdot \mathbf{n}$ †	$F\mathbf{N}_e^l \cdot \mathbf{n} = \varepsilon_e i_{tot}$ ‡ $\mathbf{J}_e^l \cdot \mathbf{n} = \varepsilon_e i_{tot}$ ‡ $\varepsilon_e i_{BV}^{SEI} = F\mathbf{N}_s^r \cdot \mathbf{n}$ ‡ $\mathbf{J}_s^l \cdot \mathbf{n} = \mathbf{J}_s^r \cdot \mathbf{n} - \varepsilon_e i_{tot}$ ‡	–	
EL	$\mathbf{N}_e^l \cdot \mathbf{n} = 0$ * $\mathbf{J}_e^l \cdot \mathbf{n} = 0$ * $0 = \mathbf{J}_s^r \cdot \mathbf{n}$ *	$F\mathbf{N}_e^l \cdot \mathbf{n} = i_{tot}$ ‡ $\mathbf{J}_e^l \cdot \mathbf{n} = i_{tot}$ ‡ $i_{BV}^{SEI} = F\mathbf{N}_s^r \cdot \mathbf{n}$ ‡ $i_{tot} = \mathbf{J}_s^r \cdot \mathbf{n}$ ‡	$\mathbf{N}_e^l \cdot \mathbf{n} = \mathbf{N}_e^r \cdot \mathbf{n}$ † $\mathbf{J}_e^l \cdot \mathbf{n} = \mathbf{J}_e^r \cdot \mathbf{n}$ † $0 = \mathbf{J}_s^r \cdot \mathbf{n}$ *	–

* Zero flux, † Perfect contact, ‡ Reaction

The entries on the top right are omitted due to symmetry. The reactive surfaces CBD-AM and EL-AM are given for the anode SEI model, for the cathode it simplifies to the intercalation model based on $i_{tot} = i_{BV}^{SEI} = i_{BV}$. The normals of the reactive surfaces are assumed to be the outer normals of AM

$$i_{BV} = 2k_s^{BV}\sqrt{c_s c_e\left(c_s^{max} - c_s\right)}\,\sinh\left(\frac{F}{2RT}\eta\right), \tag{5.4}$$

depending on the local overpotential $\eta = \phi_s - \varphi_e - U_s^{OCV}(c_s)$ and concentrations. Thus, the current density i_{BV} characterizes the exchange rate of Li-ions across the interface for a given voltage η.

For the anode-electrolyte interface $\Gamma_{AE} = \partial\Omega_A \cap (\partial\Omega_{EL} \cup \partial\Omega_{CBD})$ we account for cell aging induced by growth of the SEI by deploying a suitable interface model [22]

$$i_{tot} = i_{BV}^{SEI} + i_{SEI}, \tag{5.5}$$

$$i_{SEI} = -\frac{F D_{SEI} c_{SEI}^{Li}}{L_{SEI}} \exp\left(-\frac{FU}{RT}\right) g_{SEI} \sin^2\left(\frac{\pi}{2}\min\left\{\max\left\{0, 25g_{SEI}\right\}, 1\right\}\right), \tag{5.6}$$

$$g_{SEI} = \left(1 - \frac{F}{RT}\frac{L_{SEI}\omega_{SEI}}{\kappa_{SEI}} i_{tot}\right), \tag{5.7}$$

$$U = \phi_s - \varphi_e - \frac{L_{SEI}}{\kappa_{SEI}} i_{tot}, \tag{5.8}$$

where i_{BV}^{SEI} is the same as in equation (5.4) but depends on a slightly modified overpotential $\eta_{SEI} = \eta - \frac{L_{SEI}}{\kappa_{SEI}} i_{tot}$. Similar to the cathode interface, i_{BV}^{SEI} models the intercalation and deintercalation of Li-ions through the SEI, but it also captures an additional potential drop due the SEI interface resistance $\frac{L_{SEI}}{\kappa_{SEI}}$ causing power fade of the battery cell as the layer continuous to grow. The current density i_{SEI} characterizes the SEI growth rate and acts as lithium sink term at the interface, allowing to quantify the amount of Li-ions consumed by the SEI reactions and the resulting capacity fade of the battery cell. The expression for i_{SEI} is derived by modeling the rate limiting process through the SEI, which is assumed to be a combination of neutral lithium interstitial diffusion and electron hopping from the active material through the SEI. We refer to the original model for additional details [22]. The $\sin^2$ term in equation (5.6) acts as a heuristic factor that damps i_{SEI} as g_{SEI} approaches 0 and ensures $i_{\mathrm{SEI}} \leqslant 0$ to capture the irreversible nature of the SEI growth. The local dynamics of the SEI layer thickness L_{SEI} at a given point on the anode surface are governed by

$$\partial_t L_{\mathrm{SEI}} = -\frac{V_{\mathrm{SEI}}^{\mathrm{mol}}}{\nu_{\mathrm{SEI}} F} i_{\mathrm{SEI}}. \tag{5.9}$$

We refer to Table 5.6 for a description of the parameters.

In order to close the model formulation, we prescribe suitable boundary and initial conditions. For the lateral boundaries Γ_{lat} we assume periodic boundary conditions, matching the periodicity of the considered geometry. At the outer anode current collector face Γ_{CCA} we prescribe Dirichlet conditions $\phi_{\mathrm{s}} = \phi_{\mathrm{D}}$ for some constant ϕ_{D}, while Neumann conditions $\mathbf{J} \cdot \mathbf{n} = -i_{\mathrm{app}}(t)$ are used for the cathode current collector face Γ_{CCC} for some time-dependent applied current density $i_{\mathrm{app}}(t)$. The required initial conditions for the Li-ion concentration c and SEI thickness L_{SEI} are listed in Table 5.6.

5.3 Numerical Solver

In this section we give a brief summary of the numerical methods that are utilized to solve the previously introduced model equations from Sect. 5.2. The inclusion of the SEI model poses certain numerical challenges, since it introduces additional ordinary differential equations (5.9) and algebraic equations (5.5) characterizing the SEI growth dynamics and the interface current densities across the SEI, respectively. We address this challenge by deploying a semi-implicit numerical solution scheme which specifically takes care of the additional equations of the SEI model and has shown to outperform a monolithic solution approach solving the whole system fully-coupled. For additional information we refer to our previous publications [23, 24]. We start with the general discretization scheme in Sect. 5.3.1 and continue with the extrapolation scheme for cyclic battery operation in Sect. 5.3.2.

5.3.1 Semi-implicit Solver for the Microscale Battery Degradation Model

The spatial discretization of the considered three-dimensional battery geometry Ω is realized by a cell-centered finite volume method using a regular Cartesian grid, where the material phases are assumed to be mesh conforming [19]. At each cell center $x_i \in \Omega$, we solve for the local Li-ion concentration c_i, the Maxwell potential ϕ_i and the electrochemical potential φ_i, where the exact selection of variables depends on the material type. Fluxes between cells within the same material phase or at perfect contact interfaces are discretized using centered finite difference approximation of the gradient and harmonic averages of the potentially discontinuous or non-constant coefficients between adjacent cells. At each interface of a grid cell that coincides with the anode surface Γ_{AE} we additionally discretize equation (5.5) and (5.9) to capture the local SEI dynamics. The Butler-Volmer reactions at the cathode interface Γ_{CE} can be computed explicitly and are inserted directly into the finite volume formulation. For both reaction calculations we consider concentrations and potentials from the two adjacent computational cell centers to determine the interface current densities.

After the spatial discretization, we obtain a system consisting of algebraic equations and ordinary differential equations. We deploy a Godunov-type operator splitting of the form [23, 24]

$$\begin{pmatrix} \partial_t c \\ 0 \\ 0 \\ \partial_t L_{\text{SEI}} \end{pmatrix} = \underbrace{\begin{pmatrix} G_c(c, \phi, i_{\text{tot}}, L_{\text{SEI}}) \\ G_\phi(c, \phi, i_{\text{tot}}) \\ G_{i_{\text{tot}}}(c, \phi, i_{\text{tot}}, L_{\text{SEI}}) \\ 0 \end{pmatrix}}_{(A)} + \underbrace{\begin{pmatrix} 0 \\ 0 \\ 0 \\ G_{L_{\text{SEI}}}(\phi, i_{\text{tot}}, L_{\text{SEI}}) \end{pmatrix}}_{(B)}. \qquad (5.10)$$

Each line denotes a vector describing the spatially discretized version of one of the four continous variable types: concentration c, (Maxwell and electrochemical) potentials ϕ, SEI interface current i_{tot} or SEI thickness L_{SEI}. The functions G_* describe the corresponding discretized right hand side of the equations. For each time step, each of the two systems is solved a single time. First we solve system (A) with an implicit Euler step deploying an inexact Newton-Krylov method. Within the Newton iteration, the resulting linear systems exhibit a saddle-point structure, which is resolved by a Schur complement formulation eliminating the SEI interface currents i_{tot}. The obtained linear system is then solved by a BiCGstab Krylov solver using a single iteration of an algebraic multigrid (AMG) method as preconditioner in each iteration. The solution of system (A) provides the update for the concentrations c, potentials ϕ and SEI interface currents i_{tot}, which are then used to compute the updated SEI thickness L_{SEI} by solving system (B) with an explicit Euler method.

5.3.2 Extrapolation for Accelerated Cyclic Aging Studies

Experimentally, aging of Li-ion batteries during operation is commonly characterized by so-called cycling experiments. In these experiments, a predefined operation protocol is applied to the battery, repeating a given cycle to charge and discharge the battery. Depending on the protocol, this might result in either exactly or close to periodic boundary conditions for the battery. Due to the repetitive nature of these protocols and the overall slow growth rate of the SEI, changes between consecutive cycles are expected to be small. Hence, we propose to use a cyclic extrapolation scheme [23, 24] which alternates between continuous simulation of full cycles and extrapolation of the SEI growth and capacity fade.

In total we are concerned with three different type of time intervals: the period of a single operation cycle τ_{cyc} given by the protocol, the period of continuous simulation $\tau_{\text{sim}} = m_1 \tau_{\text{cyc}}$ to get a recent estimate on the SEI growth dynamics and the extrapolation step size $\tau_{\text{exp}} = m_2 \tau_{\text{cyc}}$. Due to the nonlinear growth behaviour during a cycle, the latter two are assumed to be conforming with the operation cycle period, i.e. $m_1, m_2 > 0$ are assumed to be integers.

For each computational cell interface $\Gamma_j \subset \Gamma_{\text{AE}}$ a linearly extrapolated layer thickness $L^{\text{exp}}_{\text{SEI},j}$ is obtained by taking into account the local growth rate over the latest simulation period

$$L^{\text{exp}}_{\text{SEI},j} = L^{\text{post}}_{\text{SEI},j} + \tau_{\text{exp}} \frac{L^{\text{post}}_{\text{SEI},j} - L^{\text{pre}}_{\text{SEI},j}}{\tau_{\text{sim}}}, \tag{5.11}$$

using the recorded SEI thickness at that specific interface at the beginning $L^{\text{pre}}_{\text{SEI},j}$ and end $L^{\text{post}}_{\text{SEI},j}$ of the latest simulation period. As a result of these localized estimates, the scheme is able to take into account the inhomogeneous growth rate along the anode surface and capture this inhomogeneity in the extrapolated values.

The extrapolated SEI growth is related to the underlying reactions and will be associated with a corresponding capacity fade. Based on the SEI model equation (5.9) the total amount of consumed Li-ions is computed and uniformly subtracted as an average from the anode concentration in each computational cell i in Ω_{A}

$$c_{\text{a},i} - \Delta c_{\text{SEI}} = c_{\text{a},i} - \frac{1}{V_{\text{A}}} \frac{\nu_{\text{SEI}}}{V^{\text{SEI}}_{\text{mol}}} \sum_{j=1}^{N_{\text{SEI}}} \varepsilon_{\text{e}} A_j \left(L^{\text{exp}}_{\text{SEI},j} - L^{\text{post}}_{\text{SEI},j} \right), \tag{5.12}$$

where V_{A} is the total anode active material volume in the considered computational domain, A_j describes the area of the computational cell interface $\Gamma_j \subset \Gamma_{\text{AE}}$ and N_{SEI} is given by the total number of cell interfaces coinciding with Γ_{AE}. The parameter ε_{e} captures the porosity of the CBD material, while the expression simplifies with $\varepsilon_{\text{e}} = 1$ for the bulk electrolyte. The capacity correction assumes that diffusion within a cycle is sufficiently fast to distribute the capacity loss evenly across the anode.

5.4 Numerical Results

In this section we are analysing the proposed extrapolation scheme in more detail. We consider a complex full cell setup of a Li-ion battery featuring strong nonlinearities, which will be cycled to study the resulting cell aging. The specific simulation setup is described in Sect. 5.4.1. In Sect. 5.4.2, we investigate different extrapolation step size strategies and compare their convergence and accuracy to a given reference solution. Finally, we show the effects of different parameters of the SEI model on the predicted aging results in Sect. 5.4.3 and summarize the overall performance of the numerical solver for the conducted simulations in Sect. 5.4.4.

5.4.1 Simulation Setup

We consider a model parameterization proposed in [23, 24] as the basis for numerical experiments in this work. The corresponding model parameters are listed in Table 5.6, while the computational grid is visualized in Fig. 5.1. The graphite anode and nickel manganese cobalt oxide (NMC) cathode exhibit a porosity of 32.63% and 32.07%, respectively. Both electrode microstructures were generated in GeoDict [15] using the GrainGeo module. Assuming 90% utilization of the anode, the areal capacity of the simulated structure amounts to around $Q_A = 2.2\,\text{mAh/cm}^2$.

The cyclic operation is realized by a profile alternating between constant current discharge and constant current charge every 1800s resulting in $\tau_{\text{cyc}} = 1\text{h}$, see Fig. 5.2. The considered time interval of 500 h covers 500 cycles. For the applied cur-

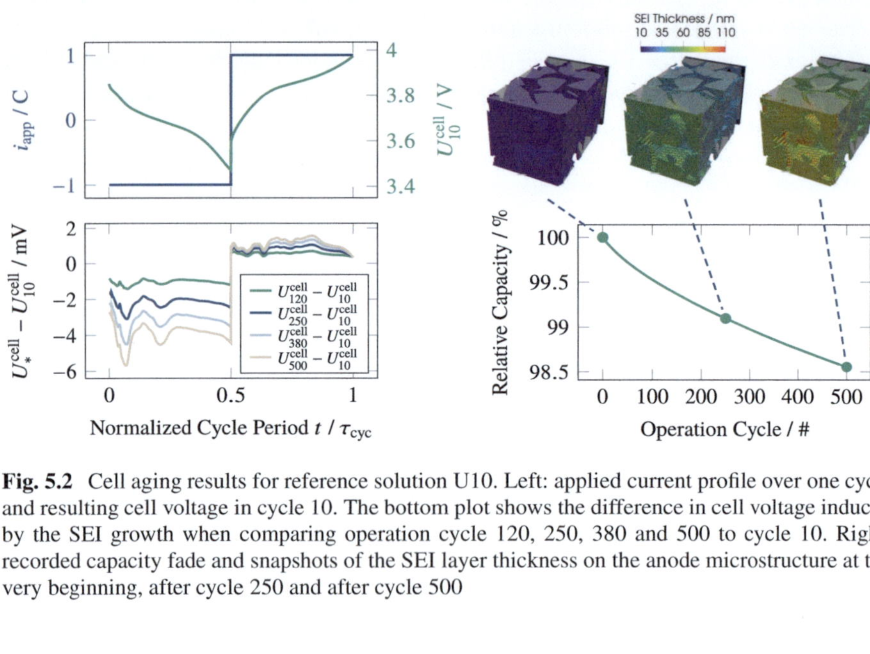

Fig. 5.2 Cell aging results for reference solution U10. Left: applied current profile over one cycle and resulting cell voltage in cycle 10. The bottom plot shows the difference in cell voltage induced by the SEI growth when comparing operation cycle 120, 250, 380 and 500 to cycle 10. Right: recorded capacity fade and snapshots of the SEI layer thickness on the anode microstructure at the very beginning, after cycle 250 and after cycle 500

rent density a value normalized by the areal capacity is considered $i_{\text{app}} = 1\text{C} = \frac{Q_A}{1\text{h}}$, such that the battery is initially cycled between 75 and 25% state of charge (SoC). All simulations are performed with our in-house simulation software Battery and Electrochemistry Simulation Tool BEST [6] using SAMG [7] as implementation for the Krylov accelerated AMG solver. Due to the strong nonlinearities in the model, we deploy an adaptive time stepper with a base step size of $\Delta t = 5\text{s}$ which halves the step size every time the numerical solver fails to solve a system of a given time step. After an adaptively chosen number of successful steps, the step size reduction will be gradually reverted.

5.4.2 Extrapolation Step Size Strategy

In this section we investigate different uniform and non-uniform step size strategies for the cyclic extrapolation scheme. Table 5.3 shows a list of the considered strategies and describes which cycles of the operation protocol are simulated. First we take a look at the uniform strategies U10, U25, U50, U100. Since there is no analytical solution known to us for the complex battery simulation at hand, we will consider our simulation with the finest uniform extrapolation step size U10 as reference solution for our error computations. The obtained cell aging results from that reference solution are shown in Fig. 5.2.

At the end of the simulation, the gradual growth of the SEI results in a total capacity fade of 1.44%. Furthermore, it causes a decrease in cell voltage during discharge and an increase during charge, due to the amplified overpotentials. This can be seen by the evolution of cell voltage difference between the later operation cycles and the 10th cycle. The asymmetric shift between charge and discharge in these voltage differences can be explained by the gradual drift in the open-circuit

Table 5.3 Summary of considered cyclic extrapolation stepping schemes, providing the exact cycles from the operation protocol which are simulated, the total number of simulated cycles and a short description of the schemes

Label	Simulated cycles	$N_{\text{cyc}}^{\text{sim}}$	Description
U10	1, 10, 20, ..., 490, 500	51	Uniform extrapolation 9 cycles each
U25	1, 25, 50, ..., 475, 500	21	Uniform extrapolation 24 cycles each
U50	1, 50, 100, ..., 450, 500	11	Uniform extrapolation 49 cycles each
U100	1, 100, 200, 300, 400, 500	6	Uniform extrapolation 99 cycles each
NU100	1, 50, 100, 200, 300, 400, 500	7	Non-uniform extrapolation from [23, 24]
PF06	1, 150, 240, 320, 380, 440, 500	7	Power function extrapolation with $p = 0.\overline{6}$
PF10	1, 80, 170, 250, 330, 420, 500	7	Power function extrapolation with $p = 1.0$
PF15	1, 30, 100, 180, 270, 380, 500	7	Power function extrapolation with $p = 1.5$
PF20	1, 10, 60, 130, 220, 350, 500	7	Power function extrapolation with $p = 2.0$
PF25	1, 10, 30, 90, 180, 320, 500	7	Power function extrapolation with $p = 2.5$

voltage based on the capacity fade. With every cycle, the anode SoC gets lower for a fixed point within the cycle, while the cathode SoC remains unchanged. As a result, cell balancing changes over time causing the aforementioned decrease in open-circuit voltage. We would like to point out, that the strength of the observed aging effects strongly depends on the considered parameters, see Sect. 5.4.3. Additionally, the aging depends on the operation protocol and will be amplified if the cell remains at a higher SoC for longer time periods, due to the low anode potential in that state. These dependencies are also observed experimentally [1, 9, 11, 21].

Next, we take a look at the SEI results obtained for the different uniform extrapolation schemes. For a discrete point in time t_k, we will consider the following scalar quantities associated with the inhomogeneous SEI profile along the computational cell interfaces $\Gamma_j \subset \Gamma_{\mathrm{AE}}$ coinciding with the anode surface

$$L_{\mathrm{SEI}}^{\mathrm{avg}}(t_k) = \frac{1}{N_{\mathrm{SEI}}} \sum_{j=1}^{N_{\mathrm{SEI}}} L_{\mathrm{SEI},j}(t_k), \tag{5.13}$$

$$L_{\mathrm{SEI}}^{\mathrm{max}}(t_k) = \max_{j\in\{1,\ldots,N_{\mathrm{SEI}}\}} L_{\mathrm{SEI},j}(t_k), \tag{5.14}$$

$$L_{\mathrm{SEI}}^{\mathrm{min}}(t_k) = \min_{j\in\{1,\ldots,N_{\mathrm{SEI}}\}} L_{\mathrm{SEI},j}(t_k), \tag{5.15}$$

which describe the arithmetic average, the maximum and the minimum of the layer thickness L_{SEI} over the total number of computational cell interfaces with SEI N_{SEI}, respectively. This gives rise to the following relative error criteria in our study

$$d_*\left(L_{\mathrm{SEI}}(t_k)\right) = \left|L_{\mathrm{SEI}}^{*}(t_k) - L_{\mathrm{SEI}}^{*,\mathrm{ref}}(t_k)\right| \cdot \left|L_{\mathrm{SEI}}^{*,\mathrm{ref}}(t_k)\right|^{-1}, \tag{5.16}$$

where $* \in \{\mathrm{avg}, \mathrm{max}, \mathrm{min}\}$. The additional superscript $\cdot^{\mathrm{ref}}$ refers to the respective quantities that are obtained by the reference solution U10, which uses the overall finest extrapolation step size in the present study. Furthermore, we consider the error criterion

$$d_2\left(L_{\mathrm{SEI}}(t_k)\right) = \left\|L_{\mathrm{SEI}}(t_k) - L_{\mathrm{SEI}}^{\mathrm{ref}}(t_k)\right\|_2 \cdot \left\|L_{\mathrm{SEI}}^{\mathrm{ref}}(t_k)\right\|_2^{-1}, \tag{5.17}$$

which utilizes the weighted L_2-norm

$$\|L_{\mathrm{SEI}}(t_k)\|_2 = \left(\sum_{j=1}^{N_{\mathrm{SEI}}} \varepsilon_{\mathrm{e},j}\right)^{-0.5} \cdot \left(\sum_{j=1}^{N_{\mathrm{SEI}}} \varepsilon_{\mathrm{e},j} \left(L_{\mathrm{SEI},j}(t_k)\right)^2\right)^{0.5}, \tag{5.18}$$

and accounts for differences along the spatially resolved SEI profile. Figure 5.3 shows the evolution of the scalar SEI quantities $L_{\mathrm{SEI}}^{\mathrm{avg}}$, $L_{\mathrm{SEI}}^{\mathrm{max}}$, $L_{\mathrm{SEI}}^{\mathrm{min}}$ and the observed convergence rate for the four different error criteria, which are evaluated at the very end of the simulation $t_k = 500\mathrm{h}$.

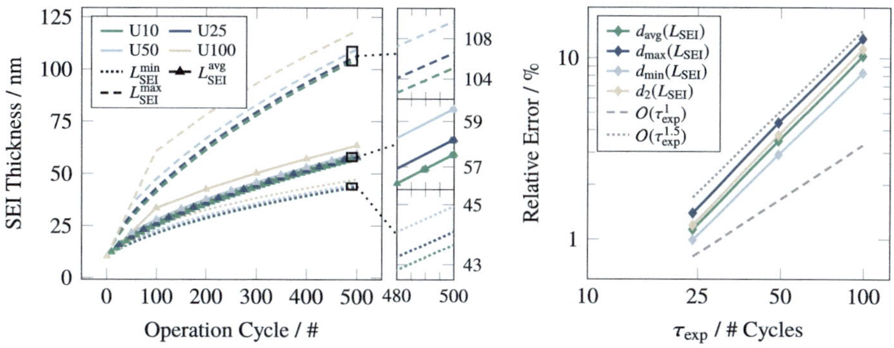

Fig. 5.3 SEI results for uniform extrapolation step sizes. Left: SEI evolution over the full simulation for different extrapolation step sizes. The three zoomed plots highlight the small differences between the three finest step sizes U50, U25, U10 at the end of the simulation. Right: observed numerical convergence rate in the final SEI thickness at $t_k = 500\,\text{h}$ for U100, U50, U25 with decreasing uniform extrapolation step size τ_{exp} when compared to the reference solution U10

Since the SEI growth rate decreases with increasing SEI thickness, we observe a general overestimation of SEI growth for larger extrapolation step sizes. However, the solutions remain stable as this overestimation will result in a reduced growth in the next simulation cycle, slowly correcting the error. When decreasing the extrapolation step size, we observe a decrease in SEI thickness, which gets smaller with every step size refinement, indicating convergence of the solutions. As the convergence plot shows, a numerical convergence order of around 1.5 in τ_{exp} is observed for all four error criteria.

In our previous study [23, 24] we deployed the almost uniform extrapolation scheme NU100 for the present simulation, where the idea was to include an additional step for the 50th operation cycle, because the SEI growth is overall the fastest at the beginning of the simulation. Here, we extend on that approach and compare different non-uniform extrapolation stepping strategies, including the original NU100 scheme. Given a fixed number of simulation cycles of $N^{sim}_{cyc} = 7$, we distribute the simulation cycles amongst the total number of operation cycles $N^{op}_{cyc} = 500$ based on normalized power functions with different rational exponents p

$$n^{op}_{cyc}\left(n^{sim}_{cyc}\right) = \left(N^{op}_{cyc} - 1\right)\left(\frac{n^{sim}_{cyc} - 1}{N^{sim}_{cyc} - 1}\right)^{p} + 1, \tag{5.19}$$

which map the simulation cycle index $n^{sim}_{cyc} \in \{1, \ldots, N^{sim}_{cyc}\}$ to the actual cycle index in the operation protocol $n^{op}_{cyc} \in [1, N^{op}_{cyc}]$. In order to ensure integer results and mesh conformity with our reference solution U10, we round the result of Eq. (5.19) to the nearest multiple of 10 when performing the evaluation. The resulting power function schemes are listed in Table 5.3.

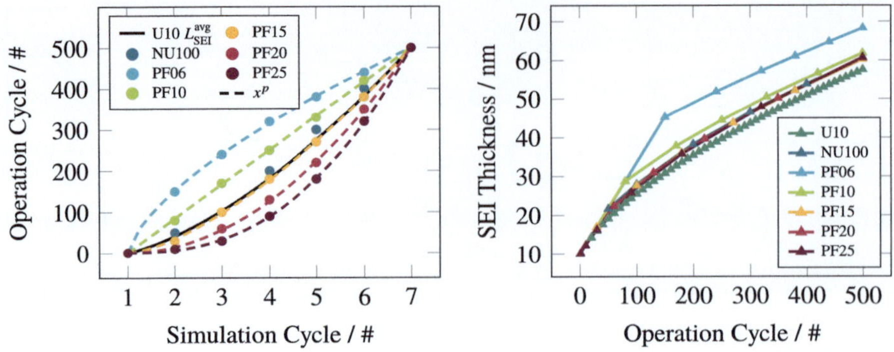

Fig. 5.4 Plots for non-uniform extrapolation step size strategies. Left: the scatter plots highlight the 7 selected simulation cycles for each strategy, while the dashed lines show the underlying continuous power functions. The black line is obtained by shifting and scaling $L_{\text{SEI}}^{\text{avg}}$ from the reference solution U10 to the interval $[1, 7]$. Right: predicted growth of $L_{\text{SEI}}^{\text{avg}}$ for different non-uniform schemes compared to the reference solution U10

Figure 5.4 visualizes the corresponding power functions and selected step sizes and shows the obtained average SEI thickness $L_{\text{SEI}}^{\text{avg}}$ for the different schemes. For the sake of visual clarity, we exclude the minimum $L_{\text{SEI}}^{\text{min}}$ and maximum $L_{\text{SEI}}^{\text{max}}$, but qualitatively they exhibit the same behaviour as the average. We observe the overall trend that the power functions using $p > 1$, which distribute more simulation cycles at the start of the operation protocol, produce more accurate results than the uniform variation PF10 and PF06, which increases the simulation cycle density towards the end of the protocol. Although PF15, PF20 and PF25 produce very similar results, looking at the exact error values in Table 5.4 shows that a minimum is reached for PF15 and the error starts to increase again when increasing p. It also beats our original strategy NU100 in terms of accuracy.

As it turns out, if one takes the result for $L_{\text{SEI}}^{\text{avg}}$ from the reference solution U10, rescales and shifts the average to match the interval $[1, N_{\text{cyc}}^{\text{sim}}]$, one obtains a nearly perfect match with the power function using $p = 1.5$, as shown by the black graph in the left plot of Fig. 5.4. Therefore, in our particular setup, the strategy PF15 is closely related to a physically motivated strategy which tries to get an even average SEI growth in each extrapolation step, which yielded optimal results amongst the tested schemes.

Table 5.4 Errors obtained at the very end of the cycling simulation $t_k = 500\text{h}$ using different non-uniform extrapolation step size strategies from Table 5.3

Error	PF06 (%)	PF10 (%)	PF15 (%)	PF20 (%)	PF25 (%)	NU100 (%)
$d_{\max}$ (L_{SEI})	22.94	9.22	**5.23**	5.86	6.25	6.10
d_{avg} (L_{SEI})	18.64	7.37	**4.63**	5.19	5.55	5.03
$d_{\min}$ (L_{SEI})	14.72	6.05	**4.51**	5.09	5.66	4.61
d_2 (L_{SEI})	20.68	8.01	**4.74**	5.29	5.61	5.26

The lowest value in each row is highlighted in bold font

5.4.3 Parameter Study for the SEI Model

In order to demonstrate the impact of different parameters of the SEI model on the obtained battery aging results, we conduct a parameter study for the present cycling simulation, while using the extrapolation step strategy PF15. We also consider its impact on the numerical solver performance which will be discussed in Sect. 5.4.4. The considered parameter variations are summarized in Table 5.5. We modify one parameter at a time, while all other model parameters from Table 5.6 remain unchanged.

Figure 5.5 shows how the cell aging characteristics change with different parameter choices, while the corresponding average SEI growth curves are displayed in Fig. 5.6. In the following, we briefly describe the effects of the individual parameters.

Increasing the initial SEI thickness $L_{\mathrm{SEI},0}$ results in a corresponding shift in the profile with slightly reduced growth dynamics, because of the longer transport paths through the SEI, causing less relative capacity fade than the base parameter set. Note, that the capacity fade calculation takes the cell capacity at $t = 0\mathrm{s}$ as reference, such that the initial SEI thickness is not accounted for in the calculation.

The fitting parameter ω_{SEI} allows to control the sensitivity of the SEI growth with respect to the local current density i_{tot} at the interface. A value of 0 restricts the SEI growth mechanism to pure diffusion and reduces overall growth, while larger values increase the influence of conductive transport for a given current density, yielding strong SEI growth for $\omega_{\mathrm{SEI}} = 100$.

Modifying the molar volume $V_{\mathrm{SEI}}^{\mathrm{mol}}$ of the SEI products changes the ratio between the thickness of the layer and the associated capacity fade. Larger molar volumes result in a stronger growth of L_{SEI} for a given amount of reactions. Since the layer thickness is directly related to its resistance, this results in increased overpotentials

Table 5.5 Overview of the performed parameter study of the SEI model, which modifies a single parameter at a time

Label	Parameter	New value	Unit	$L_{\mathrm{SEI}}^{\mathrm{min}}$ (nm)	$L_{\mathrm{SEI}}^{\mathrm{avg}}$ (nm)	$L_{\mathrm{SEI}}^{\mathrm{max}}$ (nm)
Base	–	–	–	46	60	111
L20	$L_{\mathrm{SEI},0}$	20	nm	48	63	113
L30	$L_{\mathrm{SEI},0}$	30	nm	53	68	118
w0	ω_{SEI}	0	–	41	49	66
w100	ω_{SEI}	100	–	41	182	380
K/2	κ_{SEI}	$2 \cdot 10^{-7}$	S/cm	48	73	129
Kx2	κ_{SEI}	$8 \cdot 10^{-7}$	S/cm	44	54	92
V/2	$V_{\mathrm{SEI}}^{\mathrm{mol}}$	47.93	$\mathrm{cm}^3/\mathrm{mol}$	32	40	71
Vx2	$V_{\mathrm{SEI}}^{\mathrm{mol}}$	191.72	$\mathrm{cm}^3/\mathrm{mol}$	66	93	171
D/2	D_{SEI}	$2.5 \cdot 10^{-10}$	cm^2/s	32	41	73
Dx2	D_{SEI}	10^{-9}	cm^2/s	65	91	167

The last three columns list the recorded minimum, average and maximum values of the SEI layer thickness at the very end of the respective simulations $t_k = 500\,\mathrm{h}$

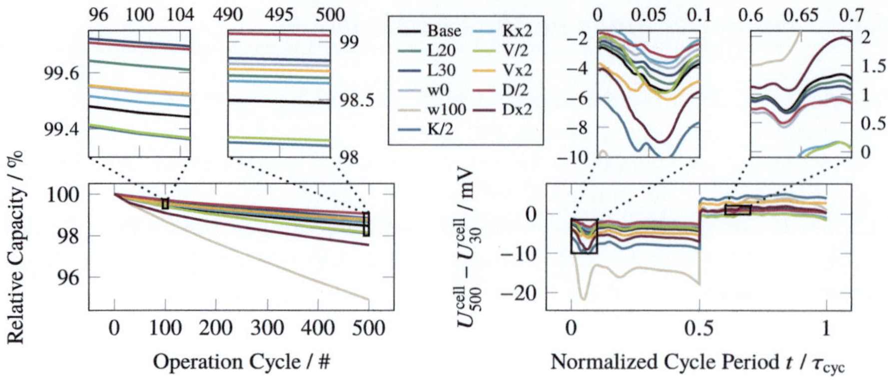

Fig. 5.5 Cell aging results for the performed parameter study of the SEI model. The left plot shows the capacity fade over the full simulation period, while the right plot visualizes the difference in cell voltage between the 500th and 30th operation cycle resulting from increased overpotentials and drift of the open-circuit voltage, due to the SEI growth

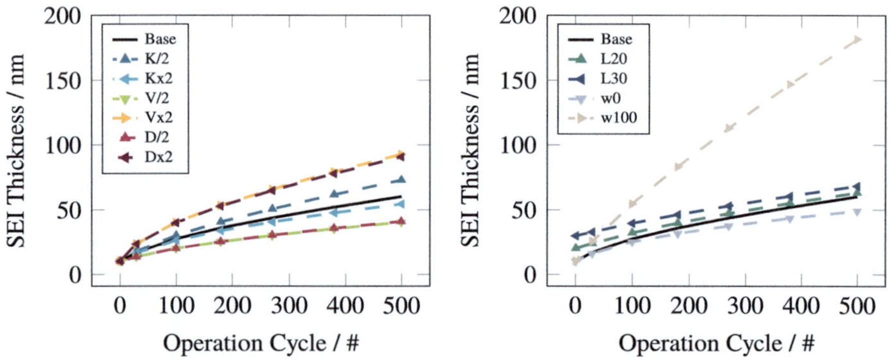

Fig. 5.6 Average SEI growth $L_{\mathrm{SEI}}^{\mathrm{avg}}$ in simulations of the SEI parameter study. Left: comparison between the base parameter set and different values for the SEI conductivity κ_{SEI}, diffusivity D_{SEI} and molar volume $V_{\mathrm{SEI}}^{\mathrm{mol}}$. Right: comparison between the base parameter set and different values for the initial SEI thickness $L_{\mathrm{SEI},0}$ and the current sensitivity parameter ω_{SEI}

across the SEI. At the same time, the reaction dynamics slow down more rapidly, due to the longer transport pathways, resulting in overall less capacity fade than lower molar volumes.

The SEI diffusion coefficient D_{SEI} directly influences the speed of transport of lithium interstitials that are consumed at the interface between SEI and electrolyte. Increasing the diffusion coefficient results in faster transport and more reactions, thus, leading to stronger capacity fade and faster growth of the layer thickness, producing larger overpotentials. The opposite is achieved by decreasing the diffusion coefficient.

The SEI conductivity κ_{SEI} controls the relation between the potential drop-off across the SEI and its layer thickness for a given current density. Reducing the conductivity results in an increased potential drop shifting the balance between intercalation and SEI growth reactions in favour of SEI growth and producing increased overpotentials on cell level.

5.4.4 *Performance*

All simulations have been executed exclusively on single nodes of our computational cluster using a 48 core dual Intel Xeon Gold 6240R processor, while no other jobs were running. For parallelization, OpenMP was deployed utilizing 16 threads for each simulation. Overall, we observe consistent performance metrics for the different simulation runs.

Since the number of simulated cycles of the operation protocol varies between simulations, we consider the runtime per simulated cycle. Averaging this runtime over all presented simulations from Sects. 5.4.2 and 5.4.3, we obtain a value of 30550 s. Regarding the performed iteration numbers of the nonlinear and linear solvers, we only consider the converged Newton iterations, which yields an average over all simulations of 5.49 Newton iterations per time step and 5.92 AMG iterations to solve a linear system for a Newton step. The same metrics can be computed for each individual simulation which allows to determine the deviation with respect to the aforementioned general mean values. These deviations are visualized in Fig. 5.7. The plot on the right features the simulations from the parameter study and demonstrates the robustness of the solver performance when the parameters of the SEI dynamics are altered. Additionally, we include the performance metrics for the different step size strategies for our base parameterization, which shows that the choice in extrapolation step size and the location of the simulated cycles in the overall operation protocol

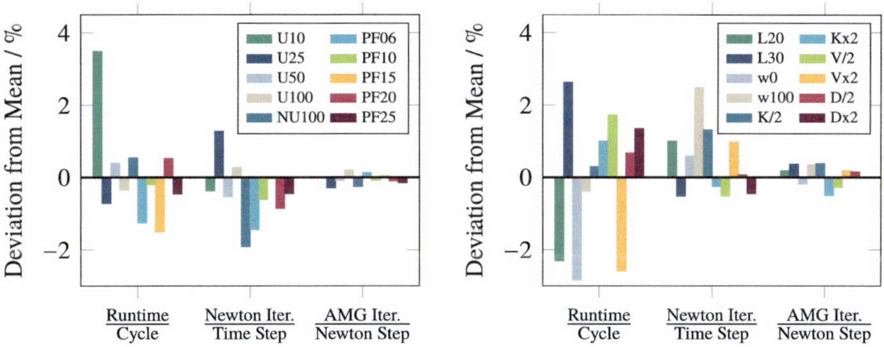

Fig. 5.7 Performance metrics for the performed microscale battery simulations including SEI growth. Each bar shows the deviation between the performance of the respective simulation relative to the average over all performed simulations displayed in the two plots. Left: simulations with different extrapolation step size strategies from Sect. 5.4.2. Right: simulations of SEI parameter study from Sect. 5.4.3

does not significantly impact the solver performance. The largest relative variations are observed in runtime per cycle, but the differences remain below 4% compared to the overall average, indicating a robust behaviour of the deployed solver for all considered scenarios. In particular, the simulations using PF15 result in a reduction of total runtime of around 86.9% compared to our reference simulation U10, while the error measures in Table 5.4 indicate deviations in predicted SEI growth of 4.51–5.23% for our base parameter set. Based on the minor deviations in performance, a linear scaling in the number of simulated cycles is to be expected which suggests a runtime improvement with the scheme PF15 of 98.6% compared to a full simulation without extrapolation, resulting from the drastically reduced number of simulated cycles.

5.5 Conclusion

In this work, we compared different step size strategies for a cyclic extrapolation scheme which performs locally linear extrapolation of the SEI thickness based on the preceding simulation period. The scheme allows to reduce the computational time of a detailed electrochemical microscale model for Li-ion batteries that captures growth of the SEI when performing repetitive cycling studies of hundreds or even thousands of cycles. Our studies for a full cell setup show a numerical order of convergence of 1.5 with respect to uniform extrapolation step sizes. Testing different non-uniform extrapolation step strategies based on power functions, we identify the power function with the closest fit to the SEI growth dynamics to yield the most accurate solution for a fixed number of simulation cycles. This result suggests that an optimal extrapolation strategy might be obtained by a physically motivated method which limits the extrapolation step size based on the growth dynamics of the previous simulation cycle, such that a predefined limit in SEI growth is not exceeded in a single extrapolation step. We would like to point out that the presented methodology is not limited to the presented microscale model class, but might be also applied to less detailed physical models to even further accelerate their computational times.

The small performance deviations between different parameters of the SEI model and extrapolation step size strategies indicate a robust behaviour of the numerical scheme which combines a semi-implicit integration of the regular time steps and the cyclic extrapolation of the SEI thickness. Furthermore, a linear scaling of runtime with respect to the number of simulated cycles in the operation protocol is to be expected.

Acknowledgements The authors would like to acknowledge funding by the German state of Rhineland-Palatinate through the project DiBaZ.

Appendix

We briefly list the required model parameters that have been considered for our simulations which are consistent with our previous studies [23, 24]. Table 5.6 provides an overview of all relevant parameters and lists values that are constant, while functional parameters are given by references to the respective figures and equations.

The ionic conductivity and the ionic diffusion coefficient of the electrolyte are given by [5]

$$\kappa_e(c_e) = \zeta_e \cdot 10^{-3} \cdot \left(2.667 \cdot \left(\frac{c_e}{c_e^{ref}}\right)^3 - 12.983 \cdot \left(\frac{c_e}{c_e^{ref}}\right)^2 + 17.919 \cdot \frac{c_e}{c_e^{ref}} + 1.726\right), \quad (5.20)$$

$$D_e(c_e) = \frac{RT}{F^2 c_e} \kappa_e(c_e), \quad (5.21)$$

using $c_e^{ref} = 1 \text{mol/l}$ and the effective ionic transport factor ζ_e to model effective transport properties resulting from the unresolved pore space of the CBD. Finally, the transference number is given by

$$t_+(c_e) = -0.19333 \cdot \left(\frac{c_e}{c_e^{ref}}\right)^3 + 0.67 \cdot \left(\frac{c_e}{c_e^{ref}}\right)^2 - 0.79667 \cdot \frac{c_e}{c_e^{ref}} + 0.58, \quad (5.22)$$

and approximates data from [17].

Table 5.6 List of model parameters utilized in the simulations

Parameter	Value	Unit	Description	References
Anode AM				
D_{a}	Figure 5.8	cm^2/s	Li-ion diffusion coefficient	[5]
κ_{a}	10^2	S/cm	Electric conductivity	[18]
$U_{\mathrm{a}}^{\mathrm{OCV}}$	Figure 5.8	V	Open-circuit voltage	[5]
$k_{\mathrm{a}}^{\mathrm{BV}}$	1.07	$\mathrm{Acm}^{2.5}/\mathrm{mol}^{1.5}$	Butler-Volmer reaction rate	[4, 5]
$c_{\mathrm{a}}^{\max}$	31.92	mol/l	Maximum Li-ion concentration	[4]
$c_{\mathrm{a},0}$	21.546	mol/l	Initial Li-ion concentration	–
Electrolyte				
D_{e}	Equation 5.21	cm^2/s	Ionic diffusion coefficient	[5]
κ_{e}	Equation 5.20	S/cm	Ionic conductivity	[5]
t_+	Equation 5.22	–	Li-ion transference number	[17]
$c_{\mathrm{e},0}$	1	mol/l	Initial Li-ion concentration	[5]
ζ_{e}	1	–	Effective ionic transport factor	–
Cathode AM				
D_{c}	Figure 5.8	cm^2/s	Li-ion diffusion coefficient	[5]
κ_{c}	0.68	S/cm	Electric conductivity	[4]
$U_{\mathrm{c}}^{\mathrm{OCV}}$	Figure 5.8	V	Open-circuit voltage	[5]
$k_{\mathrm{c}}^{\mathrm{BV}}$	0.29	$\mathrm{Acm}^{2.5}/\mathrm{mol}^{1.5}$	Butler-Volmer reaction rate	[5]
$c_{\mathrm{c}}^{\max}$	48.58	mol/l	Maximum Li-ion concentration	[4]
$c_{\mathrm{c},0}$	22.028	mol/l	Initial Li-ion concentration	–
CBD				
D_{e}	Equation 5.21	cm^2/s	Effective ionic diffusion coefficient	[5]
κ_{e}	Equation 5.20	S/cm	Effective ionic conductivity	[5]
t_+	Equation 5.22	–	Li-ion transference number	[17]
$c_{\mathrm{e},0}$	1	mol/l	Initial Li-ion concentration	[5]
ζ_{e}	0.05	–	Effective ionic transport factor	[10]
ε_{e}	0.5	–	Porosity	[10]
κ_{s}	10	S/cm	Effective electric conductivity	–
CC				
κ_{s}	10^5	S/cm	Electric conductivity	[18]
SEI				
$V_{\mathrm{SEI}}^{\mathrm{mol}}$	95.86	$\mathrm{cm}^3/\mathrm{mol}$	Average molar volume of SEI products	[22]
ν_{SEI}	2	–	Stoichiometric coeff. of Li in SEI reactions	[22]
D_{SEI}	$5 \cdot 10^{-10}$	cm^2/s	Li interstitial diffusion coefficient	[22]
$c_{\mathrm{SEI}}^{\mathrm{Li}}$	$1.5 \cdot 10^{-5}$	mol/l	Reference concentration of Li interstitials	[22]
κ_{SEI}	$4 \cdot 10^{-7}$	S/cm	Ionic conductivity	[23, 24]
ω_{SEI}	10	–	SEI growth sensitivity w.r.t. current	[22]
$L_{\mathrm{SEI},0}$	10	nm	Initial SEI layer thickness	–
Misc.				
T	296	K	Temperature	–

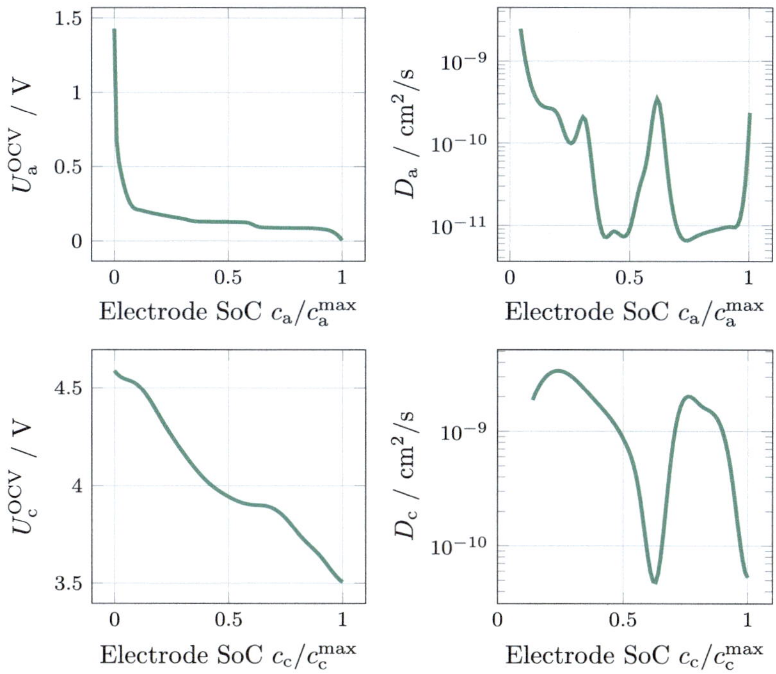

Fig. 5.8 Open-circuit voltage (left) and Li-ion diffusion coefficient (right) for anode (top) and cathode (bottom) active material, respectively

References

1. Attia, P.M., Das, S., Harris, S.J., Bazant, M.Z., Chueh, W.C.: Electrochemical kinetics of SEI growth on carbon black: part I. experiments. J. Electrochem. Soc. **166**(4), E97–E106 (2019). https://doi.org/10.1149/2.0231904jes
2. Bolay, L.J., Schmitt, T., Hein, S., Mendoza-Hernandez, O.S., Hosono, E., Asakura, D., Kinoshita, K., Matsuda, H., Umeda, M., Sone, Y., Latz, A., Horstmann, B.: Microstructure-resolved degradation simulation of lithium-ion batteries in space applications. J. Power Sources Adv. **14**, 100083 (2022). https://doi.org/10.1016/j.powera.2022.100083
3. Doyle, M., Fuller, T.F., Newman, J.: Modeling of galvanostatic charge and discharge of the lithium/polymer/insertion cell. J. Electrochem. Soc. **140**(6), 1526–1533 (1993). https://doi.org/10.1149/1.2221597
4. Ecker, M., Käbitz, S., Laresgoiti, I., Sauer, D.U.: Parameterization of a physico-chemical model of a lithium-ion battery: II. Model validation. J. Electrochem. Soc. **162**(9), A1849–A1857 (2015). https://doi.org/10.1149/2.0541509jes
5. Ecker, M., Tran, T.K.D., Dechent, P., Käbitz, S., Warnecke, A., Sauer, D.U.: Parameterization of a physico-chemical model of a lithium-ion battery: I. determination of parameters. J. Electrochem. Soc. **162**(9), A1836–A1848 (2015). https://doi.org/10.1149/2.0551509jes
6. Fraunhofer ITWM: BEST: Battery and Electrochemistry Simulation Tool (Development Version 2023). https://www.itwm.fraunhofer.de/best
7. Fraunhofer SCAI: SAMG: Algebraic Multigrid Methods for Systems (Release 2017). https://www.scai.fraunhofer.de/samg

8. Fuller, T.F., Doyle, M., Newman, J.: Simulation and Optimization of the Dual Lithium Ion Insertion Cell. Journal of The Electrochemical Society **141**(1), 1–10 (1994). https://doi.org/10.1149/1.2054684
9. Keil, P., Schuster, S.F., Wilhelm, J., Travi, J., Hauser, A., Karl, R.C., Jossen, A.: Calendar aging of lithium-ion batteries. J. Electrochem. Soc. **163**(9), A1872–A1880 (2016). https://doi.org/10.1149/2.0411609jes
10. Kremer, L.S., Hoffmann, A., Danner, T., Hein, S., Prifling, B., Westhoff, D., Dreer, C., Latz, A., Schmidt, V., Wohlfahrt-Mehrens, M.: Manufacturing process for improved ultra-thick cathodes in high-energy lithium-ion batteries. Energy Technol. **8**(2), 1900167 (2020). https://doi.org/10.1002/ente.201900167
11. Kuntz, P., Raccurt, O., Azaïs, P., Richter, K., Waldmann, T., Wohlfahrt-Mehrens, M., Bardet, M., Buzlukov, A., Genies, S.: Identification of degradation mechanisms by post-mortem analysis for high power and high energy commercial Li-Ion cells after electric vehicle aging. Batteries **7**(3) (2021). https://doi.org/10.3390/batteries7030048
12. Latz, A., Zausch, J.: Thermodynamic consistent transport theory of Li-ion batteries. J. Power Sources **196**(6), 3296–3302 (2011). https://doi.org/10.1016/j.jpowsour.2010.11.088
13. Loew, P.J., Poh, L.H., Peters, B., Beex, L.A.: Accelerating fatigue simulations of a phase-field damage model for rubber. Comput. Methods Appl. Mech. Eng. **370**, 113247 (2020). https://doi.org/10.1016/j.cma.2020.113247
14. Lüders, C., Sinapius, M., Krause, D.: Adaptive cycle jump and limits of degradation in micromechanical fatigue simulations of fibre-reinforced plastics. Int. J. Damage Mech. **28**(10), 1523–1555 (2019). https://doi.org/10.1177/1056789519833728
15. Math2Market GmbH: GeoDict Simulation Software (Release 2020). https://www.geodict.com
16. Newman, J., Thomas-Alyea, K.: Electrochemical Systems, 3rd edn. Wiley, The ECS Series of Texts and Monographs (2004)
17. Nyman, A.: An experimental and theoretical study of the mass transport in lithium-ion battery electrolytes. Ph.D. thesis, KTH, Applied Electrochemistry (2011). QC 20110128
18. Park, M., Zhang, X., Chung, M., Less, G.B., Sastry, A.M.: A review of conduction phenomena in Li-ion batteries. J. Power Sources **195**(24), 7904–7929 (2010). https://doi.org/10.1016/j.jpowsour.2010.06.060
19. Popov, P., Vutov, Y., Margenov, S., Iliev, O.: Finite volume discretization of equations describing nonlinear diffusion in Li-Ion batteries. Berichte des Fraunhofer-Instituts für Techno- und Wirtschaftsmathematik (ITWM Report) (191) (2010). http://nbn-resolving.de/urn:nbn:de:hbz:386-kluedo-16598
20. Prada, E., Domenico, D.D., Creff, Y., Bernard, J., Sauvant-Moynot, V., Huet, F.: A simplified electrochemical and thermal aging model of $LiFePO_4$-graphite Li-ion batteries: power and capacity fade simulations. J. Electrochem. Soc. **160**(4), A616–A628 (2013). https://doi.org/10.1149/2.053304jes
21. Preger, Y., Barkholtz, H.M., Fresquez, A., Campbell, D.L., Juba, B.W., Romàn-Kustas, J., Ferreira, S.R., Chalamala, B.: Degradation of commercial lithium-ion cells as a function of chemistry and cycling conditions. J. Electrochem. Soc. **167**(12), 120532 (2020). https://doi.org/10.1149/1945-7111/abae37
22. Schmitt, T.: Degradation models and simulation tools for lithium and zinc batteries. Ph.D. thesis, Universität Ulm (2020). https://doi.org/10.18725/OPARU-29529. https://oparu.uni-ulm.de/xmlui/handle/123456789/29590
23. Schneider, F.: Development and analysis of numerical simulation methods for lithium-ion battery degradation. Ph.D. thesis, TU Kaiserslautern (2023). https://doi.org/10.24406/publica-783. https://publica.fraunhofer.de/handle/publica/434276
24. Schneider, F., Zausch, J., Lammel, J., Andrä, H.: An efficient semi-implicit solver for solid electrolyte interphase growth in Li-ion batteries. Appl. Math. Model. **109**, 741–759 (2022). https://doi.org/10.1016/j.apm.2022.05.009
25. Vermeer, W., Chandra Mouli, G.R., Bauer, P.: A comprehensive review on the characteristics and modeling of lithium-ion battery aging. IEEE Trans. Transp. Electrif. **8**(2), 2205–2232 (2022). https://doi.org/10.1109/TTE.2021.3138357

26. Waldmann, T., Iturrondobeitia, A., Kasper, M., Ghanbari, N., Aguesse, F., Bekaert, E., Daniel, L., Genies, S., Gordon, I.J., Löble, M.W., Vito, E.D., Wohlfahrt-Mehrens, M.: Review-post-mortem analysis of aged lithium-ion batteries: disassembly methodology and physico-chemical analysis techniques. J. Electrochem. Soc. **163**(10), A2149–A2164 (2016). https://doi.org/10.1149/2.1211609jes
27. Wang, P., Cui, L., Lyschik, M., Scholz, A., Berger, C., Oechsner, M.: A local extrapolation based calculation reduction method for the application of constitutive material models for creep fatigue assessment. Int. J. Fatigue **44**, 253–259 (2012). https://doi.org/10.1016/j.ijfatigue.2012.04.018

Chapter 6
Compression and Simulation of Large Insurance Portfolios with New Business

Maximilian Diehl and Jörn Sass

Abstract We develop compression methods for large life insurance portfolios, where the insured collective is grouped into cohorts based on selected contract-related criteria. This allows us to simulate an extremely reduced number of representative contracts. We also show how to efficiently integrate new contracts into the existing insurance portfolio. Furthermore, we investigate the efficiency of the compression methods and their quality in approximating the uncompressed life insurance portfolio. For the simulation of the insurance business, we devise a stochastic asset-liability management model. The incorporated balance sheet model is in line with the principle of double-entry bookkeeping as required in accounting. In extensive simulation studies, we illustrate the short- and long-term behavior of our model and show impacts of different compression methods regarding performance and reliability.

Keywords Life insurance · Compression methods · Cohorts · Model points · Asset-liability management · Balance sheet

6.1 Introduction

Despite the fact that advances in technology with increasing computing capacities allow for faster computations, a contract-specific simulation of a life insurance portfolio is often an impossible task. Reasons for this are the size and heterogeneity of insurance portfolios, long time horizons, and the number of necessary Monte Carlo simulations. Instead, suitable approximation techniques or improved Monte Carlo methods are required.

Several methods are proposed in the literature, [5], e.g., use the least squares Monte Carlo method for the calculation of the solvency capital requirement (SCR) under Solvency II. For this, they need only a small number of simulations, but have to further process the output. In the recommendations of the German Association of Actuaries, an exemplary procedure for the compression of an insurance portfo-

M. Diehl · J. Sass (✉)
RPTU Kaiserslautern-Landau, Kaiserslautern, Germany
e-mail: joern.sass@rptu.de

A. Ott et al. (eds.), *Applications of Mathematics in Sciences, Engineering, and Economics*, Springer Proceedings in Mathematics & Statistics 515,
https://doi.org/10.1007/978-3-032-01279-1_6

lio is presented, which is based on solving a restricted minimization problem for chosen target figures, cf. [2]. This approach is related to cluster-based compression methods like the K-means clustering proposed by [10]. Here, the insurance portfolio is divided into a predefined number of clusters and then a K-means algorithm is applied to improve their location. The corresponding centers then form a compressed insurance portfolio. A K-means algorithm is also used by [4]. Typical drawbacks are the fact that one has to choose K in advance and the algorithm's run time. Furthermore, [6] criticize the absence of an active control of the involved risk features and the implicitly assumed linear dependence. Accordingly, they present a grouping method using neural networks, thereby capturing non-linear dependencies. Numerical investigations for term life insurance contracts and defined contribution plans within a run-off scenario (without new business) indicate that their approach is superior compared to K-means clustering. Both, K-means clustering and methods using neural networks, are not directly applicable to an ongoing insurance business with new business. Indeed, repeating the procedure in every time step where new customers arrive would not be efficient. Neural networks are known to potentially out-perform classic approaches in detecting and re-creating structures in a given data set such as an insurance portfolio. However, a poor performance is possible if the given structure substantially changes in the future and external events occur that cannot be predicted by historical developments. Examples for the latter are crashes in the capital market or extreme new business scenarios. Furthermore, neural networks often yield pointwise forecasts only. Getting information regarding the (empirical) prediction uncertainty may require an appropriate choice of the network architecture or the applied training method. In contrast, Monte Carlo-based simulation methods always provide this information directly through the corresponding quantiles.

In [3], a procedure for the generation of insurers' compressed contract portfolios is presented, in order to be able to simulate at all. Their compression and simulation method allows to illustrate the expected structure of the balance sheet in the case of different business forms of a life insurer, such as a run-off scenario (without new customers) and an ongoing business with stationary new business. Furthermore, due to manageable run times, [3] perform extensive simulation studies and investigate, e.g., the impact of possible capital market crashes on the balance sheet positions and the declared interest rate like in Fig. 6.1. Here, the stock and bond markets crash after 25 years causing stock prices to instantly drop by 40% (top) and 10% of all held bonds to default completely (bottom), respectively.

In this paper, we generalize the compression method of [3] by introducing two parameters for the grouping procedure and investigate the effects of coarser schemes. We illustrate that the introduced model suffers from unacceptable run times, partly even for single simulation runs, if we perform naive contract-specific simulations. This shows that, in particular if considering an ongoing insurance business, where the existing insurance portfolio is permanently augmented with new policies, we need flexible, approximating simulation methods.

The performance of our compression and simulation methods is investigated within a stochastic asset-liability management (ALM) framework that generalizes the version of [3] by allowing for more realistic trading strategies, e.g. prior selling of

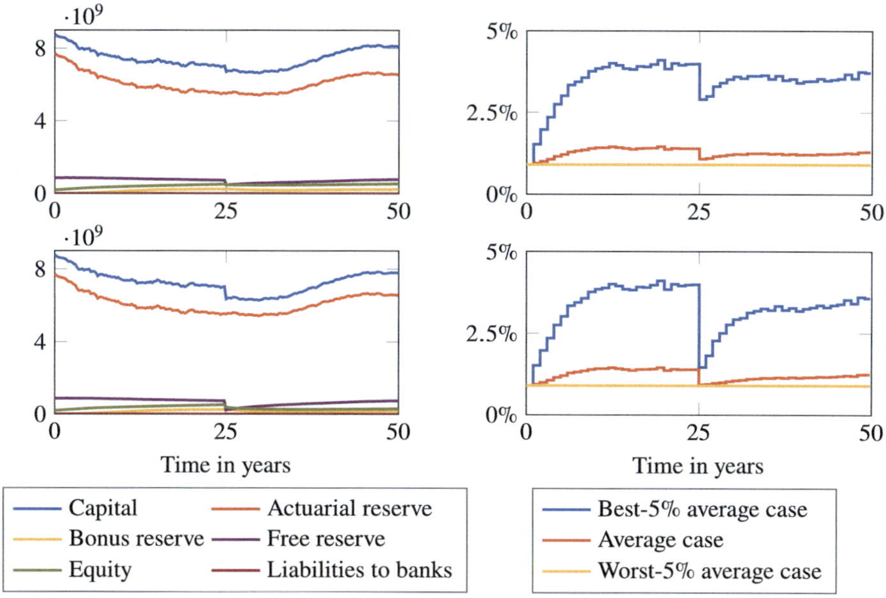

Fig. 6.1 Expected balance sheet positions (left) and declared interest rate $\widehat{i_k}$ (right) in the case of a stock market crash $\left(t^{\mathrm{C}}, z^{\mathrm{C}}\right)^s = (25, 0.4)^s$ (top) and in the case of a bond market crash $\left(t^{\mathrm{C}}, z^{\mathrm{C}}\right)^b = (25, 0.1)^b$ (bottom)

held bonds and by incorporating a minimum stock ratio. We discuss the efficiency of the simulation methods and their quality in approximating the uncompressed insurance portfolio.

6.2 Compression and Simulation Methods

In this section, we develop a class of compression and simulation methods that can be applied to an ongoing insurance business with new business. The basic and standard idea is to divide the insured collective into cohorts and then to simulate only one representative insurance contract per cohort. At first, we describe the characteristics of the considered life insurance portfolio.

6.2.1 Structure of the Considered Life Insurance Portfolio

We denote the considered time horizon by $[0, \mathcal{T}]$, where $\mathcal{T}$ is the number of years to consider and 0 represents the current time. Throughout this paper, we use an equidistant time discretization of that interval, i.e.

$$0 = t_0 < t_1 < \cdots < t_K = \mathcal{T}$$

with $t_k = k\Delta t, k = 0, \ldots, K$, and $\Delta t = \frac{\mathcal{T}}{K} = t_k - t_{k-1}$. Inventory figures like previously signed contracts are linked to past times $t_{-k} < 0$. The initial insurance portfolio at time t_0 consists of δ_0 policies and is characterized by the policyholders'

- gender $g^{\text{real},i} \in \{0, 1\}$,
- age when signing the contract $\underline{x}^{\text{real},i} \in \left[\underline{x}^{\min}, \underline{x}^{\max}\right]$,
- age at maximum contract term $\overline{x}^{\text{real},i} \in \left[\overline{x}^{\min}, \overline{x}^{\max}\right]$,
- current age $x_0^{\text{real},i} \in \left(\underline{x}^{\text{real},i}, \overline{x}^{\text{real},i}\right)$,
- current premium payment $P_0^{\text{real},i} \in \left[P^{\min}, P^{\max}\right]$,
- current contract value $V_0^{\text{real},i} \in \mathbb{R}_{>0}$.

We assume that each policy belongs to one insured person indicated by i. The auxiliary term *real* emphasizes that the corresponding quantity belongs to an actual, *real* policyholder or contract. In contrast, later we will consider *representative* customers or policies (indicated by m) to generate approximating, compressed insurance portfolios.

6.2.2 A Class of Compression and Simulation Methods

According to the compression and simulation method $\mathcal{P}^{b_1,b_2}$, with

$$(b_1, b_2) \in \left\{1, \ldots, \underline{x}^{\max} - \underline{x}^{\min}\right\} \times \left\{1, \ldots, \overline{x}^{\max} - \overline{x}^{\min}\right\}, \tag{6.1}$$

we group those policyholders of the same gender into the same cohort whose current ages and exit ages lie within intervals of length b_1 and b_2, respectively. Here, the insured collective is divided into M_0 non-empty cohorts due to the three grouping criteria

1. gender $g^{\text{real},i}$,
2. transformed current age $\langle x_0^{\text{real},i} \rangle \in \mathbb{N}$, and
3. transformed exit age $\langle \overline{x}^{\text{real},i} \rangle \in \mathbb{N}$.

For given (b_1, b_2) from (6.1), the transformation of the real-valued ages of the individual policyholders requires the construction of intervals

$$\underline{I_l} = \left[\underline{x}^{\min} + (l-1)b_1, \underline{x}^{\min} + lb_1\right),$$
$$\overline{I_l} = \left[\overline{x}^{\min} + (l-1)b_2, \overline{x}^{\min} + lb_2\right),$$

for $l = 1, 2, \ldots,$ until an interval $\underline{I_l}$ and $\overline{I_l}$ contains the maximum entry and exit age $\underline{x}^{\max}$ and $\overline{x}^{\max}$, respectively. The transformed ages are then defined by the lower boundaries of the intervals that contain the corresponding real ages, i.e.,

$$\langle \underline{x}_0^{\text{real},i} \rangle := \underline{x}^{\min} + (l_1 - 1)b_1,$$
$$\langle \overline{x}^{\text{real},i} \rangle := \overline{x}^{\min} + (l_2 - 1)b_2$$

if $x_0^{\text{real},i} \in \underline{I_{l_1}}$, $\overline{x}^{\text{real},i} \in \overline{I_{l_2}}$.

The representative policyholders

After the grouping, we assign numbers and select randomly a representative policyholder from each cohort $m \in \{1, \dots, M_0\}$. In doing so, we take fractions of the year into account. We denote the representative's

- gender by g^m,
- its current age by $x_0^m \in \mathbb{R}^+$, and
- its exit age by $\overline{x}^m \in \mathbb{R}^+$.

The remaining contract characteristics are determined by arithmetic means. The initial size of the insurance portfolio δ_0 can be written as $\delta_0 = \sum_{m=1}^{M_0} \delta_0^m$ with δ_0^m denoting the cohorts' sizes.

The updated current age x_k^m of the representative policyholder from cohort m at time t_k, $k \geq 1$, which is needed to assign the corresponding death probability, and the remaining contract terms in periods d_k^m are described recursively, i.e.

$$x_k^m = x_{k-1}^m + \Delta t \qquad \text{and} \qquad d_k^m = d_{k-1}^m - 1,$$

with $d_0^m = \left\lceil \Delta t^{-1} \left(\overline{x}^m - x_0^m \right) \right\rceil \in \mathbb{N}$ and $\lceil x \rceil = \min \{ z \in \mathbb{Z} \mid z \geq x \}$.

Remark 6.1 (*Special cases $\mathcal{P}^{1,1}$ and $\mathcal{P}^{0,0}$*) According to the compression method $\mathcal{P}^{1,1}$, two policyholders are assigned to the same cohort if they have the same gender, integer current age, and integer exit age. This special case is applied in [3]. The method of simulating the uncompressed insurance portfolio will later be denoted by $\mathcal{P}^{0,0}$ and corresponds to a naive contract-specific simulation.

The development of the cohorts

To model the new business, we assume that a random number $\delta_{k-1}^{\text{new}}$ of new customers arrives in every period k, independent of other periods. The policies are signed at the beginning of a period $[t_{k-1}, t_k]$ and associated with individual characteristics as described in Sect. 6.2.1. The collective of new customers is first divided into N_{k-1} cohorts based on the established criteria and the remaining contract characteristics are determined by arithmetic means as before. Then, we integrate new customers into existing cohorts if the grouping criteria of a new cohort $n \in \{1, \dots, N_{k-1}\}$ coincide with those of one of the existing cohorts $m \in \{1, \dots, M_{k-1}\}$, i.e.

- $g^{\text{new},n} = g^m$,
- $\langle \underline{x}^{\text{new},n} \rangle = \langle x_{k-1}^m \rangle$, and
- $\langle \overline{x}^{\text{new},n} \rangle = \langle \overline{x}^m \rangle$.

Here, the updated transformed current age $\langle x_k^m \rangle$ for $k \geq 1$ is defined by

$$\langle x_k^m \rangle = \begin{cases} \langle x_{k-1}^m \rangle + 1, & \text{if } k \equiv 0 \left(\text{mod } \frac{1}{\Delta t}\right) \\ \langle x_{k-1}^m \rangle, & \text{else.} \end{cases} \tag{6.2}$$

The representative premium payments and the contract values are then adjusted according to weighted averages, i.e.

$$\begin{aligned} P_{(k-1)^+}^m &= \frac{1}{\delta_{k-1}^m + \delta_{k-1}^{\text{new},m}} \left(\delta_{k-1}^m P_{k-1}^m + \delta_{k-1}^{\text{new},m} P_{k-1}^{new,m} \right), \\ V_{(k-1)^+}^m &= \frac{\delta_{k-1}^m}{\delta_{k-1}^m + \delta_{k-1}^{\text{new},m}} V_{k-1}^m + P_{(k-1)^+}^m, \end{aligned}$$

where $\delta_{k-1}^{\text{new},m}$ denotes the number of additional customers in cohort m. Those cohorts who could not be merged augment the set of existing model points by M_{k-1}^{add}, i.e. we then have $M_k = M_{k-1} + M_{k-1}^{\text{add}}$.

Decrements of a cohort m are assumed to occur at the end of a period $[t_{k-1}, t_k]$. The corresponding periodic death and surrender probabilities are assumed to be independent and denoted by q_k^m and u_k^m, respectively. They are associated with the representatives of every cohort m for time period $\left[t_{k-1}, t_k\right]$ and defined by

$$\begin{aligned} q_k^m &= 1 - \left(1 - q\left(\min\left\{\left\lfloor x_{k-1}^m \right\rfloor, x_{\max}\right\}, g^m, y^m\right)\right)^{\Delta t} \quad \text{and} \\ u_k^m &= \mathbb{1}_{\{d_k^m > 0\}} \cdot \left(1 - \mathrm{e}^{-\kappa \Delta t}\right). \end{aligned}$$

Here, the annual death probabilities $q = q(x, g, y)$, depending on the age x, on the gender g, and on the year of birth y, are taken from the cohort life tables for Germany provided by the Federal Office of Statistics.[1] The parameter $x_{\max}$ denotes the maximum age in the life table[2] and $y^m = \left\lfloor \mathcal{Y} - x_0^m \right\rfloor$ is the representative's year of birth with $\mathcal{Y}$ being the current calendar year. Note that we assumed a constant force of mortality throughout the year. The assumption of constant surrender probabilities might be justified by the averaging of policies including newly signed ones to form the updated cohort throughout the simulation which balances the effects of individual contracts. The indicator function $\mathbb{1}_{\{\}}$ ensures that surrender is only possible if the insurance policy is still in force in the following period.

Taking the remaining contract period d_k^m into account, the number of policyholders in model point m at time t_k which remain in the insured collective is given by

$$\delta_k^m = \mathbb{1}_{\{d_k^m > 0\}} \cdot \left(1 - u_k^m\right)\left(1 - q_k^m\right)\left(\delta_{k-1}^m + \delta_{k-1}^{\text{new},m}\right).$$

---

[1] We use the cohort life table Variant 1, which is based on a trend in the development of mortality since 2011, see [8]. More details on the derivation of the death probabilities can be found in [9].

[2] In our case, we have $x_{\max} = 100$.

In our simulation studies, we delete cohorts becoming empty and denote the number of cohorts that are actually modeled by $M_k^{\mathrm{sim}} \leq M_k$.

6.3 Application of the Compression and Simulation Methods

In this section, we provide a framework for the application of our compression and simulation methods. We introduce a stochastic balance sheet model being embedded in an asset-liability management (ALM) framework for life insurers.

6.3.1 *Structure of the ALM Model*

The model in this paper generalizes the version presented in [3]. It consists of a capital market model, a management model, a balance sheet model, and a liability model. The capital market is modeled by short rate and stock price processes and the insurer's financial assets are presented in simplified form by stocks and bonds. On the counterpart, we have liabilities formed by the insurance contracts comprising existing and new business. The management model describes the insurance company's decisions regarding the asset allocation and the declaration of the bonus payments and of the annual interest return. In contrast to the model in [3], we relax some trading constraints, e.g. by allowing for prior selling of held bonds, and we incorporate a minimum stock ratio. Due to diverging assumptions, we provide new results on the consistency of our model (cf. Theorem 6.2). Each of these four modules can be adjusted or extended further, without affecting the overall structure of our model. Corresponding modifications such as alternative modeling approaches of the capital market, different investment strategies or surplus participation schemes can be found in [7].

6.3.2 *Some Details and Results of the ALM Model*

Table 6.1 displays the balance sheet forming the reference for the simulation of the asset-liability management.

The total capital C_k of the assets is allocated to bonds of different times to maturity, stocks, and a cash position with market values C_k^b, C_k^s, and C_k^c, respectively. On the opposite side of the balance sheet the liabilities comprise the equity Q_k, the free reserve F_k, the technical reserve V_k, and the liabilities to banks L_k. The technical reserve represents both the obligations towards the policyholders arising from the guarantees embedded in the life insurance contracts and surpluses that have been

Table 6.1 Considered balance sheet at time t_k

Assets		Liabilities	
Bonds	C_k^b	Equity	Q_k
Stocks	C_k^s	Free reserve	F_k
Cash	C_k^c	Technical reserve	V_k
		Liabilities to banks	L_k
Total	C_k	Total	C_k

credited to individual contracts.[3] In contrast, policyholders are not entitled to unappropriated and unallocated surpluses being registered in the free reserve. Finally, the insurance company can take loans which must be registered as liabilities to banks.

Note that it is required that the sum of all assets equals the sum of all liabilities. This balance sheet equation is already indicated in Table 6.1 and explicitly proven for our model (cf. Theorem 6.2).

Capital market model

To model the bond market, we choose a discretized version of the Vasiček model. The short rate r_k at time t_k follows the dynamics

$$r_k = r_{k-1} + (b - ar_{k-1})\,\Delta t + \sigma_r \Delta_k W_r,$$

where $\Delta_k W_r := W_r(t_k) - W_r(t_{k-1})$ for a Brownian motion W_r under the real-world measure P and with $b = \tilde{b} + \lambda\sigma_r \in \mathbb{R}$, $a = \tilde{a} > 0$, and $\sigma_r > 0$. Here, $\tilde{b}$ and $\tilde{a}$ denote the short rate parameters under the risk-neutral measure Q defined by choosing a market price of risk for interest rates of the form $\lambda(t_k) = \lambda$ with $\lambda \in \mathbb{R}$. The corresponding price of the zero-coupon bond with maturity T is then given by

$$p\,(t_k, T) \doteq A(t_k, T)\exp\left(-r_k B(t_k, T)\right)$$

with

$$B(t_k, T) = \frac{1 - e^{-\tilde{a}(T-t_k)}}{\tilde{a}},$$

$$A(t_k, T) = \exp\left(\left(\frac{\tilde{b}}{\tilde{a}} - \frac{\sigma_r^2}{2\tilde{a}^2}\right)(B(t_k, T) - (T - t_k)) - \frac{\sigma_r^2}{4\tilde{a}} B(t_k, T)^2\right).$$

The stock prices follow a discretized geometric Brownian motion model,

[3] In [3], it corresponds to the sum of actuarial reserve and bonus reserve.

$$s_k = s_{k-1}\exp\left\{\left(\mu_s - \frac{1}{2}\sigma_s^2\right)\Delta t + \sigma_s \Delta_k W_s\right\},$$

with drift $\mu_s \in \mathbb{R}$, volatility $\sigma_s > 0$, and $W_s(t) = \rho W_r(t) + \sqrt{1-\rho^2}Z(t)$ for a Brownian motion Z independent of W_r. Thus, $W_s(t)$ and $W_r(t)$ have correlation ρ.

Liability and management model

We consider an insurance portfolio consisting of classic endowment insurances that are equipped with a surrender option, time-depending death benefits, and a lump-sum benefit payment at the expiration of the contract. Here, the individual survival, death, and surrender benefit payments resulting from cohort m at time t_k are given by

$$E_k^m = \mathbb{1}_{\{d_k^m=0\}} V_k^m, \quad T_k^m = \mathbb{1}_{\{d_k^m \geq 0\}} V_k^m, \quad \text{and } S_k^m = \mathbb{1}_{\{d_k^m > 0\}} \vartheta V_k^m,$$

where $\vartheta \in [0, 1]$ denotes the surrender factor. The policyholders' contract values are adjusted at the end of every period according to

$$V_k^m = (1 + i_k)\, V_{(k-1)^+}^m,$$

where i_k denotes the periodic interest rate derived from the annually declared interest rate $\widehat{i_k}$ (note that we may have more periods than years), with

$$\widehat{i_k} = \begin{cases} \max\left\{\widehat{i_G}, \omega \cdot (\gamma_{k-1} - \gamma)\right\} & \text{if } k \equiv 1 \left(\operatorname{mod} \frac{1}{\Delta t}\right), \\ \widehat{i_{k-1}} & \text{else}. \end{cases}$$

Here, $\widehat{i_G}$ denotes the guaranteed interest rate and ω the distribution ratio weighting the deviation between the current reserve rate

$$\gamma_{k-1} = \frac{F_{k-1}}{F_{k-1} + V_{k-1}}, \tag{6.3}$$

and a target value $\gamma \in [0, 1]$.

The life insurer's total obligations consist of expiring (long-term) credits L_k^- and due benefit payments

$$\mathcal{B}_k = E_k + T_k + S_k,$$

where the survival, death, and surrender benefit payments E_k, T_k, and S_k are calculated by

$$E_k = \sum_{m=1}^{M_k} (1 - u_k^m)(1 - q_k^m)(\delta_{k-1}^m + \delta_{k-1}^{\text{new},m}) E_k^m,$$

$$T_k = \sum_{m=1}^{M_k} q_k^m (\delta_{k-1}^m + \delta_{k-1}^{\text{new},m}) T_k^m,$$

$$S_k = \sum_{m=1}^{M_k} u_k^m (1 - q_k^m)(\delta_{k-1}^m + \delta_{k-1}^{\text{new},m}) S_k^m.$$

As investment strategy for the periodic asset allocation we consider a constant mix strategy, where the insurance company strives to have a fixed share quota $\pi_k^{s,\text{tar}} \in [\pi_k^{s,\min}, \pi_k^{s,\max}]$, with the maximum depending on regulatory requirements.

Balance sheet model

The balance sheet position cash C_k^c equals the amount of expired bonds that are not needed to meet the due obligations. The positions C_k^s and C_k^b are defined by the market values of held bonds and stocks. The policyholders' accounts V_k^m are linked to the technical reserve V_k via

$$V_k = \sum_{m=1}^{M_k} \delta_k^m V_k^m. \tag{6.4}$$

The following proposition shows the relation to the previous balance sheet at time t_{k-1}.

Proposition 6.1 (Recursive schemes of the technical reserve) *Consider the endowment insurance with surrender factor $\vartheta > 0$. Then, it holds for all $k = 1, \ldots, K$:*

$$V_k = (1 + i_k)\left(V_{k-1} + P_{k-1}\right) - \left(E_k + T_k + \frac{1}{\vartheta} S_k\right),$$

where

$$P_{k-1} = \sum_{m=1}^{M_k} \mathbb{1}_{\{d_{k-1}^m > 0\}} \left(\delta_{k-1}^m + \delta_{k-1}^{\text{new},m}\right) \cdot P_{(k-1)^+}^m$$

is the total premium income at the beginning of period $[t_{k-1}, t_k]$.

Proof See Proposition 2.1 in [3]. □

The balance sheet position L_k (liabilities to banks) consists of bridging loans, i.e. interest-free over-night credits, and long-term credits with variable interest rates, i.e. emitted bonds with different times to maturity. The last two positions, the free reserve F_k and the equity Q_k, depend on the generated surpluses G_k and the method of how they are distributed between the insured collective and the shareholders. This is done by introducing a participation rate α specifying the share of the risk surplus that has to be credited to the policyholders' accounts. In particular, F_k and Q_k satisfy

$$F_k + Q_k = F_{k-1} + Q_{k-1} + G_k.$$

The following theorem shows that the life insurer's business activities lead to equal sums of assets and liabilities as displayed in Table 6.1.

Theorem 6.2 (Verification of the model) *Consider the endowment insurance with surrender factor $\vartheta > 0$ and suppose that the sum of all assets equals the sum of all liabilities at the start of the simulation, i.e., $C_0 = V_0 + F_0 + Q_0 + L_0$. Then, the fundamental balance sheet equation is fulfilled at any time, i.e. it holds*

$$C_k = V_k + F_k + Q_k + L_k$$

for all $k = 0, \ldots, K$.

Proof Similar to the proof of Theorem 2.2 in [3]. □

6.4 Simulation Studies

In this section, we perform several simulation studies to investigate the performance of some of the compression and simulation methods $\mathcal{P}^{b_1,b_2}$ introduced in Sect. 6.2.2. We analyze their efficiency and quality in approximating the involved distributions and the development of the uncompressed insurance portfolio and the corresponding balance sheets.

6.4.1 Introduction

We consider a portfolio consisting of classic endowment insurances with characteristics as described in Sect. 6.2.1. The initial contract values are calculated according to

$$V_0^{\text{real},j} = \sum_{k=1}^{c_0^{\text{real},j}} \left(1 + \widehat{i}_G\right)^{k\Delta t} P_{-k}^{\text{real},j}, \qquad j = 1, \ldots, \delta_0.$$

Here, $P_{-k}^{\text{real},j} = P_0^{\text{real},j}$ are the past premium payments and $c_0^{\text{real},j}$ denotes the elapsed contract term in periods defined by

$$c_0^{\mathrm{real},j} = \left\lfloor \Delta t^{-1} \left(x_0^{\mathrm{real},j} - \underline{x}^{\mathrm{real},j} \right) \right\rfloor \qquad \text{with} \qquad \lfloor x \rfloor = \max \{ z \in \mathbb{Z} \mid z \leq x \}.$$

The initial number of model points M_0 and the sizes of the cohorts δ_0^m are determined by the applied grouping method and depend on the distribution of the actual biometric parameters of the insured collective. The representative contract values V_0^m and premium payments P_0^m are determined by arithmetic means.

The initial values of the technical reserve V_0 and the free reserve F_0 are given and implied by Eqs. (6.4) and (6.3), respectively. The amount of equity Q_0 is determined by the initial fraction of own funds

$$\psi_0 = \frac{F_0 + Q_0}{V_0 + F_0 + Q_0 + L_0}.$$

Respecting the fundamental balance sheet equation at time t_0 (cf. Theorem 6.2) yields

$$C_0^b + C_0^s + C_0^c = V_0 + Q_0 + F_0 + L_0,$$

where the initial values of the stock and cash position are specified by fractions of the total assets $C_0 = C_0^b + C_0^s + C_0^c$.

Table 6.2 specifies the remaining parameters and the distributional assumptions used for the following numerical investigations if not stated otherwise. The size of the insurance portfolio is chosen such that a contract specific simulation is still possible, since we aim at analyzing the error compared to the uncompressed insurance portfolio. Indeed, the compression and simulation methods can be applied to much larger insurance portfolios, see Table 6.4. Regarding the initial balance sheet, there is no liquidity gap at the beginning of the first period, i.e. $L_0^+ = L_0 = 0$, and the initial amount of cash equals the average excessive value of expiring bonds obtained by prior simulations. Using the beta distribution for the average number Λ_k of new customers per period k is motivated by the high flexibility of this distribution family. Alternative new business scenarios can be considered by choosing different shape parameters α_k and β_k as done in Table 6.3. The remaining two parameters represent the minimum and maximum amount of newly issued contracts per period. Their values are inspired by observations from annual business reports of Allianz Life[4] being currently the largest life insurer in Germany.[5]

We consider three new business scenarios (NBS) being parameterized in Table 6.3 and illustrated in Fig. 6.2. NBS 1 reflects the scenario of a stationary new business, while positive and negative shocks on the expected future number of new customers are considered in the other two scenarios.

In Fig. 6.3, we illustrate the expected structure of the balance sheet as displayed in Table 6.1 in the case of an ongoing insurance business with NBS 3. In particular, the graphs visualize the fulfillment of the fundamental balance sheet equation (cf. Theo-

[4] The annual reports from 2013 to 2023 are available at www.allianz.de.

[5] According to the statistics regarding direct insurers provided by the German Federal Financial Supervisory Authority (BaFin), the market share in 2022 was 23.51%, see [1].

Table 6.2 Exemplary scenario for the following simulation studies. We denote by $\mathcal{B}(p)$ the Bernoulli distribution with parameter p, by $\mathcal{TN}(\mu, \sigma^2, a, b)$ the truncated normal distribution with mean μ, variance σ^2, and support on $[a, b]$, by $\mathcal{U}(a, b)$ the uniform distribution on $[a, b]$, by $\mathrm{Poi}(\lambda)$ the Poisson distribution with parameter λ, and by $\mathrm{Beta}(\alpha, \beta, a, b)$ the four parameter Beta distribution with shape parameters α and β and support on $[a, b]$. For a random variable $X \sim \mathrm{Beta}(\alpha, \beta)$, we have $Y := a + (b - a)X \sim \mathrm{Beta}(\alpha, \beta, a, b)$

Parameter	Description	Value
General setting and balance sheet		
$\mathcal{Y}$	Current calendar year	2022
$\mathcal{T}$	Time horizon	40 years
Δt	Period length	0.25 years
γ_0	Initial reserve rate	10%
ψ_0	Initial fraction of own funds	12%
π_0^s	Initial ratio of stocks	10%
π_0^c	Initial ratio of cash	6.03%
L_0^+	Initial bridging credit	0
L_0	Initial value of liabilities to banks	0
$\pi_k^{s,\max}$	Maximum stock ratio	35%
Capital market model		
a	Mean-reversion speed of the short rate process	0.5
b/a	Long-term mean of the short rate process	0.7%
λ_0	Market price of interest rate risk parameter	2%
σ_r	Volatility of the short rate process	3%
r_0	Initial value of the short rate process	0.1%
μ_S	Drift of the stock price process	4%
σ_S	Volatility of the stock price process	20%
s_0	Initial value of the stock price process	100
ρ	Correlation between short rate and stock	−10%
Management model		
$\pi_k^{s,\min}$	Minimum stock ratio	5%
$\pi_k^{s,\mathrm{tar}}$	Target stock ratio in the CM strategy	10%
τ	Maturity of bonds	3 years
$\widehat{i}_G$	Annual guaranteed interest rate	0.25%
γ	Target reserve rate	10%
ω	Distribution ratio	0.3
α	Participation rate	0.9
ϑ	Surrender factor	0.9
Structure and dynamics of insurance portfolio		
$g^{\mathrm{real},i}$	Gender	$\sim \mathcal{B}(0.5)$
$\underline{x}^{\mathrm{real},i}$	Entry age	$\sim \mathcal{TN}(38, 64, 15, 65)$
$\overline{x}^{\mathrm{real},i}$	Exit age	$\sim \mathcal{TN}(73, 25, 65, 80)$
$x_0^{\mathrm{real},i}$	Current age	$\sim \mathcal{U}\left(\underline{x}^{\mathrm{real},i}, \overline{x}^{\mathrm{real},i}\right)$
κ	Surrender probability parameter	3%
$p^{\mathrm{real},i}$	Constant, periodic premium payments	$\sim \mathcal{U}(50, 500)$
δ_0	Initial size of the insurance portfolio	10,000
δ_k^{new}	Number of new customers in period k	$\sim \mathrm{Poi}(\Lambda_k)$
Λ_k	Poisson parameter in δ_k^{new}	$\sim \mathrm{Beta}\left(\alpha_k, \beta_k, 0.5\% \cdot \delta_0, 2.2\% \cdot \delta_0\right)$

Table 6.3 Parameterization of the considered new business scenarios (NBS)

NBS	Shape parameters α_k and β_k
1	$(\alpha_k, \beta_k) = (1, 1)$ for $k = 1, \ldots, K$
2	$(\alpha_k, \beta_k) = (2, 20)$ for $k = 1, \ldots, \frac{K}{2}$ and $(\alpha_k, \beta_k) = (20, 2)$ for $k = \frac{K}{2} + 1, \ldots, K$
3	$(\alpha_k, \beta_k) = (20, 2)$ for $k = 1, \ldots, \frac{K}{2}$ and $(\alpha_k, \beta_k) = (2, 20)$ for $k = \frac{K}{2} + 1, \ldots, K$

Table 6.4 Average required run times in seconds for simulating a single Monte Carlo path applying $\mathcal{P}^{0,0}$ and $\mathcal{P}^{1,1}$ for insurance portfolios of different sizes δ_0 in the case of an ongoing insurance business with NBS 1

δ_0	Run time [s] $\mathcal{P}^{0,0}$	Run time [s] $\mathcal{P}^{1,1}$
10,000	8.5	1.2
50,000	33	2.8
100,000	63	3.8
250,000	157	5.0
500,000	300	6.1
1,000,000	600 (estimated)	7.3
10,000,000	6000 (estimated)	16.0

rem 6.2). For the simulation, we used the same compression as [3], i.e. method $\mathcal{P}^{1,1}$ in our notation. In the following sections, we study the reliability of the compression and the effects of applying coarser simulation methods $\mathcal{P}^{b_1,b_2}$ with $\max\{b_1, b_2\} > 1$.

6.4.2 *Efficiency of the Compression Methods*

Before performing the simulation studies, where each Monte Carlo simulation will consist of $N = 10,000$ simulated paths, we investigate the dependence of the required run time for a single simulation run on the size δ_0 of the initial insurance portfolio (note that the number δ_k^{new} of new customers in each period also depends on δ_0). We use the parameters from Table 6.2 with a time horizon $\mathcal{T} = 50$ years and $\Delta t = 0.25$ yielding a total number of $K = 200$ considered periods. We consider an ongoing insurance business with stationary new business (NBS 1). The required run times are displayed in Table 6.4.

The contract-specific simulation $\mathcal{P}^{0,0}$ requires 300 seconds per path, i.e. 833 hours for a Monte Carlo simulation if the initial insurance portfolio consists of $\delta_0 = 500,000$ policies. The average run time of $\mathcal{P}^{1,1}$ is 6.1 and thus nearly 50 times

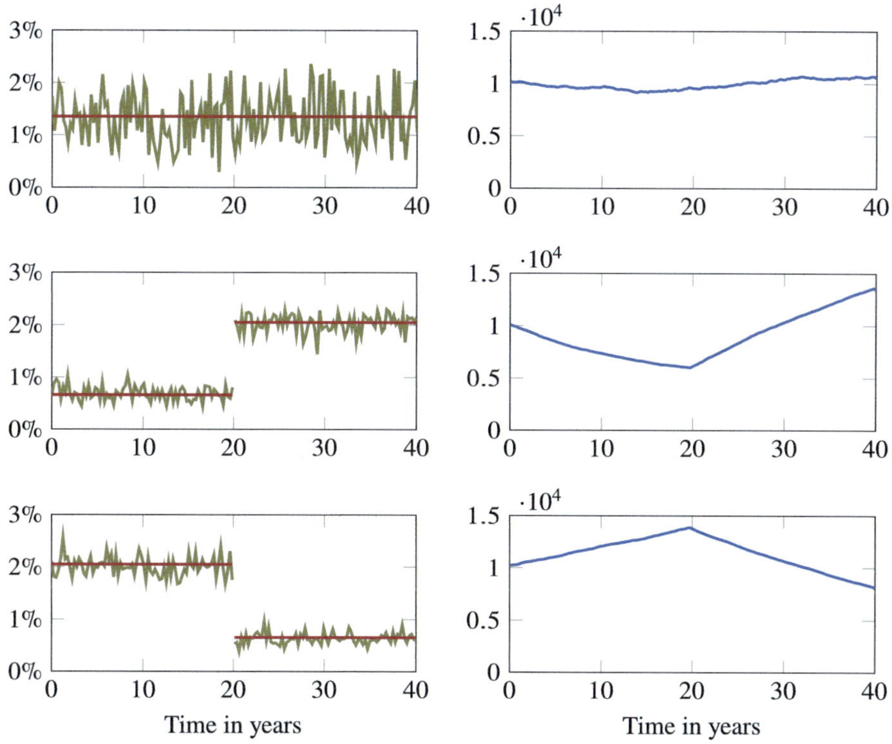

Fig. 6.2 New business scenarios NBS 1 (top) to NBS 3 (bottom). Left: number of new customers per period (expected number in red, one path in green), right: corresponding path of the (uncompressed) insurance portfolio size

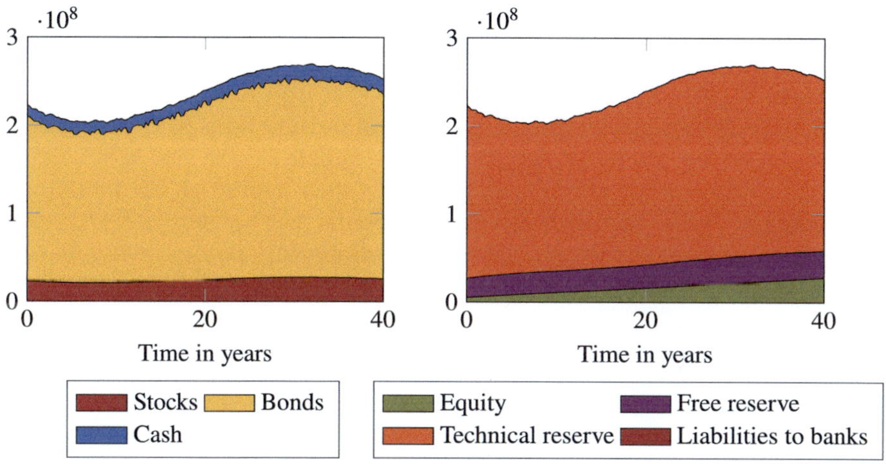

Fig. 6.3 Expected balance sheet structure in the case of an ongoing insurance business with NBS 3. Left: assets, right: liabilities

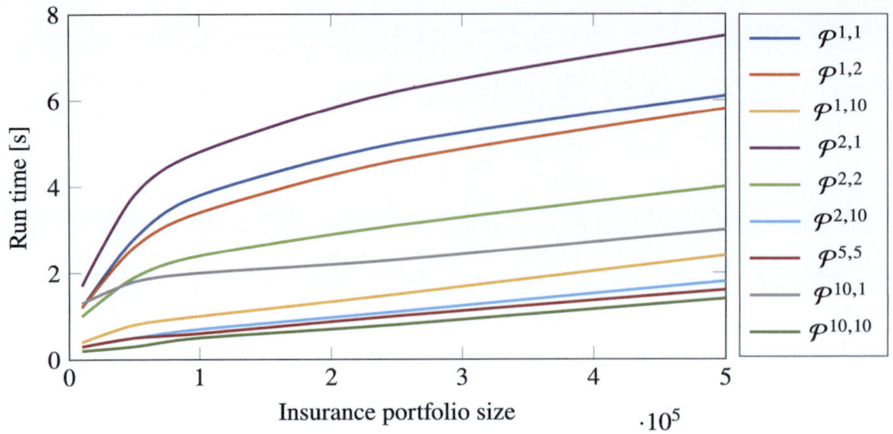

Fig. 6.4 Average required run times in seconds for simulating a single Monte Carlo path applying different simulation methods $\mathcal{P}^{b_1,b_2}$ for insurance portfolios of different sizes δ_0 in the case of an ongoing insurance business with NBS 1

smaller.[6] We also see that the dependence on δ_0 is stronger if we apply $\mathcal{P}^{0,0}$, i.e. here the run time is more exposed to the size of the insurance portfolio.

If we apply coarser simulation methods, we may even increase the time saving as illustrated in Fig. 6.4, where the corresponding run times are plotted against the sizes of the initial insurance portfolios. However, the additional amount of saved time gets smaller so that the run times of $\mathcal{P}^{2,10}$, $\mathcal{P}^{5,5}$, and $\mathcal{P}^{10,10}$ are nearly the same. Note that for $\mathcal{P}^{1,1}$, we obtain the transformed ages for grouping by the corresponding integer values. This is faster than following the procedure described in Sect. 6.2.2 for $b_1 = b_2 = 1$ and yields the same compressed insurance portfolio. As a consequence, the run time of $\mathcal{P}^{1,1}$ may be smaller than for coarser simulation methods, here $\mathcal{P}^{2,1}$. In most cases, less additional run time is required for larger δ_0. This is due to the fact that the number of generated cohorts increases slower for large insurance portfolios and is bounded from above due to the finite number of possible ages. More importantly, new policies are merged into existing cohorts if possible.

Next we elaborate on the efficiency in terms of the compression factor reflecting the proportion of the simulated (non-empty) cohorts $M_k^{\text{sim}} = M_k^{\text{sim}}\left(\mathcal{P}^{b_1,b_2}\right)$ to the actual size δ_k of the uncompressed insurance portfolio $\mathcal{P}^{0,0}$, i.e.

$$CF_k = \frac{M_k^{\text{sim}}}{\delta_k}.$$

[6] If we fix a path of new customers, the required run time is only 0.02 s per path even for $\delta_0 = 10,000,000$, i.e. about three minutes for a Monte Carlo simulation. A contract-specific simulation is eight times slower for $\delta_0 = 500,000$ and may cause memory problems for larger insurance portfolios.

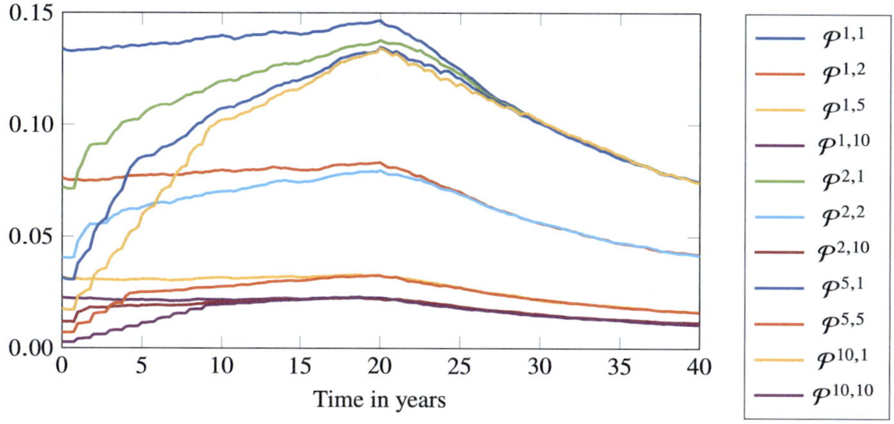

Fig. 6.5 Compression factors applying different simulation methods $\mathcal{P}^{b_1,b_2}$ in the scenario NBS 2

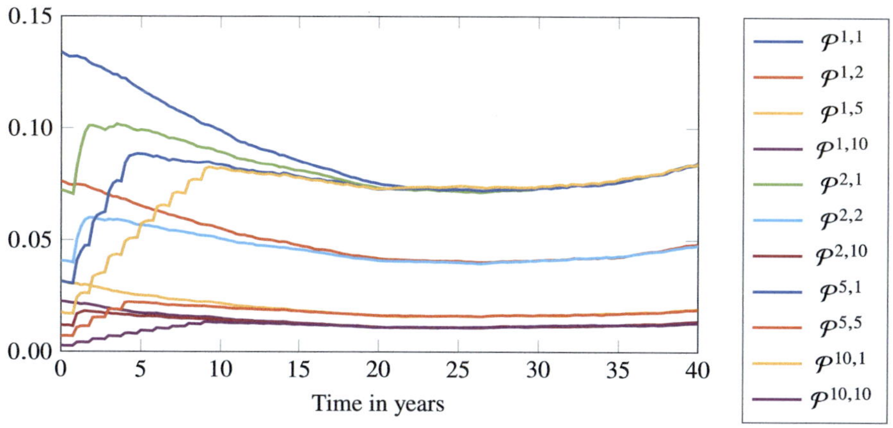

Fig. 6.6 Compression factors applying different simulation methods $\mathcal{P}^{b_1,b_2}$ in the scenario NBS 3

In particular, smaller values are associated with a more efficient compression and simulation method $\mathcal{P}^{b_1,b_2}$. In contrast to run times, the compression factor is an efficiency measure that is independent of the capability of computers.

Figures 6.5 and 6.6 display the development of the compression factor CF_k applying different methods $\mathcal{P}^{b_1,b_2}$ in the scenario with a positive and negative shock on the expected future number of new customers after 20 years (NBS 2 and NBS 3), respectively. We see that $\mathcal{P}^{1,1}$ has the worst efficiency which is due to the large number of generated cohorts. However, its efficiency remains quite stable or even improves throughout the simulation. This is not the case for e.g. $\mathcal{P}^{10,1}$ being very efficient in the short term but then the efficiency gets substantially worse. In the long term, it even has the same efficiency as $\mathcal{P}^{1,1}$. More generally, we observe that all methods $\mathcal{P}^{b_1,b_2}$ with $b_1 > 1$ suffer from a decline in efficiency throughout the simulation. This can be explained by the annual adjustment of the current ages of existing

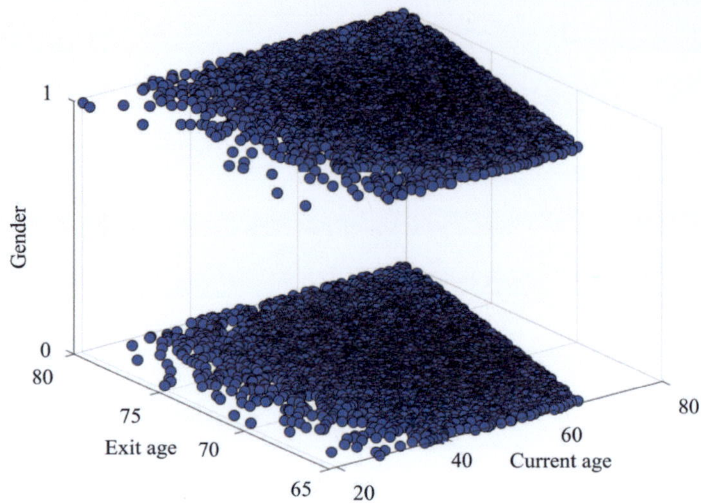

Fig. 6.7 Illustration of the uncompressed insurance portfolio regarding the grouping criteria. Each point corresponds to a contract, here characterized by the policyholder's current age $x_0^{\mathrm{real},i}$, exit age $\overline{x}^{\mathrm{real},i}$, and gender $g^{\mathrm{real},i}$

policyholders (cf. Eq. (6.2)) and the fact that the transformed ages of representative new customers are of the form $\underline{x}^{\min} + lb_1$ for a $l \geq 0$, cf. Sect. 6.2.2. Therefore, a merging of new cohorts and existing ones is not possible during the first $b_1 - 1$ years implying that M_k^{sim} increases (much) stronger than δ_k. After b_1 years, new cohorts can be merged again implying a slower increase in the cohorts. Furthermore, we observe that the efficiency of a simulation method $\mathcal{P}^{b_1,b_2}$ with $b_1 > 1$ converges to the efficiency of $\mathcal{P}^{1,b_2}$ as $b_1 \downarrow 1$. Since the exit ages remain constant, the methods $\mathcal{P}^{1,b_2}$ do not suffer from a decline in efficiency throughout the simulation.

6.4.3 *Approximation Quality of the Compression Methods*

In this section, we investigate the approximation quality of our compression methods $\mathcal{P}^{b_1,b_2}$, here exemplary for

$$(b_1, b_2) \in \{(1, 1), (2, 2), (5, 5), (1, 10), (10, 1), (10, 10)\}. \tag{6.5}$$

We use the parameters from Table 6.2 and consider an ongoing insurance business with NBS 3.

Figure 6.7 illustrates the initial (uncompressed) insurance portfolio with respect to the policyholders' current age $x_0^{\mathrm{real},i}$, exit age $\overline{x}^{\mathrm{real},i}$, and gender $g^{\mathrm{real},i} \in \{0, 1\}$. Each point corresponds to an individual contract $i \in \{1, \ldots, \delta_0\}$.

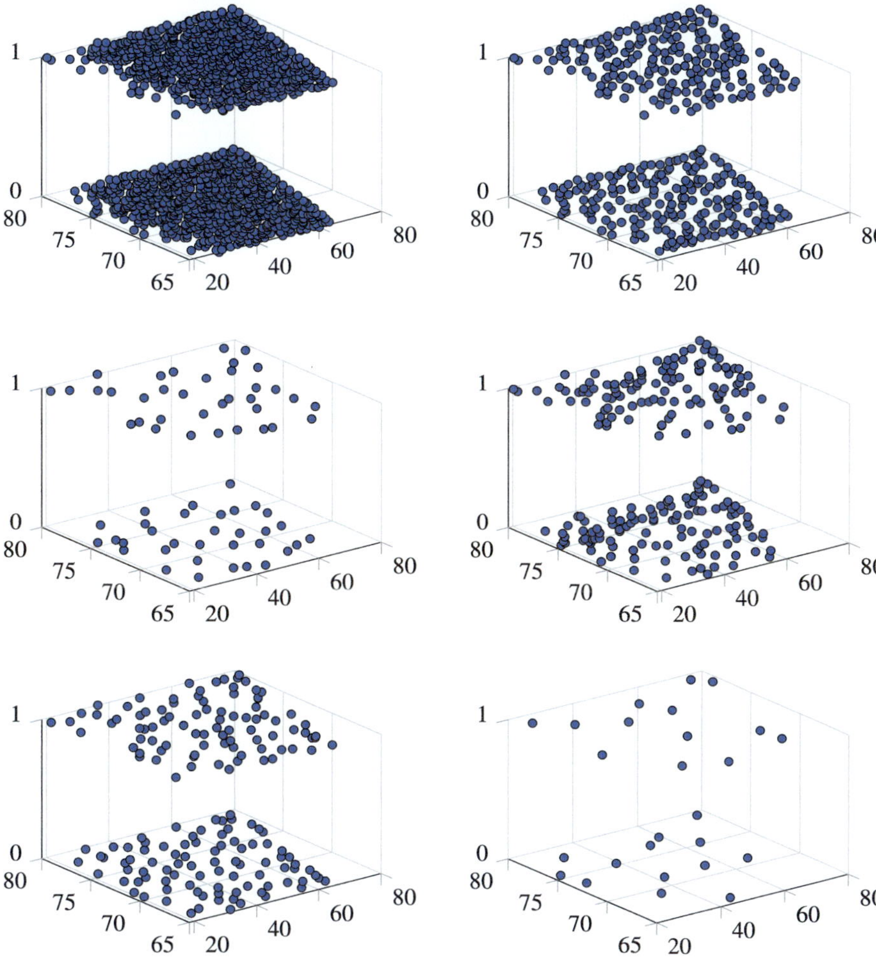

Fig. 6.8 Illustration of the compressed insurance portfolios according to different compression methods $\mathcal{P}^{b_1,b_2}$. Each point corresponds to the representative contract from a cohort $m \in \{1, \dots, M_0\}$, here characterized by the representative's current age x_0^m, exit age $\overline{x}^m$, and gender g^m. Top-left: $\mathcal{P}^{1,1}$, $M_0 = 1,339$. Top-right: $\mathcal{P}^{2,2}$, $M_0 = 406$. Middle-left: $\mathcal{P}^{5,5}$, $M_0 = 69$. Middle-right: $\mathcal{P}^{1,10}$, $M_0 = 225$. Bottom-left: $\mathcal{P}^{10,1}$, $M_0 = 176$. Bottom-right: $\mathcal{P}^{10,10}$, $M_0 = 26$

In Fig. 6.8, we illustrate the obtained compressed insurance portfolios $\mathcal{P}^{b_1,b_2}$. Each point corresponds to the representative contract from a cohort $m \in \{1, \dots, M_0\}$. The number of contracts that needs to be simulated is substantially decreased, from $\delta_0 = 10,000$ in the case of no compression to e.g. $M_0 = 1,339$ and $M_0 = 26$ in the case of $\mathcal{P}^{1,1}$ and $\mathcal{P}^{10,10}$ corresponding to a reduction of 86.61% and 99.74%, respectively. At the same time, the existing structure is maintained.

Note that for all grouping schemes $\mathcal{P}^{b_1,b_2}$, there is no approximation error regarding the aggregated premium payments and contract values at time t_0, i.e. we have

$$\sum_{m=1}^{M_0} \delta_0^m \cdot P_0^m = \sum_{i=1}^{\delta_0} P_0^{\text{real},i} \quad \text{and} \quad \sum_{m=1}^{M_0} \delta_0^m \cdot V_0^m = \sum_{i=1}^{\delta_0} V_0^{\text{real},i}.$$

Throughout the simulation, i.e. at times t_k with $k \geq 1$, this will not be the case.

Each of the compression and simulation methods $\mathcal{P}^{b_1,b_2}$ with (b_1, b_2) from (6.5) is applied $N = 10,000$ times. To investigate the direct impact of $\mathcal{P}^{b_1,b_2}$, we use the same new business scenario and dynamics of the capital market in each run, i.e. the n-th Monte Carlo paths for $\mathcal{P}^{0,0}$ and $\mathcal{P}^{b_1,b_2}$ are based on the same generated random numbers. Thereby, we allow for a pathwise comparison between compressed and uncompressed simulation methods.

The development of the capital C_k according to $\mathcal{P}^{0,0}$ and $\mathcal{P}^{b_1,b_2}$ is shown in Fig. 6.9. The corresponding 5–95% quantiles are illustrated by colored areas. $\mathcal{P}^{1,1}$ and $\mathcal{P}^{2,2}$ perform very well and also the coarser simulation methods mostly yield satisfying results. However, the approximation provided by $\mathcal{P}^{10,10}$ is quite inaccurate at most times. Looking at the size of the quantile distance illustrated by the width of the colored areas, the uncertainty regarding the capital C_k remains stable in the long term. Similar observations also hold for the insurance portfolio size δ_k, the death and surrender benefit payments T_k and S_k, and the declared interest rate $\widehat{i}_k$, see Fig. 6.10 for the latter.

To investigate the robustness of the compression and simulation methods, we look at the worst-case approximation corresponding to that simulation path $n^* = n^*\left(\mathcal{P}^{b_1,b_2}\right) \in \{1, \ldots, N\}$ for which the sum of absolute errors is maximized, i.e.

$$n^* = \operatorname{argmax}\left\{\sum_{k=0}^{K} \left|C_k^{b_1,b_2,(n)} - C_k^{0,0,(n)}\right| : n \in \{1, \ldots, N\}\right\}$$

in case of the capital C_k. Here, $C_k^{b_1,b_2,(n)}$ and $C_k^{0,0,(n)}$ denote the n-th simulated value of C_k applying $\mathcal{P}^{b_1,b_2}$ and $\mathcal{P}^{0,0}$, respectively. The results are shown in Fig. 6.11. Again, especially $\mathcal{P}^{1,1}$ yields very good approximations at all times but also the coarser methods provide mostly a good performance indicating a high reliability. Larger deviations are visible in the long term applying $\mathcal{P}^{5,5}$ and $\mathcal{P}^{10,10}$.

We conclude the simulation studies by investigating to which extent the sample distributions according to $\mathcal{P}^{0,0}$ are affected by the compression and simulation methods $\mathcal{P}^{b_1,b_2}$, here exemplary for the equity Q_k at times t_k. The results are displayed in Fig. 6.12. For all methods, even for the coarse ones, we observe small values of the total variation distance. This shows that despite the uncertainty induced by the new business, the sample distributions of the equity Q_k according to $\mathcal{P}^{0,0}$ are well approximated by the compression and simulation methods. Correspondingly, accurate approximations of the default probabilities can be expected, too.

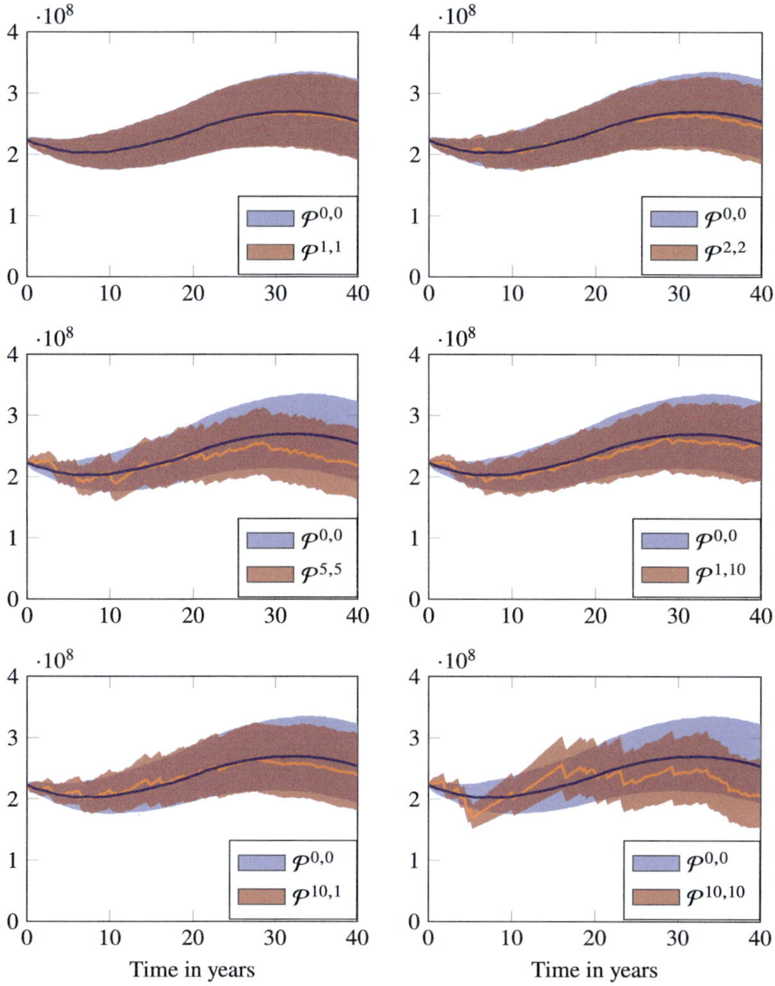

Fig. 6.9 Median with corresponding 5–95% quantiles of the capital C_k according to $\mathcal{P}^{0,0}$ and $\mathcal{P}^{b_1,b_2}$

6.5 Conclusion

We developed a class of methods for the compression and simulation of large insurance portfolios with new business based on contract-related criteria. Here, two parameters allowed to control the coarseness of the obtained compressed insurance portfolio. The performance was investigated in terms of both efficiency of different methods and their quality in approximating the uncompressed insurance portfolio. For this, we presented a stochastic asset-liability management model for life insurers that accounts for many relevant factors and yet is capable of simulating an ongoing business. Its flexibility and modular structure allows to consider different new business

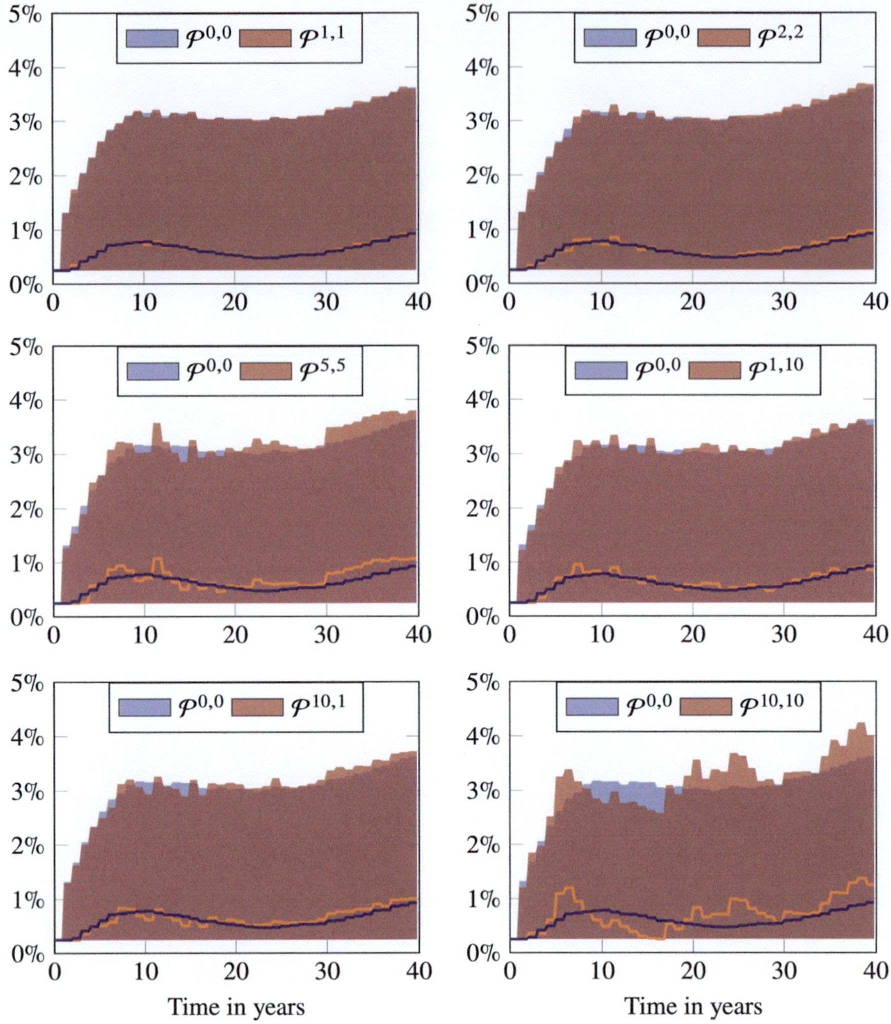

Fig. 6.10 Median with corresponding 5–95% quantiles of the declared interest rate $\widehat{i}_k$ according to $\mathcal{P}^{0,0}$ and $\mathcal{P}^{b_1,b_2}$

scenarios and alternative modeling approaches of individual parts, respectively. In addition, the model is consistent in the sense that it respects the fundamental balance sheet equation at the end of every period according to the principles of double-entry bookkeeping as required in accounting. In various simulation studies, we saw that naive contract-specific simulations require unacceptable run times while our grouping schemes provide efficient computations even when considering large insurance portfolios with new business. Furthermore, we found that especially $\mathcal{P}^{1,1}$ provides throughout very accurate and robust approximations. Coarser schemes can be more

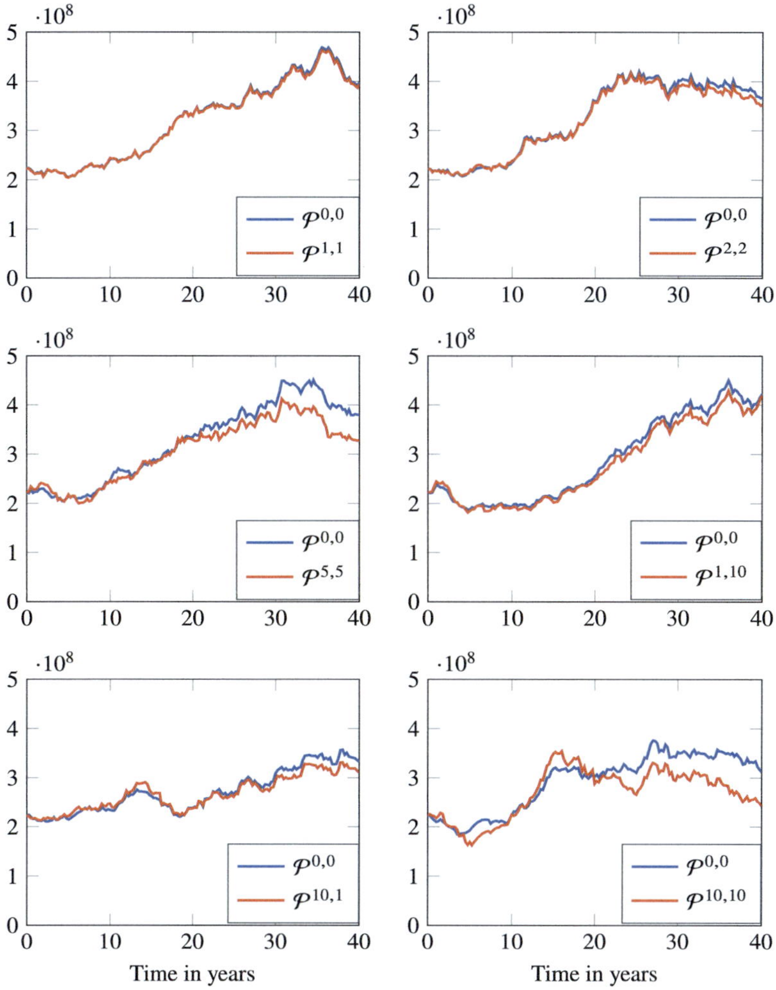

Fig. 6.11 Worst-case approximation paths n^* of the capital C_k

efficient but they may suffer heavily from approximation errors. Therefore, the compression and simulation method has to be chosen carefully in accordance with the specific application. More details about the quality in approximating several quantities of interest that may motivate the choice of specific tuning parameters b_1 and b_2 can be found in [7], Chap. 4.

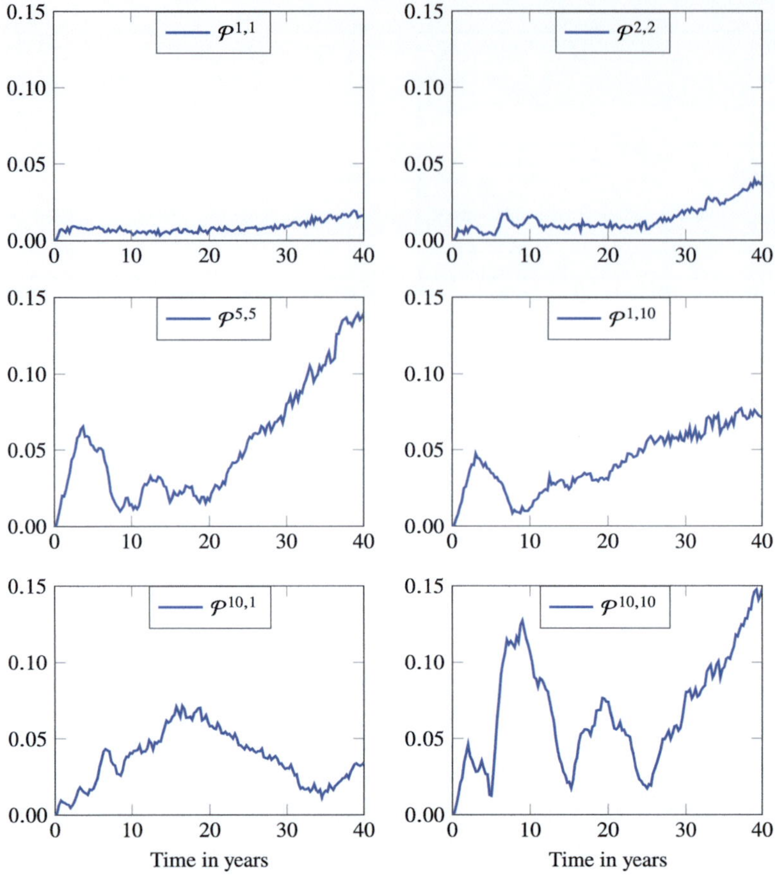

Fig. 6.12 Total variation distances of the sample distributions of the equity Q_k according to $\mathcal{P}^{0,0}$ and $\mathcal{P}^{b_1,b_2}$

References

1. Statistik der BaFin – Erstversicherungsunternehmen 2022, Bundesanstalt für Finanzdienstleistungsaufsicht (2022)
2. Deutsche Aktuarvereinigung Best Estimate in der Lebensversicherung: Hinweis (2019)
3. Diehl, M., Horsky, R., Reetz, S., et al.: Long-term stability of a life insurer's balance sheet. Eur. Actuar. J. **13**, 147–182 (2023)
4. Goffard, P., Guerrault, X.: Is it optimal to group policyholders by age, gender, and seniority for BEL computations based on model points? Eur. Actuar. J. **5**, 165–180 (2015)
5. Krah, A., Nikolić, Z., Korn, R.: Machine learning in least-squares Monte Carlo proxy modeling of life insurance companies. Risks **8** (2020)
6. Kiermayer, M., Weiß, C.: Grouping of contracts in insurance using neural networks. Scand. Actuar. J. **4**, 295–322 (2021)
7. Diehl, M.: Large Insurance Portfolios - Simulation, Strategy Decisions, Asset-Liability Management, PhD thesis, RPTU Kaiserslautern-Landau (2023)

8. Kohortensterbetafeln für Deutschland. Ergebnisse aus den Modellrechnungen für Sterbetafeln nach Geburtsjahrgang. 1920–2020, Statistisches Bundesamt (2020)
9. Kohortensterbetafeln für Deutschland. Methoden- und Ergebnisbericht zu den Modellrechnungen für Sterbetafeln der Geburtsjahrgänge 1920–2020, Statistisches Bundesamt (2020)
10. Seehafer, M., Nörtemann, S., Offtermatt, J., et al.: Actuarial Data Science: Maschinelles Lernen in der Versicherung. De Gruyter (2021)

Chapter 7
A PAC-Bayes Oracle Inequality for Sparse Neural Networks

Maximilian F. Steffen and Mathias Trabs

Abstract We study the Gibbs posterior distribution for sparse deep neural nets in a nonparametric regression setting. The posterior can be accessed via Metropolis-adjusted Langevin algorithms. Using a mixture over uniform priors on sparse sets of network weights, we prove an oracle inequality which shows that the method adapts to the unknown regularity and hierarchical structure of the regression function. The estimator achieves the minimax-optimal rate of convergence (up to a logarithmic factor).

Keywords Adaptive nonparametric estimation · Stochastic neural network · Optimal contraction rate · Oracle inequality

7.1 Introduction

Driven by the enormous success of neural networks in a broad spectrum of machine learning applications, see [18, 31] for an introduction, the theoretical understanding of network based methods is a dynamic and flourishing research area at the intersection of mathematical statistics, optimization and approximation theory. In addition to theoretical guarantees, uncertainty quantification is an important and challenging problem for neural networks and has motivated the introduction of Bayesian neural networks, where for each network weight a distribution is learned, see e.g. [10, 19] and numerous subsequent articles. The mathematical analysis of uncertainty estimates based on such stochastic neural networks has only recently been studied, see [8, 17]. In this work we study the Gibbs posterior distribution for a stochastic neural network. In a nonparametric regression problem, we show that the corresponding

M. F. Steffen (✉) · M. Trabs
Karlsruhe Institute of Technology, Institute of Stochastics, Karlsruhe, Germany
e-mail: maximilian.steffen@kit.edu

M. Trabs
e-mail: mathias.trabs@kit.edu

A. Ott et al. (eds.), *Applications of Mathematics in Sciences, Engineering, and Economics*, Springer Proceedings in Mathematics & Statistics 515,
https://doi.org/10.1007/978-3-032-01279-1_7

estimator achieves a minimax-optimal prediction risk bound up to a logarithmic factor.

Basic theoretical foundations for neural networks are summarized by [4]. The approximation properties of deep neural networks with the ReLU activation function have been revealed by [37], see also the review paper [11] and [15] for sparse deep neural networks. [32] as well as [5] have shown that deep neural network estimators can profit from a hierarchical structure of the regression function allowing for a mitigation of the curse of dimensionality. We focus on sparse neural networks with *rectified linear unit* (ReLU) activation function as studied by the aforementioned [32]. Sparsity will allow us to handle high-dimensional data which are ubiquitous in modern applications. To achieve a sparse neural network, a standard approach is to use penalized empirical risk minimization which has been analyzed for neural nets by [35].

Instead of penalized empirical risk minimization, we train a neural network by learning the Gibbs posterior distribution. We prove a *probably approximately correct* (PAC-)Bayes oracle inequality for the resulting sparse stochastic neural network which implies an adaptive version of the upper bound by [32], i.e., the network estimator achieves this upper bound without prior knowledge of the hierarchical structure and the regularity of the regression function. An analogous oracle inequality for fully connected stochastic neural networks is given in [8].

PAC-Bayes inequalities go back to [24, 25, 33]. In a quite general setting it is possible to derive upper bounds for the prediction error either in terms of the empirical risk or in terms of an oracle inequality. We refer to the review papers by [2, 21] as well as [3, 34] for applications in sparse nonparametric regression. Empirical PAC-Bayes bounds are intensively studied for (deep) neural networks, see [16, 26] and further references in [2, Section 3.3]. PAC-Bayes oracle inequalities in this setting are less well studied. Exceptions are [14], who has proved a (non-adaptive) oracle inequality for the variational approximation of the PAC-Bayes procedure based on the theory by [13], and [36] who have used the PAC-Bayes bound for an aggregation of (shallow) neural networks.

In order to sample from the Gibbs posterior, a Metropolis-adjusted Langevin algorithm (MALA) can be applied, see [6, 29]. This approach can exploit the well-established and efficient gradient descent algorithms and is well suited to navigating through the complex posterior landscape in large parameter spaces. More precisely, MALA relies on a Metropolis-Hastings algorithm where the proposal density is centered around a gradient descent step. The Markov chain Monte Carlo (MCMC) approach is different from the majority of literature on Bayesian neural networks which mainly focuses on variational Bayesian inference, see [9] for a review on variational inference and [10, 19] for early applications to neural networks. A main reason is that with a growing sample size the calculation of the acceptance probability in the Metropolis-Hastings step becomes expensive which however can be circumvented via a stochastic Metropolis-Hastings approach as demonstrated by [8]. To adapt to sparsity, we can choose a prior that prefers networks with sparse weights.

It should be noted that due to the non-convex dependence on the network weights, the computational analysis of such an MCMC algorithm is as challenging as the convergence analysis of gradient-based methods towards a global risk minimizer.

The paper is organized as follows: In Sect. 7.2, we introduce our estimation method and discuss its implementation. In Sect. 7.3, we state and discuss our main results for the proposed method. All proofs have been postponed to Sect. 7.4 and some further details on algorithmic aspects are contained in Sect. 7.5.

7.2 Estimation Method

We consider a training sample $\mathcal{D}_n := (\mathbf{X}_i, Y_i)_{i=1,\ldots,n} \subseteq \mathbb{R}^p \times \mathbb{R}$ given by $n \in \mathbb{N}$ i.i.d. copies of generic random variables $(\mathbf{X}, Y) \in \mathbb{R}^p \times \mathbb{R}$ on some probability space $(\Omega, \mathcal{A}, \mathbb{P})$. The aim is to estimate the regression function $f : \mathbb{R}^p \to \mathbb{R}$ given by $Y = f(\mathbf{X}) + \varepsilon$ with observation error ε satisfying $\mathbb{E}[\varepsilon \mid \mathbf{X}] = 0$ almost surely (a.s.). Equivalently, $f(\mathbf{X}) = \mathbb{E}[Y \mid \mathbf{X}]$ a.s. For any estimator $\widehat{f}$, the prediction risk and its empirical counterpart are given by

$$R(\widehat{f}) := \mathbb{E}_{(\mathbf{X},Y)}\big[(Y - \widehat{f}(\mathbf{X}))^2\big] \quad \text{and} \quad R_n(\widehat{f}) = \frac{1}{n}\sum_{i=1}^{n}\big(Y_i - \widehat{f}(\mathbf{X}_i)\big)^2, \tag{7.1}$$

respectively, where $\mathbb{E}$ denotes the expectation under $\mathbb{P}$ and $\mathbb{E}_Z$ is the (conditional) expectation only with respect to a random variable Z. The accuracy of the estimation procedure will be quantified in terms of the excess risk

$$\mathcal{E}(\widehat{f}) := R(\widehat{f}) - R(f) = \mathbb{E}_{\mathbf{X}}\big[(\widehat{f}(\mathbf{X}) - f(\mathbf{X}))^2\big] = \|\widehat{f} - f\|^2_{L^2(\mathbb{P}^{\mathbf{X}})}.$$

In the following we first introduce the considered class of sparse neural networks. Afterwards, we introduce the estimation method and discuss its implementation. Throughout, $|x|_q$ denotes the ℓ^q-norm of a vector $x \in \mathbb{R}^p$, $q \in [1, \infty]$. In particular, $|\cdot| := |\cdot|_2$ is the Euclidean norm. For $a, b \in \mathbb{R}$ we write $a \vee b := \max\{a, b\}$ and $a \wedge b := \min\{a, b\}$. The cardinality of a set $\mathcal{I}$ is denoted by $|\mathcal{I}|$.

7.2.1 *Sparse Neural Networks*

We consider a *feedforward multilayer perceptron* with $L \in \mathbb{N}$ hidden layers of constant width $r \in \mathbb{N}$. The latter restriction can be omitted, but simplifies our notation. The ReLU activation function $\varphi(x) := x \vee 0$, $x \in \mathbb{R}$, is used. We write $\varphi_v x := \big(\varphi(x_i + v_i)\big)_{i=1,\ldots,d}$ for vectors $x, v \in \mathbb{R}^d$. With this notation we can represent such neural networks as

$$g_\vartheta(\mathbf{x}) := W^{(L+1)}\varphi_{v^{(L)}}W^{(L)}\varphi_{v^{(L-1)}}\cdots W^{(2)}\varphi_{v^{(1)}}W^{(1)}\mathbf{x} + v^{(L+1)}, \qquad \mathbf{x} \in \mathbb{R}^p,$$

where the entries of the weight matrices $W^{(1)} \in \mathbb{R}^{r\times p}, W^{(2)}, \ldots, W^{(L)} \in \mathbb{R}^{r\times r}$, $W^{(L+1)} \in \mathbb{R}^{1\times r}$ and shift vectors $v^{(1)}, \ldots, v^{(L)} \in \mathbb{R}^r$, $v^{(L+1)} \in \mathbb{R}$ are all collected in one parameter vector ϑ. The total number of network parameters is

$$P := (p+1)r + (L-1)(r+1)r + r + 1.$$

A possibly more intuitive layer-wise representation of g_ϑ is given by

$$\begin{aligned}
\mathbf{x}^{(0)} &:= \mathbf{x} \in \mathbb{R}^p,\\
\mathbf{x}^{(l)} &:= \varphi(W^{(l)}\mathbf{x}^{(l-1)} + v^{(l)}),\ l = 1, \ldots, L,\\
g_\vartheta(\mathbf{x}) := \mathbf{x}^{(L+1)} &:= W^{(L+1)}\mathbf{x}^{(L)} + v^{(L+1)},
\end{aligned}$$

where the activation function is applied coordinate-wise. We denote the class of all such functions g_ϑ by $\mathcal{G}(p, L, r)$. A network is sparse, or more precisely *connection sparse*, if many weights in the network are zero and thus some links between nodes are inactive. For some active set $\mathcal{I} \subseteq \{1, \ldots, P\}$, the corresponding class of sparse networks is defined by

$$\mathcal{G}(p, L, r, \mathcal{I}) := \{g_\vartheta \in \mathcal{G}(p, L, r) : \vartheta_i = 0 \text{ if } i \notin \mathcal{I}\}.$$

For some fixed $C \geqslant 1$ and any $g_\vartheta \in \mathcal{G}(p, L, r)$, we denote the corresponding clipped version of g_ϑ by $f_\vartheta := (-C) \vee (g_\vartheta \wedge C)$. We abbreviate $R(\vartheta) := R(f_\vartheta)$ and $R_n(\vartheta) := R_n(f_\vartheta)$.

7.2.2 Prior and Posterior Distribution

To define our estimator based on the Gibbs posterior, we first choose a prior on the weights in the network class $\mathcal{G}(p, L, r)$. For a given active set $\mathcal{I}$ the prior $\Pi_\mathcal{I}$ on the parameter set of the class $\mathcal{G}(p, L, r, \mathcal{I})$ is defined as the uniform distribution on

$$\mathcal{S}_\mathcal{I} := \{\vartheta \in [-B, B]^P \mid \vartheta_i = 0 \text{ if } i \notin \mathcal{I}\} \tag{7.2}$$

for some $B \geqslant 1$. In the special case $\mathcal{I}^\bullet := \{1, \ldots, P\}$ we obtain a prior $\Pi^\bullet := \Pi_{\mathcal{I}^\bullet}$ on the non-sparse network class $\mathcal{G}(p, L, r)$. To allow for a data-driven choice of the active set, we define the prior Π as a mixture of the uniform priors $\Pi_\mathcal{I}$:

$$\Pi := \sum_{i=1}^{P} 2^{-i} \sum_{\substack{\mathcal{I}\subseteq\{1,\ldots,P\},\\ |\mathcal{I}|=i}} \binom{P}{i}^{-1} \Pi_\mathcal{I} \Big/ C_P \quad \text{with} \quad C_P := (1 - 2^{-P}). \tag{7.3}$$

The basis 2 of the geometric weights is arbitrary and can be replaced by a larger constant leading to a stronger preference of sparse networks. The theoretical results remain unchanged up to constants. The prior Π can be understood as a hierarchical prior, where we first draw a geometrically distributed integer $i \in \{1, \dots, P\}$ of non-zero weights, given i we uniformly choose an active set $\mathcal{I} \subseteq \{1, \dots, P\}$ with $|\mathcal{I}| = i$ and on $\mathcal{S}_\mathcal{I}$ the uniform prior $\Pi_\mathcal{I}$ is applied.

Based on Π, the *Gibbs posterior* probability distribution $\Pi_\lambda(\cdot \mid \mathcal{D}_n)$ is defined via

$$\frac{\mathrm{d}\Pi_\lambda(\vartheta \mid \mathcal{D}_n)}{\mathrm{d}\Pi} \propto \exp(-\lambda R_n(\vartheta)) \tag{7.4}$$

with the so-called *inverse temperature parameter* $\lambda > 0$ and empirical prediction risk from (7.1). While (7.4) coincides with the classical Bayesian posterior distribution if $Y_i = f_\vartheta(\mathbf{X}_i) + \varepsilon_i$ with i.i.d. $\varepsilon_i \sim \mathcal{N}(0, n/(2\lambda))$, the so-called tempered likelihood $\exp(-\lambda R_n(\vartheta))$ serves as a proxy for the unknown distribution of the observations given ϑ. As we will see, the method is indeed applicable under quite general assumptions on the regression model. The Gibbs posterior distribution weights each ϑ based on its empirical performance on the data, where the inverse temperature parameter λ determines the impact of $R_n(\vartheta)$ in comparison to the prior beliefs. Consequently, sparse neural networks with a strong empirical performance have a high probability under the Gibbs posterior. Therefore, an estimator for f can naturally be obtained by drawing from the posterior distribution, i.e.,

$$\widehat{f}_\lambda := f_{\widehat{\vartheta}_\lambda} \qquad \text{for} \qquad \widehat{\vartheta}_\lambda \mid \mathcal{D}_n \sim \Pi_\lambda(\cdot \mid \mathcal{D}_n) \tag{7.5}$$

or in the form of the posterior mean

$$\bar{f}_\lambda := \mathbb{E}\big[f_{\widehat{\vartheta}_\lambda} \,\big|\, \mathcal{D}_n\big] = \int f_\vartheta \, \Pi_\lambda(\mathrm{d}\vartheta \mid \mathcal{D}_n). \tag{7.6}$$

The maximum a posteriori (MAP) estimator could be used as well. It circumvents sampling from the posterior, but leads to a nontrivial optimization problem. We focus on the estimators from (7.5) and (7.6) in view of a simpler mathematical analysis compared to the MAP.

7.2.3 MCMC Algorithm

Since the normalizing constant of the posterior distribution is numerically infeasible, the posterior distribution itself is not accessible in practice. Popular Bayesian networks rely on a variational Bayes approach, cf. [9], and approximate the posterior by some simpler distribution. For instance, the *Bayes by backprop* method by [10] uses independent normal distributions for the weights such that the training of the Bayesian network reduces to the calibration of the means and variances for all weights. In contrast, the classical approach in Bayesian statistics is to sample $\widehat{\vartheta}_\lambda$ by constructing a Markov-chain $(\vartheta^{(k)})_{k\in\mathbb{N}_0}$ with stationary distribution $\Pi_\lambda(\cdot \mid \mathcal{D}_n)$ with a Markov chain Monte Carlo (MCMC) method, see [28]. MCMC methods aim for the exact posterior distribution and have been successfully applied to neural networks, see e.g. [1, 38].

For ease of presentation, we will discuss MALA for the prior $\Pi^\bullet$ on the non-sparse network class $\mathcal{G}(p, L, r)$, i.e., $\Pi^\bullet$ is the uniform distribution on $[-B, B]^P$. The corresponding posterior distribution is denoted by $\Pi^\bullet_\lambda(\vartheta \mid \mathcal{D}_n) \propto \exp(-\lambda R_n(\vartheta))\Pi^\bullet(\vartheta)$, where we use the same symbol to denote distributions and their densities.

Applying the generic Metropolis-Hastings algorithm to $\Pi^\bullet_\lambda(\mathrm{d}\vartheta \mid \mathcal{D}_n)$ and taking into account that the prior $\Pi^\bullet$ is uniform, we obtain the following iterative method: Starting with some initial choice $\vartheta^{(0)} \in \mathbb{R}^P$, we successively generate $\vartheta^{(k+1)}$ given $\vartheta^{(k)}$, $k \in \mathbb{N}_0$, by

$$\vartheta^{(k+1)} = \begin{cases} \tau^{(k)} & \text{with probability } \alpha(\tau^{(k)} \mid \vartheta^{(k)}) \\ \vartheta^{(k)} & \text{with probability } 1 - \alpha(\tau^{(k)} \mid \vartheta^{(k)}) \end{cases},$$

where $\tau^{(k)}$ is a random variable drawn from some conditional proposal density $q(\cdot \mid \vartheta^{(k)})$ and the *acceptance probability* is given by

$$\alpha(\tau^{(k)} \mid \vartheta^{(k)}) = \\ \min\Big\{1, \exp\big(-\lambda R_n(\tau^{(k)}) + \lambda R_n(\vartheta^{(k)})\big)\mathbf{1}_{[-B,B]^P}(\tau^{(k)})\tfrac{q(\vartheta^{(k)}\mid\tau^{(k)})}{q(\tau^{(k)}\mid\vartheta^{(k)})}\Big\}$$

The Metropolis-Hastings acceptance step ensures that $(\vartheta^{(k)})_{k\in\mathbb{N}_0}$ is a Markov chain with invariant distribution $\Pi^\bullet(\cdot \mid \mathcal{D}_n)$. The convergence to the invariant distribution follows from [30, Theorem 2.2].

The practical success of the Metropolis-Hastings algorithm fundamentally depends on the choice of the proposal distribution. In the construction of $q(\tau \mid \vartheta)$, MALA exploits the well-known gradient approach for an empirical risk minimizer

$$\vartheta^* \in \arg\min_{\vartheta} R_n(\vartheta)$$

with the empirical risk $R_n(\vartheta) = R_n(f_\vartheta)$ from (7.1). The gradient $\nabla_\vartheta R_n(\vartheta)$ of $R_n(\vartheta)$ with respect to ϑ can be computed efficiently using backpropagation. The gradient descent method commonly used to train neural networks would suggest to tweak a given $\vartheta^{(k)}$ in the direction of the gradient, that is

$$\widetilde{\vartheta}^{(k+1)} = \vartheta^{(k)} - \gamma\nabla_\vartheta R_n(\vartheta^{(k)})$$

with a *learning rate* $\gamma > 0$. Therefore, MALA proposes a normal distribution around $\widetilde{\vartheta}^{(k+1)}$ with some standard deviation $s > 0$. This standard deviation should not be too large as otherwise the acceptance probability might be too small. As a result the proposal would rarely be accepted, the chain might not be sufficiently randomized and the convergence to the invariant target distribution would be too slow in practice. On the other hand, s should not be smaller than the shift $\gamma \nabla_\vartheta R_n(\vartheta^{(k)})$ in the mean, since otherwise $q(\vartheta^{(k)} \mid \tau^{(k)})$ might be too small. Therefore, the probability density q of the proposal distribution is given by the density

$$q(\tau \mid \vartheta) = \frac{1}{(2\pi s^2)^{P/2}} \exp\Big(-\frac{1}{2s^2} \big|\tau - \vartheta + \gamma \nabla_\vartheta R_n(\vartheta)\big|^2 \Big). \tag{7.7}$$

To calculate the estimators $\widehat{f}_\lambda$ and $\bar{f}_\lambda$ from (7.5) and (7.6), respectively, one chooses a *burn-in* time $b \in \mathbb{N}$ to let the distribution of the Markov chain stabilize at its invariant distribution and then sets

$$\widehat{f}_\lambda := f_{\vartheta^{(b)}} \qquad \text{and} \qquad \bar{f}_\lambda := \frac{1}{N} \sum_{k=1}^{N} f_{\vartheta^{(b+ck)}}.$$

A sufficiently large *gap length* $c \in \mathbb{N}$ ensures the necessary variability and an approximate independence between $\vartheta^{(b+ck)}$ and $\vartheta^{(b+c(k+1))}$, whereas $N \in \mathbb{N}$ has to be large enough for a good approximation of the expectation by the empirical mean. For empirical results for $\Pi^\bullet(\cdot \mid \mathcal{D}_n)$, we refer to [8].

Note that the gradient has to be calculated only once in each MCMC iteration. While we use the standard gradient, the calculation of $\nabla_\vartheta R_n$ could be realized and combined with state-of-the-art gradient-based methods such as mini-batch stochastic gradient descent or Adam [22] as exploited by [7]. Compared to the training of a non-stochastic network, the additional computational price of MALA is determined by the computation of the acceptance probability $\alpha(\tau^{(k)} \mid \vartheta^{(k)})$ and a larger number of necessary iterations due to the potential rejection of proposals.

In order to extend the Metropolis-Hastings algorithm from the full prior $\Pi^\bullet$ to the mixing prior Π from (7.3), we have to take into account the hierarchical structure of Π. This could be realized via the reversible-jump Markov chain Monte Carlo (RJMCMC) algorithm first proposed by [20]. In particular, the RJMCMC algorithm has been successfully applied by [3] in a high-dimensional regression setting. Some further details on the application of the RJMCMC algorithm to sparse neural networks are given in Sect. 7.5. A computationally feasible implementation for large networks remains challenging.

7.3 Oracle Inequality

In this section we state theoretical guarantees for the Gibbs posterior distribution. As a benchmark for the performance of the method, we define the *oracle choice* on $\mathcal{S}_\mathcal{I}$ from (7.2) for some active set $\mathcal{I}$ as

$$\vartheta^*_\mathcal{I} \in \arg\min_{\vartheta \in \mathcal{S}_\mathcal{I}} R(\vartheta). \tag{7.8}$$

The oracle is not accessible to the practitioner because $R(\vartheta)$ depends on the unknown distribution of $(\mathbf{X}, Y)$. A solution to the minimization problem in (7.8) always exists since $\mathcal{S}_{\mathcal{I}}$ is compact and $\vartheta \mapsto R(f_\vartheta)$ is continuous. If there is more than one solution, we choose one of them. Our main result gives a theoretical guarantee that the estimator $\widehat{f_\lambda}$ from (7.5) is at least as good as the oracle $\vartheta^*_{\mathcal{I}}$ in terms of the excess risk up to some multiplicative constant and an additional stochastic error term. To this end, we need some mild assumptions on the regression model.

Assumption A.

1. For $K, C \geqslant 1$ we have $\mathbb{E}[|\mathbf{X}|^2] \leqslant pK$ and $\|f\|_\infty \leqslant C$.
2. ε is conditionally on $\mathbf{X}$ sub-Gaussian. More precisely, there are constants $\sigma, \Gamma > 0$ such that

$$\mathbb{E}[|\varepsilon|^k \mid \mathbf{X}] \leqslant \frac{k!}{2}\sigma^2\Gamma^{k-2}\text{a.s.}, \qquad \text{for all } k \geqslant 2.$$

Note that neither the loss function nor the data are assumed to be bounded. We obtain the following non-asymptotic oracle inequality:

Theorem 7.1 (PAC-Bayes oracle inequality) *Let* $\delta \in (0, 1)$. *Under Assumption A. there are constants* $Q_0, Q_1 > 0$ *only depending on* C, Γ, σ *such that for* $\lambda = n/Q_0$ *and sufficiently large* n*, we have with probability of at least* $1 - \delta$*, that*

$$\mathcal{E}(\widehat{f_\lambda}) \leqslant \min_{\mathcal{I}} \Big(4\mathcal{E}(f_{\vartheta^*_{\mathcal{I}}}) + \frac{Q_1}{n}\big(|\mathcal{I}|L\log(p \vee n) + \log(2/\delta)\big)\Big). \tag{7.9}$$

Remark 7.1 Q_0 can explicitly be chosen as $Q_0 = 16(C^2 + \sigma^2) + 16C(\Gamma \vee (2C))$, the dependence of Q_1 on C, Γ, σ is at most quadratic and $n \geqslant n_0 := 2 \vee r \vee b \vee K$ is sufficiently large.

The right-hand side of (7.9) can be interpreted similarly to the classical bias-variance decomposition in nonparametric statistics. The first term $\mathcal{E}(f_{\vartheta^*_{\mathcal{I}}}) = \mathbb{E}_{\mathbf{X}}[(f_{\vartheta^*_{\mathcal{I}}}(\mathbf{X}) - f(\mathbf{X}))^2]$ quantifies the approximation error while second term is an upper bound for the stochastic error. In particular, we recover $\frac{|\mathcal{I}|}{n}\log p$ as the typical error term for estimating high-dimensional vectors with sparsity $|\mathcal{I}|$. The factor 4 in the upper bound can be improved to $(1 + \eta)$ for any $\eta > 0$ at the cost of a larger constant Q_1. Theorem 7.1 is in line with classical PAC-Bayes oracle inequalities, see [2]. Reference [14] has obtained a similar oracle inequality for a variational approximation of the Gibbs posterior distribution, but without the minimum over the active sets. For penalized empirical risk minimization, [35] have obtained another oracle inequality with a different dependence on the depth L.

The $1 - \delta$ probability takes into account the randomness of the data and of the estimate. Denoting

$$r_{n,p}^2 := \min_{\mathcal{I}} \Big(4\|f_{\vartheta^*_{\mathcal{I}}} - f\|^2_{L^2(\mathbb{P}^{\mathbf{X}})} + \frac{Q_1}{n}|\mathcal{I}|L\log(p \vee n)\Big), \tag{7.10}$$

we can rewrite (7.9) as

$$\mathbb{E}\big[\Pi_\lambda\big(\|f_{\widehat{\vartheta}_\lambda} - f\|^2_{L^2(\mathbb{P}^{\mathbf{X}})} > r^2_{n,p} + t^2 \,\big|\, \mathcal{D}_n\big)\big] \leqslant 2\mathrm{e}^{-nt^2/Q_1}, \qquad t > 0,$$

which is a *contraction rate* result in terms of a frequentist analysis of the nonparametric Bayes method.

An immediate consequence of the previous theorem is an analogous oracle inequality for the posterior mean $\bar{f}_\lambda$ from (7.6).

Corollary 7.1 (Posterior mean) *Under the conditions of Theorem 7.1 we have with probability of at least* $1 - \delta$ *that*

$$\mathcal{E}(\bar{f}_\lambda) \leqslant \min_{\mathcal{I}} \Big(4\mathcal{E}(f_{\vartheta^*_{\mathcal{I}}}) + \frac{Q_2}{n}\big(|\mathcal{I}| L \log(p \vee n) + \log(2/\delta)\big)\Big)$$

with a constant Q_2 *only depending on* C, Γ, σ.

The infimum over all $\mathcal{I}$ in the oracle inequalities shows that the estimator may choose a sparse neural network for approximating f in a data-driven way. In particular, $\widehat{f}_\lambda$ as well as $\bar{f}_\lambda$ adapt to the unknown regularity and a potentially lower dimensional structure of f. Using the approximation properties of neural networks, the oracle inequality yields the optimal rate of convergence (up to a logarithmic factor) over the following class of hierarchical functions:

$$\begin{aligned}
&\mathcal{H}(q, \mathbf{d}, \mathbf{t}, \beta, C_0) \\
&:= \Big\{ g_q \circ \cdots \circ g_0 \colon [0,1]^p \to \mathbb{R} \,\Big|\, g_i = (g_{ij})_j^\top : [a_i, b_i]^{d_i} \to [a_{i+1}, b_{i+1}]^{d_{i+1}}, \\
&\qquad\qquad g_{ij} \text{ depends on at most } t_i \text{ arguments}, \\
&\qquad\qquad g_{ij} \in \mathcal{C}^{\beta_i}([a_i, b_i]^{t_i}, C_0), \text{ for some } |a_i|, |b_i| \leqslant C_0 \Big\},
\end{aligned}$$

where $\mathbf{d} := (p, d_1, \ldots, d_q, 1) \in \mathbb{N}^{q+2}$, $\mathbf{t} := (t_0, \ldots, t_q) \in \mathbb{N}^{q+1}$, $\beta := (\beta_0, \ldots, \beta_q) \in (0, \infty)^{q+1}$ and where $\mathcal{C}^{\beta_i}([a_i, b_i]^{t_i}, C_0)$ denote classical Hölder balls with Hölder regularity $\beta_i > 0$ and radius $C_0 \geqslant 1$. In the condition $g_{ij} \in \mathcal{C}^{\beta_i}([a_i, b_i]^{t_i}, C_0)$, the function g_{ij} is understood as a function of only t_i arguments. Theorem 7.1 reveals the following adaptive version of the upper bound by [32].

Proposition 7.1 (Rates of convergence) *Let* $\log p \leqslant n/(\log^2 n)$ *and* $\mathbf{X} \in [0,1]^p$. *In the situation of Theorem 7.1, there exists a network architecture* $(L, r) = (C_1 \lceil \log_2 n \rceil, C_2 n)$ *with* C_1 *and* C_2 *only depending on upper bounds for* $q, |(d_1, \ldots, d_q)|_\infty, |\mathbf{t}|_\infty, |\beta|_\infty$ *and* C_0 *such that the estimators* $\widehat{f}_\lambda$ *and* $\bar{f}_\lambda$ *yield an excess risk for sufficiently large* n *uniformly over all hierarchical functions* $f \in \mathcal{H}(q, \mathbf{d}, \mathbf{t}, \beta, C_0)$ *of at most*

$$\mathcal{E}(\widehat{f}_\lambda) \leqslant Q_3\Big(\frac{\log(p\vee n)\log^2(n)}{n}\Big)^{2\beta^*/(2\beta^*+t^*)} + Q_3\frac{\log(2/\delta)}{n} \quad \text{and}$$
$$\mathcal{E}(\bar{f}_\lambda) \leqslant Q_4\Big(\frac{\log(p\vee n)\log^2(n)}{n}\Big)^{2\beta^*/(2\beta^*+t^*)} + Q_4\frac{\log(2/\delta)}{n}$$

with probability of at least $1-\delta$, *respectively, where* β^* *and* t^* *are given by*

$$\beta^* := \beta^*_{i^*}, \quad t^* := t^*_{i^*} \quad \text{for} \quad i^* \in \operatorname*{arg\,min}_{i=0,\dots,q} \frac{2\beta_i^*}{2\beta_i^* + t_i^*} \quad \text{and} \quad \beta_i^* := \beta_i \prod_{l=i+1}^{q} (\beta_l \wedge 1).$$

The constants Q_3 *and* Q_4 *only depend on* $q, (d_1, \dots, d_q), \mathbf{t}, \beta$ *and* C_0 *and* C, Γ, σ.

For a fixed dimension p it has been proved by [32] that this rate is indeed optimal in a minimax sense up to a logarithmic factor in n. Studying classical Hölder balls $\mathcal{C}_p^\beta([0,1]^p, C_0)$, a contraction rate of order $n^{-2\beta/(2\beta+p)}$ has been derived by [14, 27], where [27] have used a hierarchical prior to obtain adaptivity with respect to β and the result by [14] is non-adaptive, but incorporates a variational Bayes method.

7.4 Proofs

We first provide some preliminary results in Sect. 7.4.1 before we prove the oracle inequality in Sect. 7.4.2. All remaining proofs can be found in the last two subsections.

7.4.1 A PAC-Bayes Bound

Let μ, ν be probability measures on a measurable space $(E, \mathscr{A})$. The *Kullback-Leibler divergence* of μ with respect to ν is defined via

$$\mathrm{KL}(\mu \mid \nu) := \begin{cases} \int \log\big(\frac{\mathrm{d}\mu}{\mathrm{d}\nu}\big)\,\mathrm{d}\mu, & \text{if } \mu \ll \nu \\ \infty, & \text{otherwise} \end{cases}.$$

The following classical lemma is a key ingredient for PAC-Bayes bounds. A proof can be found in [12, p. 159] or [2].

Lemma 7.1 *Let* $h\colon E \to \mathbb{R}$ *be a measurable function such that* $\int \exp\circ h\,\mathrm{d}\mu < \infty$. *With the convention* $\infty - \infty = -\infty$, *it then holds that*

$$\log\Big(\int \exp\circ h\,\mathrm{d}\mu\Big) = \sup_\nu\Big(\int h\,\mathrm{d}\nu - \mathrm{KL}(\nu \mid \mu)\Big), \tag{7.11}$$

where the supremum is taken over all probability measures ν *on* $(E, \mathscr{A})$. *If additionally,* h *is bounded from above on the support of* μ, *then the supremum in* (7.11) *is attained for* $\nu = g$ *with the Gibbs distribution* g, *i.e.,* $\frac{\mathrm{d}g}{\mathrm{d}\mu} \propto \exp \circ h$.

Note that no generality is lost by considering only those probability measures ν on $(E, \mathscr{A})$ such that $\nu \ll \mu$ and thus

$$\log\Big(\int \exp \circ h \,\mathrm{d}\mu\Big) = -\inf_{\nu \ll \mu}\Big(\mathrm{KL}(\nu \mid \mu) - \int h \,\mathrm{d}\nu\Big).$$

Next we state a Bernstein type inequality for the excess risk and its empirical counterpart given by

$$\mathcal{E}(\vartheta) := \mathcal{E}(f_\vartheta) = R(\vartheta) - R(f) \qquad \text{and} \qquad \mathcal{E}_n(\vartheta) := R_n(\vartheta) - R_n(f),$$

respectively, with R_n and R given in (7.1). The proof is given in Sect. 7.4.4.

Lemma 7.2 *For* $C_{n,\lambda} := \frac{\lambda}{n}\frac{8(C^2+\sigma^2)}{1-w\lambda/n}$ *and* $w := 16C(\Gamma \vee 2C)$, *we have for all* $\lambda \in [0, n/w)$ *that*

$$\max\big\{\mathbb{E}\big[\exp\big(\lambda(\mathcal{E}(\vartheta) - \mathcal{E}_n(\vartheta))\big)\big], \mathbb{E}\big[\exp\big(\lambda(\mathcal{E}_n(\vartheta) - \mathcal{E}(\vartheta))\big)\big]\big\} \leqslant \exp\big(C_{n,\lambda}\lambda\mathcal{E}(\vartheta)\big).$$

Based on these two lemmas, we can verify a PAC-Bayes bound for the excess risk. The basic proof strategy is standard in the PAC-Bayes literature, see e.g. [3].

Proposition 7.2 (PAC-Bayes bound) *For any* $\mathcal{D}_n$*-dependent (in a measurable way) probability measure* $\rho \ll \Pi$ *and any* $\lambda \in (0, n/w)$ *such that* $C_{n,\lambda} \leqslant 1/2$, *we have*

$$\mathcal{E}(\widehat{\vartheta}_\lambda) \leqslant 3\int \mathcal{E}\,\mathrm{d}\rho + \frac{4}{\lambda}\big(\mathrm{KL}(\rho \mid \Pi) + \log(2/\delta)\big) \tag{7.12}$$

with probability of at least $1 - \delta$.

Proof Lemma 7.2 yields

$$\begin{aligned}&\mathbb{E}\big[\exp\big(\lambda(1 - C_{n,\lambda})\mathcal{E}(\vartheta) - \lambda\mathcal{E}_n(\vartheta) - \log(\delta^{-1})\big)\big] \leqslant \delta \qquad \text{and}\\ &\mathbb{E}\big[\exp\big(\lambda\mathcal{E}_n(\vartheta) - \lambda(1 + C_{n,\lambda})\mathcal{E}(\vartheta) - \log(\delta^{-1})\big)\big] \leqslant \delta.\end{aligned}$$

Integrating in ϑ with respect to the prior probability measure Π and applying Fubini's theorem, we conclude

$$\begin{aligned}&\mathbb{E}\Big[\int \exp\big(\lambda(1 - C_{n,\lambda})\mathcal{E}(\vartheta) - \lambda\mathcal{E}_n(\vartheta) - \log(\delta^{-1})\big)\,\mathrm{d}\Pi(\vartheta)\Big] \leqslant \delta \qquad \text{and}\\ &\mathbb{E}\Big[\int \exp\big(\lambda\mathcal{E}_n(\vartheta) - \lambda(1 + C_{n,\lambda})\mathcal{E}(\vartheta) - \log(\delta^{-1})\big)\,\mathrm{d}\Pi(\vartheta)\Big] \leqslant \delta.\end{aligned} \tag{7.13}$$

For the posterior distribution $\Pi_\lambda(\cdot \mid \mathcal{D}_n) \ll \Pi$ with corresponding Radon-Nikodym density

$$\frac{\mathrm{d}\Pi_\lambda(\vartheta \mid \mathcal{D}_n)}{\mathrm{d}\Pi} = D_\lambda^{-1} \exp(-\lambda R_n(\vartheta)), \qquad D_\lambda := \int \exp(-\lambda R_n(\vartheta))\,\mathrm{d}\Pi(\vartheta)$$

with respect to Π, we obtain

$$\begin{aligned}
&\mathbb{E}_{\mathcal{D}_n,\widehat{\vartheta}\sim\Pi_\lambda(\cdot|\mathcal{D}_n)}\big[\exp\big(\lambda(1-C_{n,\lambda})\mathcal{E}(\widehat{\vartheta}) - \lambda\mathcal{E}_n(\widehat{\vartheta}) - \log(\delta^{-1}) + \lambda R_n(\widehat{\vartheta}) + \log D_\lambda\big)\big] \\
&= \mathbb{E}_{\mathcal{D}_n,\widehat{\vartheta}\sim\Pi_\lambda(\cdot|\mathcal{D}_n)}\Big[\exp\Big(\lambda(1-C_{n,\lambda})\mathcal{E}(\widehat{\vartheta}) - \lambda\mathcal{E}_n(\widehat{\vartheta}) - \log(\delta^{-1}) \\
&\quad - \log\Big(\frac{\mathrm{d}\Pi_\lambda(\widehat{\vartheta} \mid \mathcal{D}_n)}{\mathrm{d}\Pi}\Big)\Big)\Big] \\
&= \mathbb{E}_{\mathcal{D}_n}\Big[\int \exp\big(\lambda(1-C_{n,\lambda})\mathcal{E}(\vartheta) - \lambda\mathcal{E}_n(\vartheta) - \log(\delta^{-1})\big)\,\mathrm{d}\Pi(\vartheta)\Big] \leqslant \delta.
\end{aligned}$$

Since $\mathbf{1}_{[0,\infty)}(x) \leqslant \mathrm{e}^{\lambda x}$ for all $x \in \mathbb{R}$, we deduce for $\widehat{\vartheta} \sim \Pi_\lambda(\cdot \mid \mathcal{D}_n)$ with probability not larger than δ that

$$\big(1 - C_{n,\lambda}\big)\mathcal{E}(\widehat{\vartheta}) - \mathcal{E}_n(\widehat{\vartheta}) + R_n(\widehat{\vartheta}) - \lambda^{-1}\big(\log(\delta^{-1}) - \log D_\lambda\big) \geqslant 0.$$

Provided $\big(1 - C_{n,\lambda}\big) > 0$, we thus have with probability of at least $1 - \delta$:

$$\mathcal{E}(\widehat{\vartheta}) \leqslant \frac{1}{1 - C_{n,\lambda}}\Big(- R_n(f) + \lambda^{-1}\big(\log(\delta^{-1}) - \log D_\lambda\big)\Big).$$

Lemma 7.1 yields

$$\begin{aligned}
-\log D_\lambda &= -\log\Big(\textstyle\int \exp(-\lambda R_n(\vartheta))\,\mathrm{d}\Pi(\vartheta)\Big) \\
&= \inf_{\rho\ll\Pi}\Big(\mathrm{KL}(\rho \mid \Pi) + \textstyle\int \lambda R_n(\vartheta)\,\mathrm{d}\rho(\vartheta)\Big).
\end{aligned}$$

Therefore, we have with probability of at least $1 - \delta$:

$$\mathcal{E}(\widehat{\vartheta}) \leqslant \frac{1}{1 - C_{n,\lambda}} \inf_{\rho\ll\Pi}\Big(\int \mathcal{E}_n(\vartheta)\,\mathrm{d}\rho(\vartheta) + \lambda^{-1}\big(\log(\delta^{-1}) + \mathrm{KL}(\rho \mid \Pi)\big)\Big).$$

In order to reduce the integral $\int \mathcal{E}_n(\vartheta)\,\mathrm{d}\rho(\vartheta)$ to $\int \mathcal{E}(\vartheta)\,\mathrm{d}\rho(\vartheta)$, we use Jensen's inequality and (7.13) to obtain for any probability measure $\rho \ll \Pi$ (which may depend on $\mathcal{D}_n$)

$$
\begin{aligned}
&\mathbb{E}_{\mathcal{D}_n}\Big[\exp\Big(\int \lambda\mathcal{E}_n(\vartheta) - \lambda(1+C_{n,\lambda})\mathcal{E}(\vartheta)\,\mathrm{d}\rho(\vartheta) - \mathrm{KL}(\rho \mid \Pi) - \log(\delta^{-1})\Big)\Big] \\
&\quad = \mathbb{E}_{\mathcal{D}_n}\Big[\exp\Big(\int \lambda\mathcal{E}_n(\vartheta) - \lambda(1+C_{n,\lambda})\mathcal{E}(\vartheta) \\
&\qquad - \log\Big(\frac{\mathrm{d}\rho}{\mathrm{d}\Pi}(\vartheta)\Big) - \log(\delta^{-1})\,\mathrm{d}\rho(\vartheta)\Big)\Big] \\
&\quad \leqslant \mathbb{E}_{\mathcal{D}_n,\vartheta\sim\rho}\Big[\exp\Big(\lambda\mathcal{E}_n(\vartheta) - \lambda(1+C_{n,\lambda})\mathcal{E}(\vartheta) - \log\Big(\frac{\mathrm{d}\rho}{\mathrm{d}\Pi}(\vartheta)\Big) - \log(\delta^{-1})\Big)\Big] \\
&\quad = \mathbb{E}_{\mathcal{D}_n}\Big[\int \exp\big(\lambda\mathcal{E}_n(\vartheta) - \lambda(1+C_{n,\lambda})\mathcal{E}(\vartheta) - \log(\delta^{-1})\big)\,\mathrm{d}\Pi(\vartheta)\Big] \leqslant \delta.
\end{aligned}
$$

Using $\mathbf{1}_{[0,\infty)}(x) \leqslant \mathrm{e}^{\lambda x}$ again, we conclude with probability of at least $1-\delta$:

$$
\int \mathcal{E}_n(\vartheta)\,\mathrm{d}\rho(\vartheta) \leqslant (1+C_{n,\lambda})\int \mathcal{E}(\vartheta)\,\mathrm{d}\rho(\vartheta) + \lambda^{-1}\big(\mathrm{KL}(\rho \mid \Pi) + \log(\delta^{-1})\big).
$$

Therefore, we conclude with probability of at least $1-2\delta$

$$
\mathcal{E}(\widehat{\vartheta}) \leqslant \frac{1}{1-C_{n,\lambda}} \inf_{\rho\ll\Pi}\Big((1+C_{n,\lambda})\int \mathcal{E}(\vartheta)\,\mathrm{d}\rho(\vartheta) + \frac{2}{\lambda}\big(\log(\delta^{-1}) + \mathrm{KL}(\rho \mid \Pi)\big)\Big)
$$

which yields the claimed bound if $C_{n,\lambda} \leqslant 1/2$. □

7.4.2 *Proof of Theorem 7.1*

We fix some index set $\mathcal{I}$, a radius $\eta \in (0,1]$ and apply Proposition 7.2 with $\rho = \rho_{\mathcal{I},\eta}$ defined via

$$
\frac{\mathrm{d}\rho_{\mathcal{I},\eta}}{\mathrm{d}\Pi_{\mathcal{I}}}(\vartheta) \propto \mathbf{1}_{\{|\vartheta-\vartheta^*_{\mathcal{I}}|_\infty \leqslant \eta\}}
$$

with $\vartheta^*_{\mathcal{I}}$ from (7.8). Note that indeed $C_{n,\lambda} \leqslant 1/2$ for the given choice of λ. In order to control the integral term, we decompose

$$
\begin{aligned}
\int \mathcal{E}\,\mathrm{d}\rho &= \mathcal{E}(\vartheta^*_{\mathcal{I}}) + \int \mathbb{E}\big[(f_\vartheta(\mathbf{X}) - f(\mathbf{X}))^2 - (f_{\vartheta^*_{\mathcal{I}}}(\mathbf{X}) - f(\mathbf{X}))^2\big]\,\mathrm{d}\rho(\vartheta) \\
&= \mathcal{E}(\vartheta^*_{\mathcal{I}}) + \int \mathbb{E}[(f_{\vartheta^*_{\mathcal{I}}}(\mathbf{X}) - f_\vartheta(\mathbf{X}))^2]\,\mathrm{d}\rho(\vartheta) \\
&\quad + 2\int \mathbb{E}[(f(\mathbf{X}) - f_{\vartheta^*_{\mathcal{I}}}(\mathbf{X}))(f_{\vartheta^*_{\mathcal{I}}}(\mathbf{X}) - f_\vartheta(\mathbf{X}))]\,\mathrm{d}\rho(\vartheta) \\
&\leqslant \mathcal{E}(\vartheta^*_{\mathcal{I}}) + \int \mathbb{E}[(f_{\vartheta^*_{\mathcal{I}}}(\mathbf{X}) - f_\vartheta(\mathbf{X}))^2]\,\mathrm{d}\rho(\vartheta) \\
&\quad + 2\int \mathbb{E}\big[(f(\mathbf{X}) - f_{\vartheta^*_{\mathcal{I}}}(\mathbf{X}))^2\big]^{1/2}\mathbb{E}\big[(f_{\vartheta^*_{\mathcal{I}}}(\mathbf{X}) - f_\vartheta(\mathbf{X}))^2\big]^{1/2}\,\mathrm{d}\rho(\vartheta)
\end{aligned}
$$

$$\leqslant \frac{4}{3}\mathcal{E}(\vartheta_{\mathcal{I}}^*) + 4\int \mathbb{E}[(f_{\vartheta_{\mathcal{I}}^*}(\mathbf{X}) - f_\vartheta(\mathbf{X}))^2]\,\mathrm{d}\rho(\vartheta),$$

using $2ab \leqslant \frac{a^2}{3} + 3b^2$ in the last step. To bound the remainder, we use the well-established Lipschitz continuity of the map $\vartheta \mapsto f_\vartheta(\mathbf{x})$ for fixed $\mathbf{x}$. The explicit statement can be found in [8, Lemma 17]:

Lemma 7.3 *Let $\vartheta, \widetilde{\vartheta} \in [-B, B]^P$. Then we have for $\mathbf{x} \in \mathbb{R}^p$ that*

$$|f_\vartheta(\mathbf{x}) - f_{\widetilde{\vartheta}}(\mathbf{x})| \leqslant 4(2rB)^L(|\mathbf{x}|_1 \vee 1) \cdot |\vartheta - \widetilde{\vartheta}|_\infty.$$

Due to the support of $\rho = \rho_{\mathcal{I},\eta}$, we obtain

$$\int \mathcal{E}\,\mathrm{d}\rho \leqslant \frac{4}{3}\mathcal{E}(\vartheta_{\mathcal{I}}^*) + \frac{4}{n^2} \quad \text{for} \quad \eta = \frac{1}{8(2rB)^L pKn}. \tag{7.14}$$

It remains to bound the Kullback-Leibler term in (7.12) which can be done with the following lemma:

Lemma 7.4 *We have*

(i) $\mathrm{KL}(\rho_{\mathcal{I},\eta} \mid \Pi) = \mathrm{KL}(\rho_{\mathcal{I},\eta} \mid \Pi_{\mathcal{I}}) + \log(C_{\mathcal{I}})$ *where* $C_{\mathcal{I}} := C_P 2^{|\mathcal{I}|}\binom{P}{|\mathcal{I}|}$,

(ii) $\mathrm{KL}(\rho_{\mathcal{I},\eta} \mid \Pi_{\mathcal{I}}) \leqslant |\mathcal{I}|\log(2B/\eta)$.

In particular,

$$\mathrm{KL}(\rho_{\mathcal{I},\eta} \mid \Pi) \leqslant |\mathcal{I}|\log(2B/\eta) + \log(C_{\mathcal{I}}).$$

Together with $\binom{P}{|\mathcal{I}|} \leqslant \frac{P^{|\mathcal{I}|}}{(|\mathcal{I}|)!} \leqslant (\mathrm{e}P)^{|\mathcal{I}|}$ by Stirling's formula, the previous lemma yields

$$\mathrm{KL}(\rho \mid \Pi) \leqslant |\mathcal{I}|\log\big(4C_P BP\mathrm{e}/\eta\big). \tag{7.15}$$

Plugging (7.14) and (7.15) into the PAC-Bayes bound (7.12), we conclude

$$\mathcal{E}(\widehat{\vartheta}_\lambda) \leqslant 4\mathcal{E}(\vartheta_{\mathcal{I}}^*) + \frac{4}{n^2} + \frac{4}{\lambda}\big(|\mathcal{I}|\log\big(32BP\mathrm{e}(2rB)^L pKn\big) + \log(2/\delta)\big).$$

$$\leqslant 4\mathcal{E}(\vartheta_{\mathcal{I}}^*) + \frac{Q_1}{n}\big(|\mathcal{I}|L\log(p \vee n) + \log(2/\delta)\big) \tag{7.16}$$

for $n \geqslant n_0 := 2 \vee r \vee B \vee K$ and some constant Q_1 only depending on C, Γ, σ. Note that the upper bound in (7.16) is deterministic and $\mathcal{I}$ is arbitrary. Therefore, we can choose $\mathcal{I}$ such that this bound is minimized, which completes the proof. □

7.4.3 Remaining Proofs for Section 7.3

7.4.3.1 Proof of Corollary 7.1

Jensen's and Markov's inequality yield for $r_{n,p}^2$ from (7.10) that

$$
\begin{aligned}
&\mathbb{P}\Big(\mathcal{E}(\bar{f}_\lambda) > r_{n,p}^2 + \frac{Q_1}{n} + \frac{Q_1}{n}\log(2/\delta)\Big) \\
&\quad = \mathbb{P}\Big(\|\mathbb{E}[f_{\widehat{\vartheta}_\lambda} \mid \mathcal{D}_n] - f\|_{L^2(\mathbb{P}^{\mathbf{X}})}^2 > r_{n,p}^2 + \frac{Q_1}{n} + \frac{Q_1}{n}\log(2/\delta)\Big) \\
&\quad \leqslant \mathbb{P}\Big(\mathbb{E}\big[\|f_{\widehat{\vartheta}_\lambda} - f\|_{L^2(\mathbb{P}^{\mathbf{X}})}^2 \,\big|\, \mathcal{D}_n\big] > r_{n,p}^2 + \frac{Q_1}{n} + \frac{Q_1}{n}\log(2/\delta)\Big) \\
&\quad = \mathbb{P}\Big(\int_{\frac{Q_1}{n}\log(2/\delta)}^{\infty} \Pi_\lambda\big(\|f_{\widehat{\vartheta}_\lambda} - f\|_{L^2(\mathbb{P}^{\mathbf{X}})}^2 > r_{n,p}^2 + t \,\big|\, \mathcal{D}_n\big)\,\mathrm{d}t > \frac{Q_1}{n}\Big) \\
&\quad \leqslant \frac{n}{Q_1}\int_{\frac{Q_1}{n}\log(2/\delta)}^{\infty} \mathbb{E}\big[\Pi_\lambda\big(\|f_{\widehat{\vartheta}_\lambda} - f\|_{L^2(\mathbb{P}^{\mathbf{X}})}^2 > r_{n,p}^2 + t \,\big|\, \mathcal{D}_n\big)\big]\,\mathrm{d}t.
\end{aligned}
$$

Using Theorem 7.1, we conclude

$$
\mathbb{P}\Big(\mathcal{E}(\bar{f}_\lambda) > r_{n,p}^2 + \frac{Q_1}{n} + \frac{Q_1}{n}\log(2/\delta)\Big) \leqslant \frac{2n}{Q_1}\int_{\frac{Q_1}{n}\log(2/\delta)}^{\infty} \mathrm{e}^{-nt/Q_1}\,\mathrm{d}t = \delta.
$$

□

7.4.3.2 Proof of Proposition 7.1

Throughout, denote by $C_i, i = 1, 2, \ldots$ constants only depending on upper bounds for q, $|(d_1, \ldots, d_q)|_\infty$, $|\mathbf{t}|_\infty$, $|\beta|_\infty$ and C_0.

We will verify that for any sufficiently large n, $M \in \mathbb{N}$ there exists a sparse ReLU neural network $g = g_\vartheta \in \mathcal{G}(p, C_1\lceil\log_2 n\rceil, C_2 M, \mathcal{I})$ with $|\mathcal{I}| \leqslant C_3 M\lceil\log_2 n\rceil$ and $|\vartheta|_\infty \leqslant 1 \leqslant B$ such that

$$
\|g - f\|_{L^\infty([0,1]^p)} \leqslant C_4 M^{-\beta^*/t^*}. \tag{7.17}
$$

Careful inspection of the proof of [32, Theorem 1] reveals that there exists a sparse ReLU neural network $g \in \mathcal{G}(p, L, r, \mathcal{J})$ with weights and shifts absolutely bounded by 1 and

$$
\begin{aligned}
L &= 3(q-1) + \sum_{i=0}^{q} \big(8 + (\lceil \log_2 n \rceil + 5)(1 + \lceil \log_2(t_i \vee \beta_i) \rceil)\big), \\
r &= 6M \max_{i=0,\dots,q} d_{i+1}(t_i + \lceil \beta_i \rceil) \qquad \text{and} \\
|\mathcal{J}| &\leqslant \sum_{i=0}^{q} d_{i+1}\big(141(t_i + \beta_i + 1)^{3+t_i} M(\lceil \log_2 n \rceil + 6) + 4\big)
\end{aligned}
$$

such that (7.17) holds with

$$
M = \Big\lceil \Big(\frac{n}{\log^2(n)\log(p \vee n)} \Big)^{t^*/(2\beta^* + t^*)} \Big\rceil \leqslant n,
$$

provided $M \geqslant \max_{i=0,\dots,q}(\beta_i + 1)^{t_i} \vee (C_0(2C_0)^{\beta_i} + 1)e^{t_i}$. Hence, it remains to show that g can also be represented as a ReLU neural network in

$$
\mathcal{G}(p, C_1\lceil \log_2 n \rceil, C_2 n, \mathcal{I}). \tag{7.18}
$$

To do this, we employ the *embedding properties of network function classes* from [32, Sect. 7.1].

Note that L, r and the upper bound for $|\mathcal{J}|$ are independent of $d_0 = p$ and monotonically increasing in q, $|(d_1, \dots, d_q)|_\infty$, $|\mathbf{t}|_\infty$ and $|\beta|_\infty$. Also, r is of order M, L is of order $\lceil \log_2 n \rceil$ and the upper bound for $|\mathcal{J}|$ is of order $M\lceil \log_2 n \rceil$. Using the *enlarging* and the *depth synchronisation properties*, g can indeed be written as a ReLU neural network in (7.18). Note that to ensure the depth of the network, we added additional layers after the last hidden layer, instead of right after the input to preserve the order of the sparsity.

Theorem 7.1 together with $\mathcal{E}(f_{\vartheta^*_{\mathcal{I}}}) \leqslant \|g_\vartheta - f\|^2_{L^\infty([0,1]^p)}$ now yields

$$
\mathcal{E}(\widehat{\vartheta}_\lambda) \leqslant 4C_4 M^{-2\beta^*/t^*} + \frac{Q_1 C_3}{n} M\lceil \log_2 n \rceil^2 \log(p \vee n) + Q_1 \frac{\log(2/\delta)}{n}
$$

with probability of at least $1 - \delta$.

The convergence rate for the posterior mean is obtained analogously using Corollary 7.1. □

7.4.4 Proofs of the Auxiliary Results

7.4.4.1 Proof of Lemma 7.2

We write $\mathcal{E}_n(\vartheta) = \frac{1}{n}\sum_{i=1}^{n} Z_i$ with centered and independent random variables

$$
Z_i := (Y_i - f_\vartheta(\mathbf{X}_i))^2 - (Y_i - f(\mathbf{X}_i))^2 = -\big(2\varepsilon_i + f(\mathbf{X}_i) - f_\vartheta(\mathbf{X}_i)\big)\big(f_\vartheta(\mathbf{X}_i) - f(\mathbf{X}_i)\big)
$$

Since f and f_ϑ are bounded by C and ε_i is sub-Gaussian we have

$$\mathbb{E}[Z_i^2] = \mathbb{E}\big[\big(2\varepsilon_i + f(\mathbf{X}_i) - f_\vartheta(\mathbf{X}_i)\big)^2\big(f_\vartheta(\mathbf{X}_i) - f(\mathbf{X}_i)\big)^2\big] \leqslant 2(4\sigma^2 + 4C^2)\mathcal{E}(\vartheta) =: U$$

and for $k \geqslant 3$

$$\begin{aligned}\mathbb{E}[(Z_i)_+^k] &\leqslant \mathbb{E}\big[|2\varepsilon_i + f(\mathbf{X}_i) - f_\vartheta(\mathbf{X}_i)|^k |f_\vartheta(\mathbf{X}_i) - f(\mathbf{X}_i)|^{k-2}(f_\vartheta(\mathbf{X}_i) - f(\mathbf{X}_i))^2\big] \\ &\leqslant (2C)^{k-2}\mathbb{E}\big[|2\varepsilon_i + f(\mathbf{X}_i) - f_\vartheta(\mathbf{X}_i)|^k (f_\vartheta(\mathbf{X}) - f(\mathbf{X}))^2\big] \\ &\leqslant (2C)^{k-2}2^{k-1}\big(k!2^{k-1}\sigma^2\Gamma^{k-2} + (2C)^k\big)\mathcal{E}(\vartheta) \\ &\leqslant (2C)^{k-2}k!8^{k-2}\big(\Gamma^{k-2} \vee (2C)^{k-2}\big)v = k!U w^{k-2}.\end{aligned}$$

In view of $\mathbb{E}[\mathcal{E}_n(\vartheta)] = \mathcal{E}(\vartheta)$, Bernstein's inequality [23, inequality (2.21)] yields

$$\mathbb{E}\big[\exp\big(\lambda\big(\mathcal{E}_n(\vartheta) - \mathcal{E}(\vartheta)\big)\big)\big] \leqslant \exp\Big(\frac{U\lambda^2}{n(1 - w\lambda/n)}\Big).$$

The same bound remains true if we replace Z_i by $-Z_i$. □

7.4.4.2 Proof of Lemma 7.4

(i) We will show that

$$\frac{\mathrm{d}\rho_{\mathcal{I},\eta}}{\mathrm{d}\Pi} = C_{\mathcal{I}}\frac{\mathrm{d}\rho_{\mathcal{I},\eta}}{\mathrm{d}\Pi_{\mathcal{I}}} \tag{7.19}$$

from which we can deduce

$$\begin{aligned}\mathrm{KL}(\rho_{\mathcal{I},\eta} \mid \Pi) = \int \log\Big(\frac{\mathrm{d}\rho_{\mathcal{I},\eta}}{\mathrm{d}\Pi}\Big)\,\mathrm{d}\rho_{\mathcal{I},\eta} &= \int \log\Big(\frac{\mathrm{d}\rho_{\mathcal{I},\eta}}{\mathrm{d}\Pi_{\mathcal{I}}}\Big)\,\mathrm{d}\rho_{\mathcal{I},\eta} + \log(C_{\mathcal{I}}) \\ &= \mathrm{KL}(\rho_{\mathcal{I},\eta} \mid \Pi_{\mathcal{I}}) + \log(C_{\mathcal{I}}).\end{aligned}$$

For (7.19), we need to show

$$\rho_{\mathcal{I},\eta}(A) = \int_A C_{\mathcal{I}}^{-1}\frac{\mathrm{d}\rho_{\mathcal{I},\eta}}{\mathrm{d}\Pi}\,\mathrm{d}\Pi_{\mathcal{I}}$$

for all $A \in \mathscr{B}_{\mathbb{R}^P}$. Observe that for the sets

$$\mathcal{S}_{\mathcal{J},\Leftrightarrow} := \{\vartheta \in \mathcal{S}_{\mathcal{J}} \mid \vartheta_i \neq 0 \Leftrightarrow i \in \mathcal{J}\}$$

with $\emptyset \neq \mathcal{J} \subseteq \{1, \ldots, P\}$, we have

$$\Pi_{\mathcal{J}}(\mathcal{S}_{\mathcal{J},\Leftrightarrow}) = 1. \tag{7.20}$$

In particular, (7.20) holds for $\mathcal{J} = \mathcal{I}$. Since also $\rho_{\mathcal{I},\eta}(\mathcal{S}_{\mathcal{I},\Leftrightarrow}) = 1$, no generality is lost in additionally assuming $A \subseteq \mathcal{S}_{\mathcal{I},\Leftrightarrow}$. Note how

$$\mathcal{S}_{\mathcal{J},\Leftrightarrow} \cap \mathcal{S}_{\mathcal{I},\Leftrightarrow} = \emptyset \qquad \forall \mathcal{J} \neq \mathcal{I}. \tag{7.21}$$

Combining (7.20) with (7.21), we see that

$$\int_A \frac{\mathrm{d}\rho_{\mathcal{I},\eta}}{\mathrm{d}\Pi} \,\mathrm{d}\Pi_{\mathcal{J}} = 0 \qquad \forall \mathcal{J} \neq \mathcal{I}.$$

Therefore,

$$\rho_{\mathcal{I},\eta}(A) = \int_A \frac{\mathrm{d}\rho_{\mathcal{I},\eta}}{\mathrm{d}\Pi} \,\mathrm{d}\Pi = \int_A C_{\mathcal{I}}^{-1} \frac{\mathrm{d}\rho_{\mathcal{I},\eta}}{\mathrm{d}\Pi} \,\mathrm{d}\Pi_{\mathcal{I}}.$$

(ii) Since $\rho_{\mathcal{I},\eta}$ and $\Pi_{\mathcal{I}}$ are product measures, their KL-divergence is equal to the sum of the KL-divergences in each of the P factors. For factors with index $i \notin \mathcal{I}$, this is zero, as both factors have all their mass in 0. For factors with index $i \in \mathcal{I}$, we are comparing

$$\mathcal{U}([(\vartheta_{\mathcal{I}}^*)_i - \eta, (\vartheta_{\mathcal{I}}^*)_i + \eta] \cap [-B, B]) \qquad \text{with} \qquad \mathcal{U}([-B, B]).$$

The KL-divergence of these distributions is equal to

$$\log\left(\frac{\boldsymbol{\lambda}([-B, B])}{\boldsymbol{\lambda}([(\vartheta_{\mathcal{I}}^*)_i - \eta, (\vartheta_{\mathcal{I}}^*)_i + \eta] \cap [-B, B])}\right) \leqslant \log\left(\frac{\boldsymbol{\lambda}([-B, B])}{\boldsymbol{\lambda}([0, \eta])}\right) = \log(2B/\eta).$$

Thus,

$$\begin{aligned} \mathrm{KL}(\rho_{\mathcal{I},\eta} \mid \Pi_{\mathcal{I}}) &= \sum_{i \in \mathcal{I}} \mathrm{KL}\left(\mathcal{U}([(\vartheta_{\mathcal{I}}^*)_i - \eta, (\vartheta_{\mathcal{I}}^*)_i + \eta] \cap [-B, B]) \mid \mathcal{U}([-B, B])\right) \\ &\leqslant |\mathcal{I}| \log(2B/\eta). \end{aligned}$$

□

7.5 Implementation of the Mixing Prior Distribution

To incorporate sparse networks, the algorithm from Sect. 7.2.3 has to be extended. The following augmentations to the proposal density are inspired by [3].

For a fixed active set $\mathcal{I}$ we modify the proposal density from (7.7) to

$$\psi_{\mathcal{I}}\big((\tau)_{i \in \mathcal{I}} \mid \vartheta\big) = \frac{1}{(2\pi s^2)^{|\mathcal{I}|/2}} \exp\Big(-\frac{1}{2s^2} \sum_{i \in \mathcal{I}} \big(\tau_i - \vartheta_i + \gamma \tfrac{\partial}{\partial \vartheta_i} R_n(\vartheta)\big)^2\Big)$$

and set $\tau_i = 0$ for all $i \notin \mathcal{I}$. The mixture over sparse active sets leads to the conditional proposal density

$$q(\tau \mid \vartheta) = \frac{q_-(\tau \mid \vartheta) + 2q_=(\tau \mid \vartheta) + q_+(\tau \mid \vartheta)}{4}\mathbf{1}_{\{1<|\mathcal{I}|<P\}} + \frac{q_+(\tau \mid \vartheta) + 2q_=(\tau \mid \vartheta)}{3}\mathbf{1}_{\{|\mathcal{I}|=1\}} + \frac{q_-(\tau \mid \vartheta) + 2q_=(\tau \mid \vartheta)}{3}\mathbf{1}_{\{|\mathcal{I}|=P\}},$$

with $\mathcal{I} = \{i : \vartheta_i \neq 0\}$ and some q_- and q_+ proposing to remove or add a component, respectively, while $q_=$ leaves the active set untouched. Specifically, we choose

$$q_=(\tau \mid \vartheta) = \psi_{\mathcal{I}}(\tau \mid \vartheta), \quad q_-(\tau \mid \vartheta) = \sum_{i\in\mathcal{I}} w_i^- \psi_{\mathcal{I}\setminus\{i\}}(\tau \mid \vartheta) \quad \text{and}$$
$$q_+(\tau \mid \vartheta) = \sum_{i\in\mathcal{I}^{\mathrm{c}}} w_i^+ \psi_{\mathcal{I}\cup\{i\}}(\tau \mid \vartheta)$$

$$w_i^- = \exp(-|\vartheta_i|) \Big/ \sum_{j\in\mathcal{I}} \exp(-|\vartheta_j|) \qquad \text{and} \qquad w_i^+ = \widetilde{w}_i^+ \Big/ \sum_{j\in\mathcal{I}^{\mathrm{c}}} \widetilde{w}_j^+,$$

where $\widetilde{w}_i^+ = |\{j \in \mathcal{I}^{\mathrm{c}} : |\frac{\partial}{\partial\vartheta_j} R_n(\vartheta)| \leqslant |\frac{\partial}{\partial\vartheta_i} R_n(\vartheta)|\}|^2$. The weights $\widetilde{w}_i^-$ are chosen such that (absolutely) smaller entries have a larger probability of being removed. On the other hand, $\widetilde{w}_i^+$ are chosen such that components with a large impact (as measured by $\nabla_\vartheta R_n(\vartheta)$) on the model have a higher probability of being included. Note that this extension of the algorithm requires no additional computation of $\nabla_\vartheta R_n(\vartheta)$, as this was already done in the previous MCMC step. Finally, the mixing prior from (7.3) also has to be accounted for in the acceptance probabilities, leading to

$$\alpha(\tau \mid \vartheta) = \min\Big\{1, \exp\big(-\lambda R_n(\tau) + \lambda R_n(\vartheta)\big)\frac{\Pi(\tau)q(\vartheta \mid \tau)}{\Pi(\vartheta)q(\tau \mid \vartheta)}\Big\},$$

where with some abuse of notation $\Pi(\vartheta)$ denotes the probability density of the prior.

Acknowledgements Financial support by the Deutsche Forschungsgemeinschaft (DFG) through project TR 1349/3-1 is gratefully acknowledged.

References

1. Alexos, A., Boyd, A.J., Mandt, S.: Structured stochastic gradient MCMC. In: International Conference on Machine Learning, pp. 414–434 (2022)
2. Alquier, P.: User-friendly introduction to PAC-Bayes bounds. Found. Trends Mach. Learn. **17**(2), 174–303 (2024)
3. Alquier, P., Biau, G.: Sparse single-index model. J. Mach. Learn. Res. **14**, 243–280 (2013)
4. Anthony, M., Bartlett, P.L.: Neural Network Learning: Theoretical Foundations. Cambridge University Press (1999)

5. Bauer, B., Kohler, M.: On deep learning as a remedy for the curse of dimensionality in nonparametric regression. Ann. Stat. **47**(4), 2261–2285 (2019)
6. Besag, J.: Comments on "Representations of knowledge in complex systems" by U. Grenander, M.I. Miller. J. R. Stat. Society. Ser. B. Methodol. **56**(4), 549–581 (1994)
7. Bieringer, S., Kasieczka, G., Steffen, M.F., Trabs, M.: AdamMCMC: combining Metropolis adjusted Langevin with momentum-based optimization (2023). arXiv:2312.14027
8. Bieringer, S., Kasieczka, G., Steffen, M.F., Trabs, M.: The surrogate Gibbs-posterior of a corrected stochastic MALA: towards uncertainty quantification for neural networks (2025). arXiv:2310.09335
9. Blei, D.M., Kucukelbir, A., McAuliffe, J.D.: Variational inference: a review for statisticians. J. Am. Stat. Assoc. **112**(518), 859–877 (2017)
10. Blundell, C., Cornebise, J., Kavukcuoglu, K., Wierstra, D.: Weight uncertainty in neural networks. In: International Conference on Machine Learning, pp. 1613–1622 (2015)
11. Bolcskei, H., Grohs, P., Kutyniok, G., Petersen, P.: Optimal approximation with sparsely connected deep neural networks. SIAM J. Math. Data Sci. **1**(1), 8–45 (2019)
12. Catoni, O.: Statistical Learning Theory and Stochastic Optimization. Springer, Berlin (2004)
13. Catoni, O.: PAC-Bayesian supervised classification: the thermodynamics of statistical learning, vol. 56. Lecture Notes-Monograph Series. Institute of Mathematical Statistics (2007)
14. Chérief-Abdellatif, B.-E.: Convergence rates of variational inference in sparse deep learning. In: International Conference on Machine Learning, pp. 1831–1842 (2020)
15. DeVore, R., Hanin, B., Petrova, G.: Neural network approximation. Acta Numer. **30**, 327–444 (2021)
16. Dziugaite, G.K., Roy, D.M.: Computing nonvacuous generalization bounds for deep (stochastic) neural networks with many more parameters than training data. Proceedings of the Thirty-Third Conference on Uncertainty in Artificial Intelligence (2017)
17. Franssen, S., Szabó, B.: Uncertainty quantification for nonparametric regression using empirical Bayesian neural networks (2022). arXiv:2204.12735
18. Goodfellow, I., Bengio, Y., Courville, A.: Deep Learning. MIT Press (2016)
19. Graves, A.: Practical variational inference for neural networks. In: Advances in Neural Information Processing Systems, vol. 24 (2011)
20. Green, P.J.: Reversible jump Markov chain Monte Carlo computation and Bayesian model determination. Biometrika **82**(4), 711–732 (1995)
21. Guedj, B.: A primer on PAC-Bayesian learning. In: Proceedings of the 2nd Congress of the Société Mathématique de France, pp. 391–414 (2019)
22. Kingma, D.P., Ba, J.: Adam: a method for stochastic optimization (2014). arXiv:1412.6980
23. Massart, P.: Concentration inequalities and model selection. In: Lecture Notes in Mathematics, vol. 1896. Springer, Berlin (2007)
24. McAllester, D.A.: PAC-Bayesian model averaging. In: Proceedings of the Twelfth Annual Conference on Computational Learning Theory, pp. 164–170 (1999a)
25. McAllester, D.A.: Some PAC-Bayesian theorems. Mach. Learn. **37**(3), 355–363 (1999b)
26. Pérez-Ortiz, M., Rivasplata, O., Shawe-Taylor, J., Szepesvári, C.: Tighter risk certificates for neural networks. J. Mach. Learn. Res. **22** (2021)
27. Polson, N.G., Ročková, V.: Posterior concentration for sparse deep learning. Adv. Neural. Inf. Process. Syst. **31**, 938–949 (2018)
28. Robert, C.P., Casella, G.: Monte Carlo Statistical Methods, 2nd edn. Springer, Berlin (2004)
29. Roberts, G.O., Tweedie, R.L.: Exponential convergence of of Langevin distributions and their discrete approximations. Bernoulli. Off. J. Bernoulli Soc. Math. Stat. Probab. **2**(4), 341–363 (1996a)
30. Roberts, G.O., Tweedie, R.L.: Geometric convergence and central limit theorems for multidimensional Hastings and Metropolis algorithms. Biometrika **83**(1), 95–110 (1996b)
31. Schmidhuber, J.: Deep learning in neural networks: an overview. Neural Netw. **61**, 85–117 (2015)
32. Schmidt-Hieber, J.: Nonparametric regression using deep neural networks with ReLU activation function. Ann. Stat. **48**(4), 1875–1897 (2020)

33. Shawe-Taylor, J., Williamson, R.C.: A PAC analysis of a Bayesian estimator. In: Proceedings of the Tenth Annual Conference on Computational Learning Theory, pp. 2–9 (1997)
34. Steffen, M.F.: PAC-Bayes bounds for high-dimensional multi-index models with unknown active dimension (2023). arXiv:2303.13474
35. Taheri, M., Xie, F., Lederer, J.: Statistical guarantees for regularized neural networks. Neural Netw. **142**, 148–161 (2021)
36. Tinsi, L., Dalalyan, A.: Risk bounds for aggregated shallow neural networks using Gaussian priors. In: Conference on Learning Theory, pp. 227–253 (2022)
37. Yarotsky, D.: Error bounds for approximations with deep ReLU networks. Neural Netw. **94**, 103–114 (2017)
38. Zhang, R., Li, C., Zhang, J., Chen, C., Wilson, A.G.: Cyclical stochastic gradient MCMC for Bayesian deep learning. In: 8th International Conference on Learning Representations (2020)

Chapter 8
Comparison of Model Output Statistics and Neural Networks to Postprocess Wind Gusts

Cristina Primo, Benedikt Schulz, Sebastian Lerch, and Reinhold Hess

Abstract Wind gust prediction plays an important role in warning strategies of national meteorological services due to the high impact of its extreme values. However, forecasting wind gusts is challenging because they are influenced by small-scale processes and local characteristics. To account for the different sources of uncertainty, meteorological centers run ensembles of forecasts and derive probabilities of wind gusts exceeding a threshold. These probabilities often exhibit systematic errors and require postprocessing. Model output statistics (MOS) is a common operational postprocessing technique, although more modern methods such as neural network-based approaches have shown promising results in research studies. The transition from research to operations requires an exhaustive comparison of both techniques. Taking a first step into this direction, our study presents a comparison of a postprocessing technique based on linear and logistic regression approaches with different neural network methods proposed in the literature to improve wind gust predictions, specifically distributional regression networks and Bernstein quantile networks. We further contribute to investigating optimal design choices for neural network-based postprocessing methods regarding changes of the numerical model in the training period, the use of persistence predictors, and the temporal composition of training datasets. The performance of the different techniques is compared in terms of calibration, accuracy, reliability and resolution based on case studies of wind gust forecasts from the operational weather model of the German weather service and observations from 170 weather stations.

C. Primo (✉) · R. Hess
Deutscher Wetterdienst, Offenbach, Germany
e-mail: Cristina.Primo-Ramos@dwd.de

R. Hess
e-mail: Reinhold.Hess@dwd.de

B. Schulz · S. Lerch
Karlsruhe Institute of Technology, Karlsruhe, Germany
e-mail: Benedikt.Schulz2@kit.edu

S. Lerch
e-mail: Sebastian.Lerch@kit.edu

A. Ott et al. (eds.), *Applications of Mathematics in Sciences, Engineering, and Economics*,
Springer Proceedings in Mathematics & Statistics 515,
https://doi.org/10.1007/978-3-032-01279-1_8

Keywords Wind gusts · Postprocessing · Ensembles · Model output statistics · Neural networks · Probability forecasts/models/distribution

8.1 Introduction

Presently, the majority of weather forecasts rely on the outcomes generated by physical models, wherein systems of partial differential equations characterizing atmospheric processes are solved on a grid. In order to quantify forecast uncertainty and obtain probabilistic projections, these numerical weather prediction (NWP) models are run multiple times with diverse sets of initial conditions and/or model physics. This process yields an ensemble of predictions. Despite notable advancements in recent decades, ensemble forecasts still show systematic errors, necessitating correction through statistical postprocessing methods to obtain accurate and reliable forecasts [29].

Postprocessing has been a focus of research interest at the intersection of statistics and atmospheric sciences. Over the past decades, a large variety of methods has been developed and successfully applied in research and operations, see [28] for an overview. A particular focus has been on probabilistic forecasts that provide information about forecast uncertainty, the estimation and communication of which is especially important for forecasts of hazardous weather conditions such as severe wind gusts. Modern postprocessing methods correct systematic errors with distributional regression models for the conditional probability distribution of the weather variable of interest given predictors from the NWP system. In many of these postprocessing models, the probabilistic forecast is given by a parametric probability distribution with parameters depending on predictors through a suitably chosen link function e.g., ensemble model output statistic (EMOS) [8]. Summary statistics of ensemble predictions of the variable of interest are often used as sole covariates. Over the past years, deep learning methods based on neural networks (NNs) that were first proposed in [22] have been demonstrated to provide state-of-the-art predictive performance and have seen broad application and extensions towards other variables e.g., [7, 10, 27], types of inputs e.g., [5, 30], or time-scales e.g., [14, 25].

Despite the promising research results, hardly any of the novel machine learning methods have been used in an operational context at meteorological services thus far, and a number of methodological and practical challenges remain for a successful transfer from research to operations [11, 29]. During many years, the German weather service (*Deutscher Wetterdienst*, DWD) has postprocessed forecasts by using model output statistics (MOS) techniques based on linear regressions. The currently operational technique, MOSMIX, as well as the weather warning guidance, WarnMOS, correct errors in deterministic outputs of a combination of two different global NWP systems: the ICOsahedral Nonhydrostatic (ICON) model at DWD and the Integrated Forecasting System (IFS) at the European Centre for Medium-Range Weather Forecasts (ECMWF). Both MOS-techniques consist of a two step process. In a first step, ICON and IFS are postprocessed individually providing ICON-MOS and IFS-MOS forecasts and in a second step, the resulting optimized forecasts are combined in the best possible way. MOSMIX currently postprocesses meteorological parameters commonly measured at observation sites, e.g., temperature and dew point in 2 m height, speed and direction of wind and wind gusts in 10 m height, expected precipitation amounts, precipitation phases (rain or snow), air pressure, sunshine duration, and many more. The postprocessing aims at providing optimized forecasts that are downscaled from the NWP model grid to user-relevant points of interest (5,400 locations worldwide). Further, WarnMOS is oriented to warning events and provides threshold exceedance probabilities for several meteorological variables.

In order to provide skillful probabilistic forecasts for weather warning events, a new system was implemented at DWD, ModelMIX [12, 13], which not only combines the deterministic models, but also accounts for the uncertainty given by the ensemble prediction systems, ICON-EPS and IFS-EPS, as well as the uncertainty in convective permitting scales, the ensemble of the limited area model from the COnsortium for Small-scale MOdelling, COSMO-DE-EPS [1]. All these systems are first individually postprocessed to obtain MOS forecasts, and then statistically combined in an optimal way to produce the probability of a warning event with uniform spatial and temporal resolution. The resulting probabilities are then internally used to produce automated warning proposals to support warning management at DWD.

In an attempt to modernize postprocessing at DWD with techniques that have shown promising results in research applications (e.g., NNs), efforts are in place now in assessing the benefits of using NN techniques in comparison with the current MOS approaches. It is important to remark that our work presented here focuses on wind gust forecasts for the next 21 h, whereas MOS techniques at DWD provide postprocessing of many more meteorological parameters (approx. 150) and up to 240 h. Nevertheless, as a starting point in the comparison, we focus on postprocessing of one variable, which is particularly important for many warnings, to better understand the benefit of using new techniques, as well as to develop ways to account for potential issues in the transition from research to operations. In addition, the underlying wind gust dataset has been used in previous research [15, 20, 27], and various skillful benchmark models for postprocessing are thus available.

The overarching aim of this article thus is to contribute to closing the gap between research on NN-based postprocessing methods and operational practices at weather services. Our contributions are twofold: On the one hand, we provide a comprehensive comparison of the forecast performance of NN models for postprocessing from research [27] to a pre-operational MOS technique for ensembles at DWD, ModelMIX. Hereby, we focus on probability forecasts for exceeding warning level thresholds of wind gusts. On the other hand, we conduct experiments to investigate design choices of NN-based postprocessing models regarding their capabilities to cope with challenges that occur in the operational use of postprocessing models, such as changes in the NWP model. Changes to the resolution, model physics or components of the NWP system may invalidate the assumption of homogeneous error characteristics of the ensemble forecasts in the training data and thus represent a major challenge for the operational implementation of postprocessing models, in particular for complex models relying on the availability of large archives of past forecast and observation data for training. For EMOS models, [16] find that utilizing multiple years of training data irrespective of NWP model changes during that period show superior performance compared to time-adaptive training schemes based on sliding windows where only forecast cases from the most recent days are considered, even though the latter should be able to more quickly adapt to changes in forecast error characteristics. However, it is an open question how changes in the NWP system affect the forecast performance of NN-based postprocessing models and how those can best be adapted to deal with such updates. In addition to the temporal composition of the training datasets, we further investigate the effect of other design choices for NN-based postprocessing models motivated by analogous techniques used in operational practice, specifically the incorporation of persistence-based predictor variables, and the joint modeling of multiple lead times.

We present two case studies utilizing forecasts from the operational EPS of DWD, which underwent several significant updates since its inception in 2010. Our focus is on probabilistic forecasts of exceedances of warning level thresholds for wind gusts, and we compare different variants of statistical and NN-based postprocessing models to address the research questions outlined above.

The remainder of this article is organized as follows. Section 8.2 introduces the dataset used in the case studies, and Sect. 8.3 presents the postprocessing models. Following a short overview of verification methods in Sect. 8.4, the main results are presented in Sect. 8.5. The article closes with a discussion in Sect. 8.6.

8.2 Data

8.2.1 NWP Model Data and Observations

The forecasts are generated by the operational 20-member EPS at convection-permitting resolution of DWD. The EPS has undergone several significant changes over time, which are specified in the following and summarized in Fig. 8.1.

The EPS started as the COSMO-DE-EPS in pre-operational mode in December 2010. The ensemble members are obtained from physically perturbing the initial and boundary conditions of four different global models combined with five sets

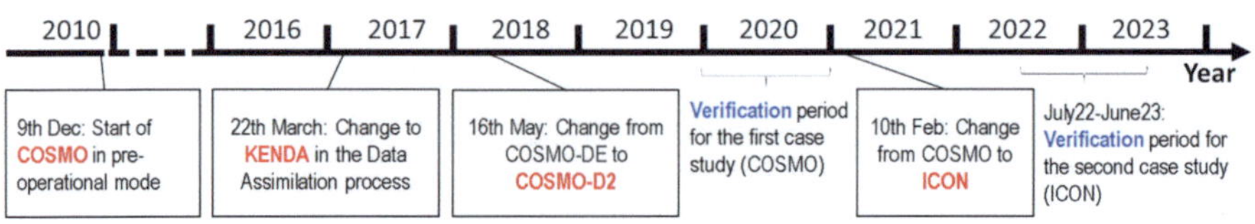

Fig. 8.1 Overview of relevant changes in the NWP model during the time range of the data considered

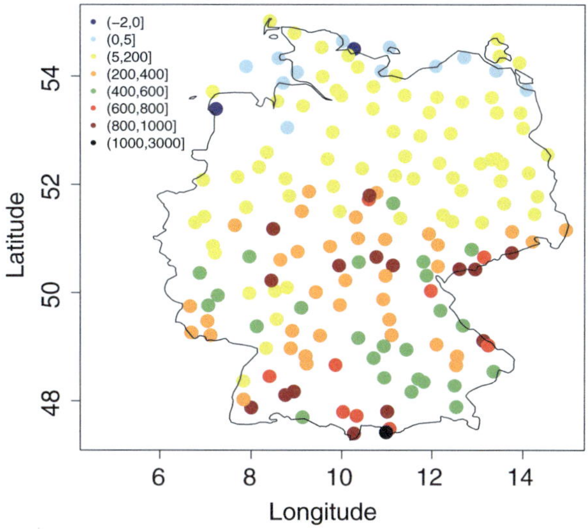

Fig. 8.2 Map of SYNOP stations. The color represents the station altitude (m)

of physical perturbations [6, 21]. The model runs at a horizontal grid spacing of 2.8km and is initialized every 3 h, with hourly forecasts produced up to 21h ahead. In 2017, the lateral boundary conditions started being driven by the global ICON-EPS system, and initial conditions were given by a kilometer-scale ensemble data assimilation system KENDA [26]. In May 2018, the model was extended to better represent convective structures, with COSMO-DE being replaced by COSMO-D2, where the horizontal and vertical resolution increased, as well as the horizontal area covered.

In February 2021, the convection-permitting model setup ICON-D2 (i.e., using the limited-area mode of the new NWP system ICON) replaced COSMO-D2 [23]. Starting values of ICON-D2-EPS are based on the local ensemble transform Kalman filter (LETKF) data assimilation of the global model ICON-EPS. Its grid consists of a set of spherical triangles that allows for a nearly constant grid spacing. Our first case study utilizes data from 2020 for verification and focuses on COSMO-D2-EPS, whereas the second case study is based on verification data from July 2022 to June 2023 and the ICON-D2-EPS model.

Observations come from 170 SYNOP stations in Germany operated by DWD, which are displayed in Fig. 8.2. Each station is paired with the nearest grid point to create pairs of observations and forecasts.

8.2.2 Predictors and Target

In this work, we use the same set of predictors for ModelMIX as in [12] (the most important predictors are listed in Table 8.2 in the Appendix) and the same for the

NNs as in [27] (see Table 8.3 in the Appendix). Those predictors include basically all meteorological variables that the model generates as well as selected temporal and spatial information. In addition, some of the postprocessing methods presented in Sect. 8.3 make use of temporal predictors that encode the model version. For each major model update (KENDA, D2, ICON), we introduce a binary variable that jumps at the time of the model change. Finally, we briefly comment on persistence predictors, which are used by some of the postprocessing methods evaluated in Sect. 8.5. Here, the persistence predictors refer to the most recently observed wind gust values at the associated station, that is, at the time the NWP model is initialized as well as one and two hours later. As discussed below, our evaluation starts at a lead time of 3h to mimic an operational setup, and incorporating persistence predictors for the first two hours thus is admissible.

Our interest lies in predicting the threshold exceedance probabilities of hourly wind gusts for a particular list of thresholds the DWD uses, namely 25, 27, 33, 40, 47, 55, 63 and 75 kt. In this work, we mostly focus on the first and lowest threshold, which is most frequently exceeded. The NWP model provides an ensemble of wind gusts as maximum wind speed values over the last hour. Of the available NWP model runs, we will focus on those initialized at 00UTC and 12UTC, with hourly outputs from 3h up to 21h. It takes around two hours to have all forecasts from the reference NWP (COSMO-DE-EPS, COSMO-D2-EPS or ICON-EPS) available and any postprocessing method used operationally could only run after having these data. Therefore, the first two lead times (1 and 2 h) are not considered here.

8.3 Postprocessing Methods

In the following, all postprocessing methods and variants thereof, which are considered in the comparisons, will be introduced. First, we present ModelMIX, the postprocessing technique used at DWD, then we present the approaches from research on probabilistic wind gust forecasting.

An overview is available in Table 8.1, divided by the two case studies. Note that the computing times stated in the table should only be seen as rough approximations and are not directly comparable due to differences in the underlying hardware and software environments used for model estimation.

8.3.1 Model Output Statistics (MOS)

ModelMIX is a classical MOS technique and is addressed here as a reference method. It is based on an operational MOS scheme at DWD that postprocesses the deterministic forecasts of ICON-D2 and IFS of ECMWF. ModelMIX has been enhanced for the ensemble forecasts of COSMO-DE/D2-EPS and ICON-D2-EPS and is tailored to extreme events of almost all relevant weather warnings of DWD.

Table 8.1 Overview of the main characteristics of the different postprocessing methods/variants. The bottom-index of the methods refers to the verification period considered (1 = 2020, 2 = 2022/2023) and the training period (C = COSMO, D = Post-D2, K = Post-Kenda, I = ICON), the top-index refers to whether persistence predictors were used (P = including persistence) and whether a joint model for all lead times and initializations was estimated (J = joint model). The column '*samples*' refers to the (average) training set size for each individual model, '*total runtime*' refers to the total time required to train and predict for all models. The total run time of $BQN_{1,D}$ and $DRN_{1,D}$ is marked with a star (*), as the model was run for only one of the two initializations. For comparison with the other variants, we double this runtime. ModelMIX is also marked (**) since it postprocesses simultaneously about 150 variables in a parallelized process, so provided numbers are estimated values from the total sample size and total run time comprising all variables

Method	Estimation and predictors			Training set			Runtime
Abbreviation	Local	Joint	Persistence	Period	Start	Samples	Total
Training until December 2019, verification on COSMO-D2-EPS until December 2020							
$ModelMIX_1$	(✓)	(✓)	✓	COSMO	09/12/2010	**500,000	**48 h
EMOS	✓	–	–	Post-D2	16/05/2018	585	3 min
EMOS-GB	✓	–	–	COSMO	09/12/2010	3,243	34.2 h
$BQN_{1,C}$	–	–	–	COSMO	09/12/2010	562,462	39.3 h
$BQN_{1,D}$	–	–	–	Post-D2	16/05/2018	101,749	*6.8 h
$DRN_{1,C}$	–	–	–	COSMO	09/12/2010	564,217	42.0 h
$DRN_{1,D}$	–	–	–	Post-D2	16/05/2018	101,749	*7.6 h
$DRN^{P}_{1,C}$	–	–	✓	COSMO	09/12/2010	564,217	41.2 h
Training until June 2022, verification on ICON-D2-EPS until July 2023							
$ModelMIX_2$	(✓)	(✓)	✓	COSMO+ICON	09/12/2010	**639,000	**48 h
$DRN^{P}_{2,I}$	–	–	✓	ICON	10/02/2021	82,701	5.6 h
$DRN^{P}_{2,K}$	–	–	✓	Post-KENDA	22/03/2017	323,675	24.2 h
$DRN^{P,J}_{2,I}$	–	✓	✓	ICON	10/02/2021	3,142,654	6.2 h
$DRN^{P,J}_{2,K}$	–	✓	✓✓	Post-KENDA	22/03/2017	12,299,665	48.4 h

Threshold exceedance probabilities are generated in two steps: First, wind gust velocities are optimized using multiple linear regressions. Ensemble mean and spread of 56 model variables are used as model input, from which altogether about 300 predictors are derived, including area means of model variables around the locations for which training is performed.

Each hourly time step is trained separately and for thresholds up to 27 kt individually for each location. For thresholds above 33 kt, locations with similar climatology are aggregated in 9 clusters and training is performed for each cluster in order to gather enough events for statistical modelling.

In a second step and based on the wind gust velocities, global logistic regressions are fitted for all locations together and parameterized for all time steps to produce probabilities, see [12] for more details (note that [12] refers to ModelMIX as Ensemble-MOS). Location-specific characteristics are considered to be taken into account already by the individual optimization of the velocities. The logistic parameterization focuses on the time step-dependent forecast errors and allows for providing meaningful statistical forecast for high thresholds up to 75kt, where the number of events is too small for training for individual stations or even time steps. Since threshold exceedance probabilities of wind gusts are based on the same deterministic wind gust velocities, consistency among thresholds is ensured.

For the first case study (verification period 2020), training was performed using forecasts of COSMO-DE-EPS and COSMO-D2-EPS together up to the end of 2019. For the second case study (July 2022 to June 2023), additionally forecasts of ICON-D2-EPS were used until June 2022. The model change from COSMO-D2-EPS to ICON-D2-EPS is taken into account by the binary model update predictor and by double weighting of ICON-D2-EPS in the training data set.

Even though ModelMIX does not run operationally, it was designed for an operational time lapse, therefore it starts to process forecasts after a time step of 3h. Moreover, the latest observations at 2h after nominal model issue are available at start of the postprocessing and are used as persistence predictors in order to boost the performance of short-term statistical forecasts.

The computing time of ModelMIX is about 15min including preparation of numerical and observation data as input. The evaluation of the statistical model itself is computed in parallel on DWD's high performance computer and requires less than 2min for an issue of ICON-D2-EPS comprising about 10^6 statistical equations altogether (predicting about 150 statistical forecasts for almost all weather warnings of DWD).

8.3.2 Postprocessing Methods Leading to Predictive Distributions

While ModelMIX generates probability forecasts tailored to the specific thresholds, another approach for the generation of threshold exceedance probabilities is to first develop a model for the entire distribution of the target variable and to then derive the desired probabilities from that distribution. This approach has not only the advantage that probabilities for any threshold can be calculated, but also that, if multiple thresholds are considered, only one model needs to be estimated. Further, this guarantees that threshold exceedance probabilities are decreasing with respect to the threshold, which could be an issue, if the threshold probabilities are forecast individually. On the other hand, models specifically tailored to certain thresholds will likely result in superior performance, as they directly target the specific event probability of interest during estimation.

Much work in research on statistical postprocessing of ensemble forecasts has focused on forecasting the entire distribution of the target variable (see e.g., [28]), and a variety of methods for that task is readily available. Reference [27] provide a systematic comparison of ensemble postprocessing methods for wind gusts and find that NN-based postprocessing methods outperform established state-of-the-art postprocessing approaches. Based on their work, we apply the NN methods to generate threshold exceedance probabilities. In addition, we also use two of the benchmark methods of this study for a broader comparison.

As [27] apply the methods to the same NWP model and observational database (although for the period 2010–2016), we directly adopt their setup of the postprocessing methods. This includes the implementation details of the postprocessing methods, the choice of hyperparameters and that of the predictor variables. Note that we did not tune the hyperparameters specifically to the application at hand. Hence, we refer to [27] for a detailed description of the postprocessing methods and accompanying code, and only describe changes with respect to this setup in the following.

8.3.2.1 Ensemble Model Output Statistics (EMOS)

Before introducing the NN methods, we briefly comment on the two benchmark methods considered in the comparison. The EMOS [8] approach assumes that the target variable follows a parametric distribution, here, a truncated logistic distribution, and links the parameters of the distribution to summary statistics of the ensemble forecasts, where the link function is estimated minimizing a proper scoring rule. For the incorporation of additional predictor variables, the classic EMOS approach can be extended using ideas from machine learning, namely, gradient boosting (EMOS-GB) [18], which has shown to produce more accurate forecasts [22, 27].

Here, we follow the standard approach of estimating both variants locally and for each lead time separately. Since preliminary tests showed that using only data since the last model update was superior to a longer training period including binary predictors on the model version, we decided to use the former training scheme for the EMOS model. This reduces the amount of training data such that we do not utilize the seasonal training approach of [27]. For the gradient boosting extension of EMOS, incorporating the binary predictors on the model version is feasible. Therefore, we include those and use the entire COSMO period for training.

8.3.2.2 Neural Network Methods

The distributional regression network (DRN) approach [22] extends the EMOS framework towards the use of methods from artificial intelligence, in that a NN links the predictor variables with the distribution parameters. Analogous to the EMOS framework, we employ a truncated logistic distribution as forecast distribution. NNs are data-driven models that typically work the best the more data is available to feed the model. For station-based postprocessing, similar observations have been made, in that a single model is estimated jointly for all stations, but made locally adaptive via a technique originally proposed in natural language processing, so-called embeddings, which notably improves the predictive performance. For details and comparisons to alterative setups, see [22].

The change from ICON to COSMO represents a substantial change in the NWP model and, as only one year of training data for the prediction in 2022/2023 is available, the question of the ideal composition of the training set arises. While [27] fit a separate model for each lead time (for a fixed model run), we investigate, in case of the ICON data, the performance of a joint model for all lead times and the two model runs at 00UTC and 12UTC, thereby increasing the training set size by a factor of 38. For the joint model, we change the target variable from the variable itself to the bias of the ensemble mean. This centers the target variable and is intended to help the model handle changes in the diurnal cycle, which is not present in the data when considering each lead time separately, as the forecasts then all validate at the same time of the day. Further, we also standardize the target variable in this case (for training). Note that we now use a non-truncated logistic distribution for estimation and do not account for a possibly non-positive support of the forecast distribution of the wind gusts. The forecast, however, which is the sum of the ensemble mean and the forecast distribution of the bias, is truncated in zero.

While the joint model is only considered for the ICON data in 2022/2023, one general modification with respect to the setup in [27] concerns all NN models. Instead of using the last year within the training period for validation, we sample randomly 20% of the training data to obtain a validation subset. We changed the selection in order to have a balanced portion of all model versions both within the data sets used for estimating and validating the NNs.

For predicting exceedance probabilities at high thresholds, the type of forecast distribution might be an influential factor, hence we consider one of the two more flexible alternatives to DRN presented in [27]. The Bernstein quantile network (BQN) [3] approach differs from DRN in the choice of the forecast distribution and the associated loss function. BQN models the quantile function as a linear combination of Bernstein basis polynomials, a class of flexible function approximators. Note that the change in the DRN model regarding the choice of the validation set also applies to the BQN models.

8.4 Verification Methods

Forecast quality has many different attributes such as accuracy, reliability, resolution or sharpness, but no verification measure is able to summarize all of them. In this paper, we use standard tools tailored to the evaluation of probability forecasts, which we present in the following. Reliability and sharpness diagrams are used to account for calibration and sharpness, and the Brier score (with its decomposition into three terms) to analyze the accuracy.

8.4.1 Reliability Diagrams

The reliability diagram [24] is a graphical tool to assess whether the postprocessing methods appropriately quantify the uncertainty, i.e., if they are calibrated. In a calibrated system, forecast probabilities correspond to the observed frequencies. To construct the reliability diagram, forecasts probabilities of a selected event (a threshold being exceeded, in our case) are distributed into bins, and compared to the observed frequency for each bin. A perfectly calibrated system would lie on the diagonal, and the deviation from the diagonal gives the conditional bias depending on the forecast probability. A line close to the horizontal line given by the climatology indicates that the system is not able to distinguish situations with different frequencies of occurrence (no resolution). A system is skillful if the reliability diagram is closer to the diagonal than the intermediate line that lies between the diagonal and the climatology.

On the other hand, each point of the reliability diagram refers to a different number of cases. Since we are dealing with extreme events which are comparatively rare, small probability forecasts tend to be much more frequent than higher values. The sharpness diagram depicts the relative frequency of the event in each forecast probability bin. In the case of a perfectly sharp probabilistic system, the sharpness diagram would consist of a histogram with two bars, one over the value zero for those cases when the model is sure no event would happen, and another one over the value one, when the model forecast the event. Sharpness is a property of the forecasts only, and—subject to calibration—sharper forecasts are preferable [9].

8.4.2 Brier Score Decomposition

To measure the accuracy, i.e., the level of agreement between probabilities and the binary observation, we use the Brier score (BS) [4], defined as the mean squared error of the probability forecast. In a set of n forecast-observation pairs, each probability forecast f_i is compared to the binary observation o_i, that is 1 if the event occurs (i.e., the wind is above the threshold), and 0 otherwise,

$$BS = \frac{1}{n} \sum_{i=1}^{n} (f_i - o_i)^2. \tag{8.1}$$

The BS always takes values between 0 and 1, and it can be decomposed into three terms, namely, the reliability, the resolution and the uncertainty term [19]:

$$BS = \underbrace{\frac{1}{n}\sum_{k=1}^{K} n_k(f_k - \bar{o}_k)^2}_{\text{reliability}} - \underbrace{\frac{1}{n}\sum_{k=1}^{K} n_k(\bar{o}_k - \bar{o})^2}_{\text{resolution}} + \underbrace{\bar{o}(1-\bar{o})}_{\text{uncertainty}} = \text{Rel} - \text{Res} + \text{Unc}, \tag{8.2}$$

with n being the total number of forecasts issued, K the number of unique forecasts issued, $\bar{o}$ the observed climatological base rate for the event to occur (observed average), n_k the number of forecasts with the same probability category and $\bar{o}_k$ the observed frequency, given forecasts of probability f_k. The reliability term refers to the agreement between forecast probability and mean observed frequency, which is also visualized in the reliability diagrams. The resolution shows if the observed outcome changes as the forecast changes. The uncertainty only depends on the variability of the observations. Hence, one should not compare the BS of events with different frequencies, because by definition, the lower the frequency of the event, the lower the contribution of the uncertainty term to the total BS, without necessarily implying a better accuracy.

To analyze the degree of improvement of the postprocessed forecasts with respect to a reference system (with BS BS_{ref}), we can calculate the Brier skill score (BSS) via

$$BSS = \frac{BS_{ref} - BS}{BS_{ref}}. \tag{8.3}$$

If the BSS is negative, the reference system is superior to the postprocessed forecasts, if it is equal to zero, they have the same skill, and if it is positive the postprocessed forecasts outperform the reference with a maximum skill score of 1 indicating optimal predictive performance.

In addition, we can specifically analyze the improvement in terms of the resolution term. While the optimal BS is equal to 0 (together with a reliability term of 0), the associated resolution term is equal to the uncertainty term. Hence, the (positively oriented) resolution term is bounded by the uncertainty term, which we need to take into account when calculating a skill score analogous to the BSS. To do this, we use the difference of the uncertainty term and the resolution term as underlying score in Eq. (8.3), that is,

$$\frac{(\text{Unc} - \text{Res}_{ref}) - (\text{Unc} - \text{Res})}{(\text{Unc} - \text{Res}_{ref})} = \frac{\text{Res}_{ref} - \text{Res}}{\text{Res}_{ref} - \text{Unc}}. \tag{8.4}$$

This term can be interpreted analogously to the BSS. Note that the uncertainty component is identical for all forecasting systems.

8.5 Results

This section shows a comparison of the different postprocessing techniques, namely (ensemble) MOS techniques and NNs. Since in an operational weather center, improvements in the initial conditions of the models, as well as in the numerical model itself, are frequently introduced, the relevance of these changes within the training period of the NNs is analyzed and discussed. For example, this includes the major changes in the data assimilation process of DWD (introduction of KENDA and later LETKF) and the change from COSMO-D2 to ICON-D2. This section also presents results regarding other important aspects in an operational center, e.g., ensuring the timely generation of postprocessed forecasts as well as design choices regarding the training of the networks.

8.5.1 Comparison of MOS and NN Methods

We start with the first case study, where we compare the five different postprocessing methods introduced in Sect. 8.3 with the direct model output of COSMO-D2-EPS. Note that all model abbreviations in the plots are listed in Table 8.1.

To assess calibration, Fig. 8.3 shows the reliability diagrams of hourly wind gusts exceeding 25kt, for the model run at run 00UTC, for two different lead times, 3 h (**a**) and 15 h (**b**), as well as the corresponding sharpness diagrams (**c** and **d**, respectively). All postprocessing methods calibrate the direct model output and lie close to the diagonal, although EMOS and EMOS-GB tend to lie above the diagonal (underforecasting) whereas the other models below (overforecasting), especially ModelMIX for longer lead times.

The histograms in the sharpness diagrams are dominated by values close to zero, as expected. For small lead times, COSMO-D2 and ModelMIX are sharper (with more zero and one values than the other models) but this changes for higher lead times. Classic EMOS shows the lowest sharpness.

Regarding accuracy, Fig. 8.4 shows the BS (a) and its decomposition in reliability (b), resolution (c) and uncertainty (d) terms. Postprocessing improves accuracy, and apart from the first few hours, where ModelMIX shows the highest accuracy, the NN methods show the best results. In line with previous research results e.g., [27] and other comparisons not shown here, there are only very minor differences between the two NN methods, DRN and BQN. This holds in particular also for high thresholds. We will therefore focus on DRN in the following. ModelMIX exhibits a notably worse reliability term, but the contribution of this term to the total BS is small. The improvement in the BS of NN compared to ModelMIX is mainly due to an improvement in resolution. Since all methods are verified against the same observations, the uncertainty term is equal for all of them.

The main difference regarding the predictors used by ModelMIX and the NNs, is that the NNs do not use any persistence predictors, whereas ModelMIX uses all

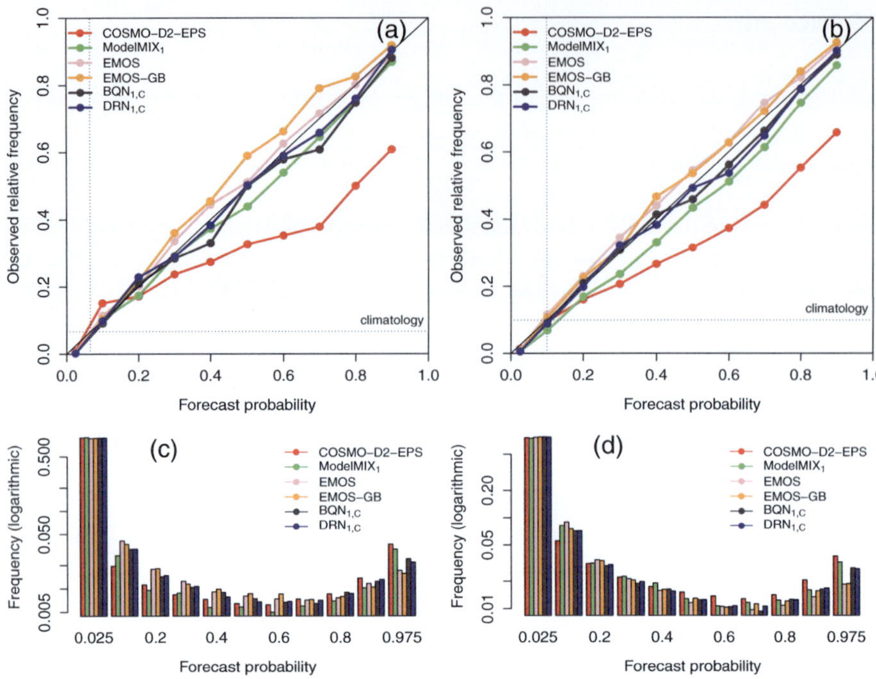

Fig. 8.3 Reliability diagrams (**a**, **b**) and sharpness histograms (**c**, **d**) of the five postprocessing techniques and the direct model output of COSMO-D2-EPS, for two different lead times: 3h (left) and 15h (right). The evaluation is based on data from year 2020, wind threshold 25kt, and the model run at 00UTC

available observations up to the time it is run operationally, i.e., two hours later than the run of the NWP model. To evaluate whether this affects the forecast performance during the first forecast hours, we run an experiment where DRN was trained using three additional persistence predictors: the wind gust observation at initialization time (of the NWP model), at initialization time plus 1h and at initialization time plus 2h. This setup mimics the use of persistence predictors by ModelMIX. Figure 8.5 shows a comparison of previous models before and after using persistence predictors. These predictors notably improve the BS within the first hours, and the NN model with persistence predictors now slightly outperforms ModelMIX. In particular, the improvement is related to an increase in the resolution term. Therefore, we will keep this setup for all further comparisons below.

Since the model performance is usually affected by the daily cycle, we further analyzed raw and postprocessed forecasts from the model run initialized at 12UTC to account for differences between day and night. Figure 8.6 shows the reliability diagrams and sharpness histograms of hourly wind gusts exceeding 25 kt for lead time 3h (a, c) and 21h (b, d), respectively. COSMO-D2-EPS notably less well calibrated in this run, as well as ModelMIX. On the contrary, the EMOS techniques are better calibrated than for the model run initialized at 00UTC. Regarding accuracy, for which

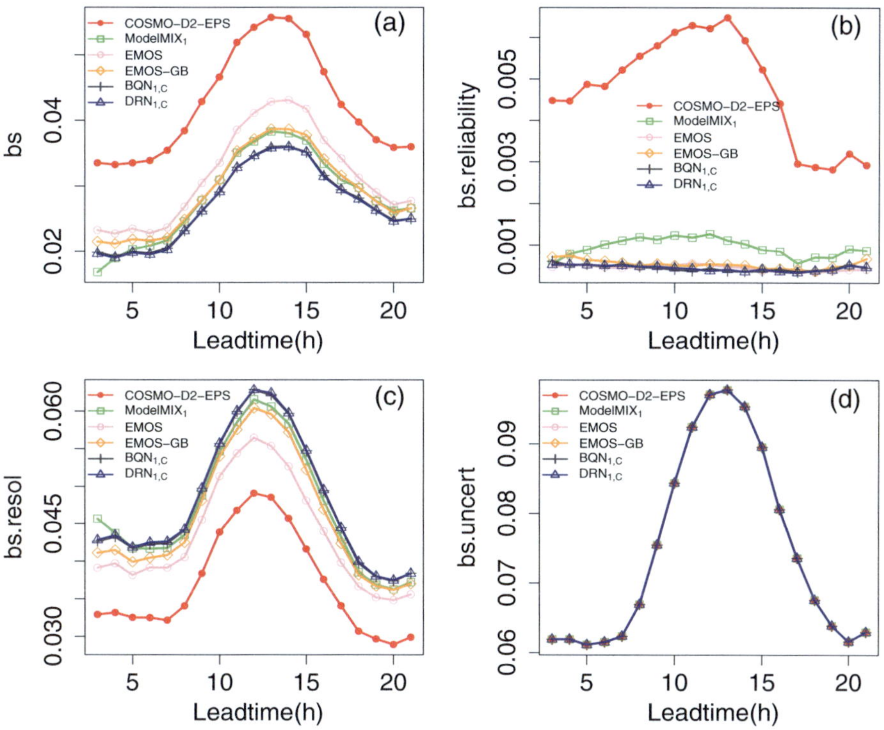

Fig. 8.4 Temporal evolution of the BS (**a**) and its reliability (**b**), resolution (**c**) and uncertainty (**d**) terms for the different postprocessing methods dependent on the lead time. The evaluation is based on data from year 2020, wind threshold 25 kt, and the model run at 00UTC

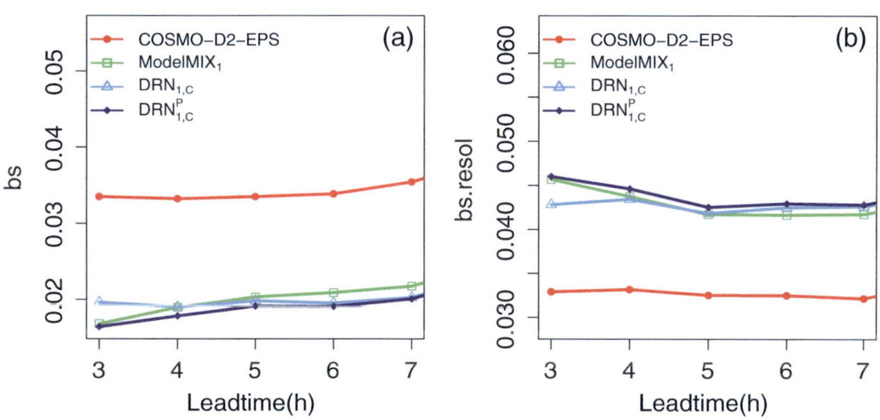

Fig. 8.5 Temporal evolution of **a** the BS and **b** the resolution term during the first hours when using persistence predictors. The evaluation is based on data from year 2020, wind threshold 25 kt, and the model run at 00UTC

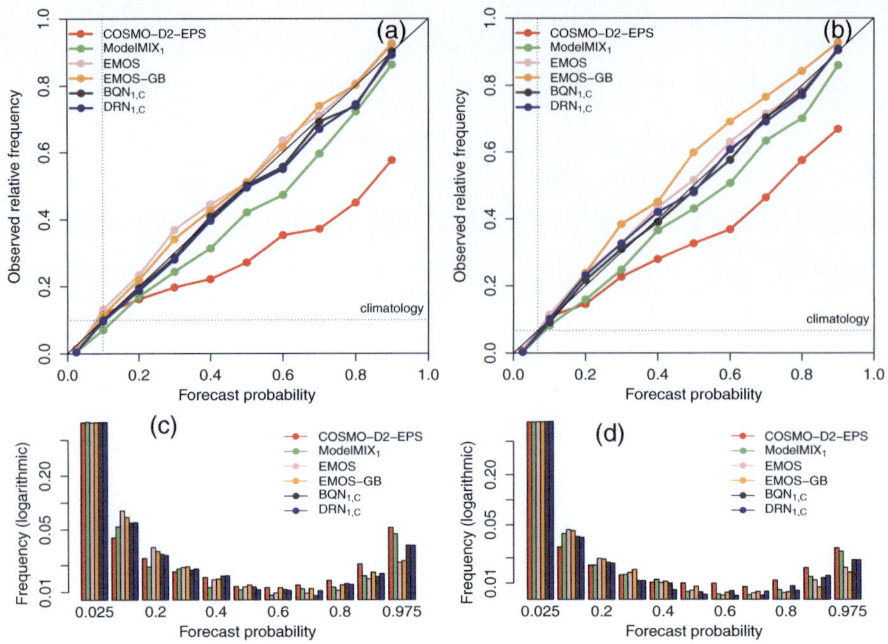

Fig. 8.6 Reliability diagrams (**a**, **b**) and sharpness histograms in logarithmic scale (**c**, **d**) of the five postprocessing techniques and the direct model output of COSMO-D2-EPS, for two different lead times: 3 h (left) and 15 h (right). The evaluation is based on data from year 2020, wind threshold 25 kt, and the model run at 12UTC

results are shown in Fig. 8.7, both runs show similar results. The EMOS techniques improve over the direct model output, and the NN methods outperform EMOS, with the improvements mainly being due to an improvement in resolution. The event of interest, i.e., the wind gusts exceeding 25kt, is observed mainly during daytime hours, and on those hours, forecasts show higher BS than during nighttime hours, where the event occurs less frequently. Therefore a direct comparison of BS for run 00UTC and run 12UTC should be avoided.

Figure 8.8 shows the BSS evolution with respect to the wind threshold for the model run initialized at 00UTC. Note that even though the BS depends on the frequency of the event, so the lower this frequency, the lower its value, and therefore both BS and BS_{ref} will by definition tend to zero for extreme events, the BSS is based on the division of both quantities so it behaves differently. According to the figure, higher differences are expected to happen for lower thresholds. For all methods, the improvements over the direct model output decrease for larger threshold values, but the ranking between the postprocessing methods mostly remains consistent. In line with the results of [27], we find that the NN methods provide the best forecasts also for higher threshold values. However, the evaluation might be less conclusive for very high threshold values due to the low event frequency, which implies large uncertainties in the score values, and the forecast distribution assigning smaller

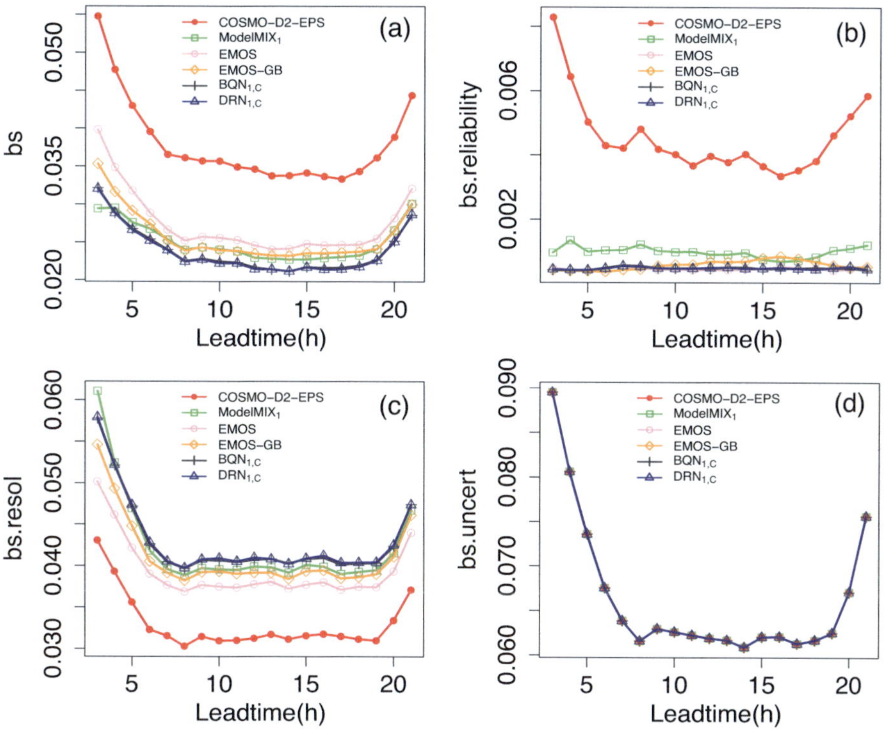

Fig. 8.7 Temporal evolution of the BS (**a**) and its reliability (**b**), resolution (**c**) and uncertainty (**d**) terms for different postprocessing methods dependent on the lead time. The evaluation is based on data from year 2020, wind threshold 25 kt, and the model run at 12UTC

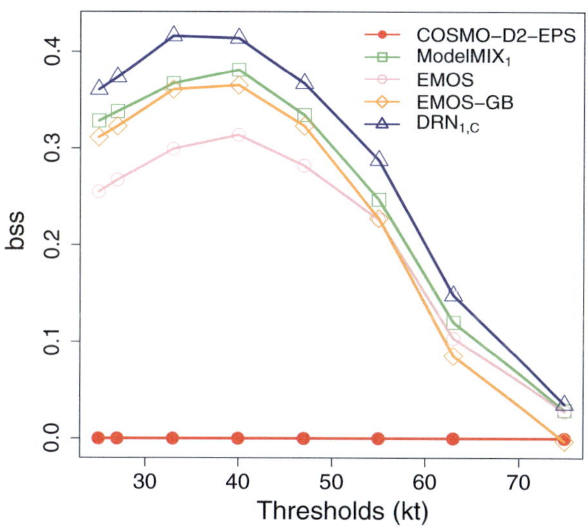

Fig. 8.8 Evolution of the BSSs of the postprocessing methods dependent on the wind threshold, where the reference is the BS of COSMO-D2-EPS. The evaluation is based on data from year 2020, all lead times, and the model run at 00UTC

probability to this region will be preferred (for details, see [17, Sect. 3.4]). Further, a detailed evaluation of individual case studies such as selected winter storms [20] could potentially highlight additional advantages or disadvantages of the different postprocessing methods, but is beyond the scope of this paper.

8.5.2 Influence of NWP Model Changes

The NN methods considered in the previous section were trained with the complete available dataset, i.e., data from 2010 to 2019. However, in March 2017 a major improvement occurred in the data assimilation method with the introduction of the KENDA technique. To evaluate the effect of these changes on the performance of the NN methods, we trained the same DRN model using two different training periods: from 2010 up to 2019, and only from March 2017 up to 2019. Note that the models include binary predictors to indicate the introduction of KENDA and COSMO-D2. These networks were trained without persistence predictors, so they are also compared to a DRN trained using the complete dataset with persistence predictors. These three NN methods are used to produce forecasts for the year 2020.

Further, in February 2021, the NWP model changed, as ICON-D2 replaced COSMO-D2. This was a major change in the NWP setup, so we have added another comparison with two more DRN variants, one trained with data from February 2021 up to June 2022, i.e., only ICON data, and another trained with data from March 2017 up to June 2022 (after KENDA was introduced). These models forecast the one year period from July 2022 to June 2023.

Note that the frequency of the event differs from one verification year to the other. The uncertainty term of the BS is directly affected by the observed event, so changes in the frequency imply changes in the BS due to the contribution of the uncertainty term, even though the performance of the model might not change. Therefore, the verification results for both periods (2020 and 2022/2023) should not be directly compared.

Figure 8.9 shows the BS (a, c) and its resolution term (b, d) of the direct output of the NWP model and two postprocessing techniques, ModelMIX and DRN. The first row shows the verification for the year 2020 and compares performance when changes in the assimilation process (KENDA) were introduced. The second row shows the corresponding verification for the years 2022/2023 and compares two different underlying NWP models, COSMO and ICON, used in the training period. The reliability diagrams are similar to previous results, showing good calibration of the forecasts and thus are omitted. Major changes occur in the accuracy during the first hours whenever persistence predictors are used. These changes are more relevant than the training period length. Generally, longer training periods lead to slightly better results. For example, the BS improves mainly due to an improvement of the resolution, especially when forecasting midday. The difference in the BSs between the year 2020 and 2022/2023 are mainly due to the differences in the uncertainty term, i.e., due to the inter-annual variability of the observed frequency of the event.

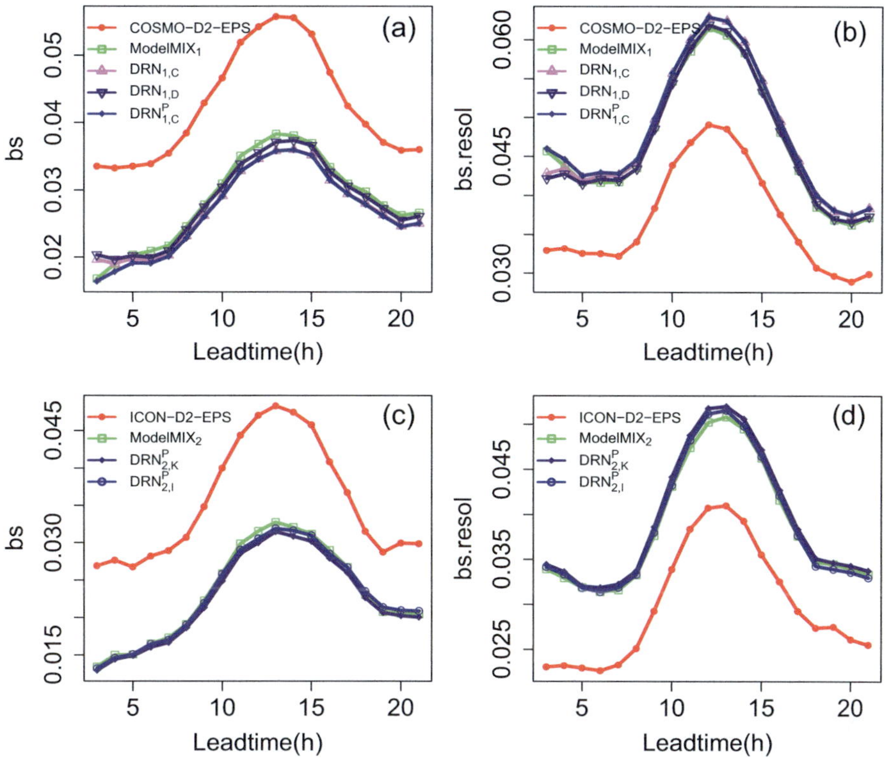

Fig. 8.9 Relevance of changes in the training period, due to: **a**, **b** changes in the assimilation process (KENDA; verified year: 2020) and **c**, **d** changes in the NWP model (ICON; verified year: 2022/2023). **a**, **c** BS; **b**, **d** resolution. The evaluation is based on wind threshold 25 kt, and the model run at 00UTC

In a nutshell, adding binary variables indicating the model version seems to work well, and the longer the training period, the better the results.

8.5.3 *Temporal Composition of Training Datasets*

When postprocessing ensemble forecasts in an operational context at large scales, computational complexity and technical requirements are more critical than in research studies. Hence, model configurations should aim at providing sufficiently good forecasts while keeping the computational complexity as low as possible. Due to the flexibility of NNs, several design choices can be considered when building the model.

One choice is whether to build joint models (over multiple spatial locations, and in a temporal context, i.e., over multiple lead times and/or model runs), or to use separate

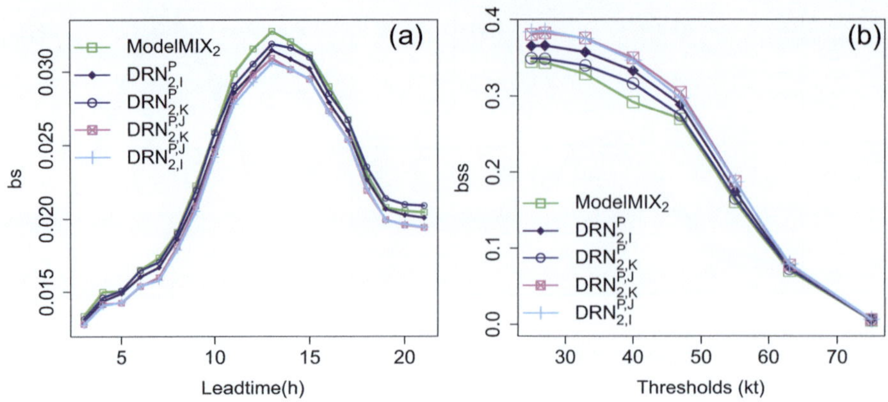

Fig. 8.10 Comparison of two NNs trained with all lead times and runs (indicated by a J in the exponent) and two NNs trained for each lead time and run. The evaluation is based on data from year 2022/2023 and the model run at 00UTC. **a** BS for wind threshold 25 kt, **b** BSS for all lead times

models in each individual case. While it has already been shown that building one NN model for all stations is feasible and improves the predictive performance, we briefly investigate how a joint model for all lead times and model runs compares to the previous practice of building separate models. As mentioned in Sect. 8.3.2.2, such a joint model increases the training set size by a factor of 38 in our case and vice versa reduces the number of model instances that need to be estimated by the same factor. This trade-off is then connected with the question of the length of the training period. Individual models have less data available for the same period of time and might thus require training periods that range back longer in time, and vice versa. Here, we therefore compare NNs trained for each lead time and model runs with NNs that are trained jointly for all lead times and runs. An added benefit of joint models is that fewer individual models need to be tuned and optimized. In our case, however, this is not relevant as we use the same architecture without any additional tuning for all variants.

Figure 8.10 compares two DRN variants trained separately for each lead time and run for the period from March 2017 to June 2022 ($DRN^{P}_{2,K}$) and from February 2021 to June 2022 ($DRN^{P}_{2,I}$) with two networks trained jointly for all lead times and the two runs from 00UTC and 12UTC, one for the period from March 2017 to June 2022 ($DRN^{P,J}_{2,D}$) and another from February 2021 to June 2022 ($DRN^{P,J}_{2,I}$). All networks are trained with persistence predictors and forecast hourly probabilities of wind gusts above 25 kt for the NWP model run initialized at 00UTC from July 2022 to June 2023. All models are well calibrated. This increase of the training data set size leads to some improvement in performance, although it also implies additional costs for the more complex training scheme. Overall, we can conclude that more data leads to better results, but when a sufficient amount of training data is available, no more training data is beneficial in an operational context. That said, further improvements might be possible with specifically targeted hyperparameter tuning for those settings.

8.5.4 Score Cards

To support the decision process of implementing changes in the operational postprocessing of a meteorological center, it is useful to have diagrams summarizing the relevant verification results, for example showing traffic light colors (green if there is an improvement and red if there is none), that compare the new method with the method considered as reference. In this study, we have analyzed the benefit of applying postprocessing methods compared to the direct model output of the NWP system, and we have compared the performance of different postprocessing methods, namely EMOS and NN. Therefore, we utilize *score cards*, i.e., diagrams showing the improvement by using different postprocessing techniques compared to our reference model, COSMO-D2-EPS for year 2020 and ICON-EPS for year 2023.

Results in this work show that all postprocessing methods always improve the direct model output and the performance of the different postprocessing methods mainly differs in the BS due to changes in the resolution. The temporal evolution of the differences in BS and resolution compared to the reference model are summarized in Figs. 8.11 and 8.12 for two years (2020 and 2023) and two runs (00UTC and 12UTC). The point size indicates the difference between the score of the postprocessing method and the score of the corresponding reference forecast. The color scale represents the skill of the score, as described in Eqs. (8.3) and (8.4).

The benefit of applying ModelMIX or NN is much bigger than using EMOS or EMOS-GB. However, the difference between ModelMIX and NN scores are not relevant compared to the differences with the reference, and the circle sizes of ModelMIX and NN thus are similar for a particular lead time. The improvement is affected by the daily cycle, since the biggest circles for run 00UTC are around lead time 14h and smaller for the later lead times, whereas for run 12UTC bigger circles are shown for the earlier lead times and the differences are not much smaller for the later lead times. Note that a circle in case of the resolution term is smaller than its counterpart in case of the BS, as the reliability term is additionally present in the BS difference (and becomes smaller by postprocessing).

8.6 Summary and Conclusions

We compare different techniques to postprocess the probability of hourly wind gust exceedance, namely ensemble-MOS techniques and NNs. The classical EMOS and EMOS with gradient boosting are compared to the EMOS technique applied at the German weather service (DWD), ModelMIX, the distributional regression network (DRN) and the Bernstein quantile network (BQN). All methods are applied to the

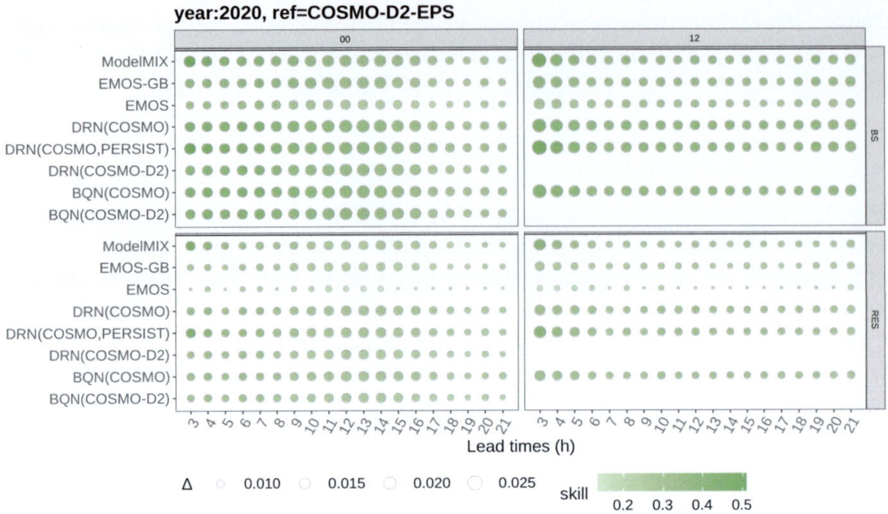

Fig. 8.11 Score cards comparing the benefit in BS and resolution of different postprocessing methods for wind threshold 25 kt, year 2020 and two model runs (00UTC and 12UTC). Green color indicates an improvement after the postprocessing compared to the reference model, COSMO-D2-EPS. Size and color scale show how big the improvement is

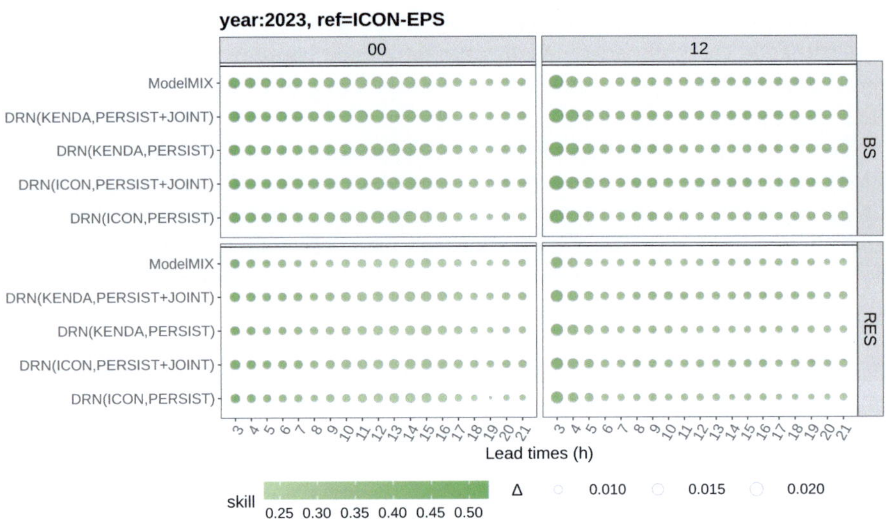

Fig. 8.12 Score cards comparing the benefit in BS and resolution of different postprocessing methods for wind threshold 25 kt, year 2023 and two model runs (00UTC and 12UTC). Green color indicates an improvement after the postprocessing compared to the reference model, ICON-EPS. Size and color scale show how big the improvement is

operational NWP ensemble prediction system of DWD, COSMO-D2-EPS (and later ICON-D2-EPS), over 170 SYNOP stations in Germany, for two years (2020 and 2022/2023).

Forecast performance is measured in terms of calibration, sharpness and accuracy. Overall, we find that ModelMIX outperforms classical EMOS techniques, and NNs outperform ModelMIX. Both NN methods perform similarly. On the one hand, ModelMIX has the advantage of using all possible and more recent observations, and it contains persistence predictors. This leads to a better performance, in particular during the first hours. We further investigate the relevance of including persistence predictors in the NNs, showing that improvements over the first hours can be achieved this way, but there is no effect on later lead times. On the other hand, NNs predict a forecast distribution (instead of just the probability of exceeding a threshold), therefore, by default yield consistent probability forecasts at arbitrary thresholds. This work utilizes score cards summarizing the verification results to help in the decision process of changing the operational postprocessing technique.

In operational centers, improvements in the numerical model occur frequently, which leads to model changes during the training period of a NN. This issue is investigated, and forecasts produced by NNs trained on data including no model change are compared to NNs trained on data including relevant model changes. In particular, changes in data assimilation approach (KENDA) and the new numerical weather prediction model of the German weather service (ICON) are analyzed. Results show that the longer data sets are beneficial even if there are model changes. We here use binary variables to account for model changes. More sophisticated approaches such as transfer learning [31] might offer superior alternatives for the NN-based methods, and would be an interesting starting point for future comparisons.

Another important aspect for the operational use of postprocessing models at weather centers is the time efficiency (systems need to have a limited complexity to produce forecasts on time) and time consistency (forecasts should not jump from one time step to the next one or from one model run to the next one). While models estimated jointly over all lead times and model runs might have advantages in this regard, a detailed evaluation of potential forecast jumps is required to assess this aspect, and represents an interesting avenue for future research.

While we focused on the use of operational forecast datasets, some meteorological centers produce reforecasts by (re-)running the current model version for past forecast cases. This generates a larger training archive of cases with the current model setup, however, comes at high computational costs. In light of the improvements obtained via longer training datasets it would be interesting to investigate the effects of using reforecast data (without model changes) versus forecast data (with model changes) in more detail. From an operational perspective, this aspect is particularly relevant since the computational costs of generating reforecasts need to be carefully balanced against the costs of improvements to the NWP system (such as increases in the spatial or temporal resolution, or the number of ensemble members), see [29] for a more detailed discussion of this aspect.

Evidently, our study only constitutes a first step and more detailed comparisons are needed in future research to comprehensively assess the respective advantages and disadvantages of the various postprocessing approaches. In particular, with a potential operational use of these methods in mind, several other aspects are of importance beyond the predictive performance assessed via standard verification metrics. For example, ModelMIX is already applied to postprocess many other variables. While DRN has the advantage of providing full distribution forecasts, this comes at the cost of having to adapt the model for different target variables, e.g., by choosing an appropriate parametric forecast distribution. On the other hand, it might be possible to further improve the predictive performance of the NN models by tailoring them better to the task at hand, for example by moving towards categorical forecasting or classification models with suitably adapted loss function (see e.g., [2]). Another important aspect not considered in our study is the interpretability of the different approaches, which might play an important role in an operational setting. While ModelMIX in principle enables a straightforward interpretation via the estimated regression coefficients, the large number of regression equations poses practical challenges. NN methods are inherently less interpretable, but techniques such as the evaluation of permutation-based feature importance conducted in [27] can provide some insight into what the models have learned. In addition, detailed meteorological case studies of individual events can also help to improve the assessment of advantages and disadvantages of the different approaches. Finally, our study further illustrates that fair and direct comparisons between operational and new postprocessing techniques are challenging due to differences in terms of the specific setup, available inputs, and training dataset choices. We here focused on the ModelMIX approach since it was possible to make the setups more comparable, even though ModelMIX does not run operationally at DWD. Instead, the operational technique is based on MOS applied to the global model (instead of to the limited area model) and in its deterministic version (instead of the EPS).

Acknowledgements Benedikt Schulz and Sebastian Lerch gratefully acknowledge funding within the project C5 "Dynamical feature-based ensemble postprocessing of wind gusts within European winter storms" of the Transregional Collaborative Research Center SFB/TRR 165 "Waves to Weather" funded by the German Research Foundation (DFG). Sebastian Lerch further gratefully acknowledges support by the Vector Stiftung through the Young Investigator Group "Artificial Intelligence for Probabilistic Weather Forecasting." We thank Sebastian Trepte for providing the observation data, and Robert Redl for assisting in data handling. We further thank an anonymous reviewer for constructive comments.

8.7 Appendix

See Tables 8.2 and 8.3.

Table 8.2 Table 1 from [12]: Predictors for statistical forecasts of the maximal speed of wind gusts within 1h that have relative weights higher than 1%. The relative weights are aggregated for all stations, seasons, forecast runs, and forecast lead times. The ensemble mean of COSMO-D2-EPS is denoted by DMO (direct model output). Parentheses within the predictor names denote time shifts. For a time shift of −30 min, denoted as (−0:30), the predictors are interpolated based on values for the previous and the current forecast hour. The required statistical forecasts have to be evaluated in advance

Predictor name	Rel. weight (%)	Description
`FF(-0:30)StF`	35.7	Statistical forecast of mean wind speed in 10 m height
`FX1(-1)StF`	12.1	Statistical forecast of speed of wind gusts of the previous hour
`VMAX_10M_LS`	8.9	DMO of speed of wind gusts for a large surrounding area (mean of 11×11 grid points)
`VMAX_10M`	8.6	DMO of speed of wind gusts for next model grid point
`FF_850(-0:30)`	4.8	DMO of wind speed in 800 hPa height
`VMAX_10M_MS`	4.1	DMO of speed of wind gusts for a medium surrounding area (mean of 6×6 grid points)
`Oa_D_0.5`	3.2	Latest observation of the speed of wind gusts
`FF_10m(-0:30)`	2.8	DMO of mean wind speed in 10 m height
`StFT2m_T950`	1.9	Statistical forecast of temperature difference between 2 m and 950 hPa height (stability index)
`Location-height`	1.6	Height of station
`FF_1000(-0:30)`	1.1	DMO of wind speed in 1000 hPa height

Table 8.3 Table 1 from [27]: Overview of available predictors [for the NN models]. For the meteorological variables, ensemble forecasts are available, with the term '500–1,000 hPa' referring to the specific model levels at 500, 700, 850, 950 and 1,000 hPa. Spatial predictors contain station-specific information

Meteorological variables

Feature	Unit	Description
VMAX	m/s	Maximum wind, i.e. wind gusts (10 m)
U	m/s	U-component of wind (10 m, 500–1,000 hPa)
V	m/s	V-component of wind (10 m, 500–1,000 hPa)
WIND	m/s	Wind speed, derived from U and V via $\sqrt{U^2+V^2}$ (10 m, 500–1,000 hPa)
OMEGA	Pa/s	Vertical velocity (Pressure) (500–1,000 hPa)
T	K	Temperature (Ground-level, 2 m, 500–1,000 hPa)
TD	K	Dew point temperature (2 m)
RELHUM	%	Relative humidity (500–1,000 hPa)
TOT_PREC	kg/m^2	Total precipitation (Accumulation)
RAIN_GSP	kg/m^2	Large scale rain (Accumulation)
SNOW_GSP	kg/m^2	Large scale snowfall—water equivalent (Accumulation)
W_SNOW	kg/m^2	Snow depth water equivalent
W_SO	kg/m^2	Column integrated soil moisture (multilayers; 1, 2, 6, 18, 54)
CLCT	%	Total cloud cover
CLCL	%	Cloud cover (800–soil)
CLCM	%	Cloud cover (400–800 hPa)
CLCH	%	Cloud cover (000–400 hPa)
HBAS_SC	m	Cloud base above mean sea level, shallow connection
HTOP_SC	m	Cloud top above mean sea level, shallow connection
ASOB_S	W/m^2	Net short wave radiation flux (at the surface)
ATHB_S	W/m^2	Net long wave radiation flux (m) (at the surface)
ALB_RAD	%	Albedo (in short-wave)
PMSL	Pa	Pressure reduced to mean sea level
FI	m^2/s^2	Geopotential (500–1,000 hPa)

Other predictors

Feature	Type	Description
yday	Temporal	Cosine transformed day of the year: $\cos\left(2\pi\frac{t-1}{365}\right)$, where t is the day of the year
lat	Spatial	Latitude of the station
lon	Spatial	Longitude of the station
alt	Spatial	Altitude of the station
orog	Spatial	Difference of station altitude and model surface height of nearest grid point
loc_bias	Spatial	Mean bias of wind gust ensemble forecasts from 2010–2015 at the station
loc_cover	Spatial	Mean coverage of the wind gust ensemble forecasts from 2010–2015 at the station

References

1. Baldauf, M., Seifert, A., Förstner, J., Majewski, D., Raschendorfer, M., Reinhardt, T.: Operational convective scale numerical weather prediction with the COSMO model description and sensitivities. Mon. Weather Rev. **139**(12), 3887–3905 (2011). https://doi.org/10.1175/MWR-D-10-05013.1
2. Baran, A., Lerch, S., El Ayari, M., Baran, S.: Machine learning for total cloud cover prediction. Neural Comput. Appl. **33**(7), 2605–2620 (2021). https://doi.org/10.1007/s00521-020-05139-4
3. Bremnes, J.B.: Ensemble postprocessing using quantile function regression based on neural networks and Bernstein polynomials. Mon. Weather Rev. **148**(1), 403–414 (2020). https://doi.org/10.1175/mwr-d-19-0227.1
4. Brier, G.W.: Verification of forecasts expressed in terms of probability. Mon. Weather Rev. **78**(1), 1–3 (1950). https://doi.org/10.1175/1520-0493(1950)078<0001:vofeit>2.0.co;2
5. Chapman, W.E., Monache, L.D., Alessandrini, S., Subramanian, A.C., Ralph, F.M., Xie, S.P., Lerch, S., Hayatbini, N.: Probabilistic predictions from deterministic atmospheric river forecasts with deep learning. Mon. Weather Rev. **150**(1), 215–234 (2022). https://doi.org/10.1175/MWR-D-21-0106.1
6. Gebhardt, C., Theis, S., Krahe, P., Renner, V.: Experimental ensemble forecasts of precipitation based on a convection-resolving model. Atmos. Sci. Lett. **9**(2), 67–72 (2008). https://doi.org/10.1002/asl.177
7. Ghazvinian, M., Zhang, Y., Seo, D.J., He, M., Fernando, N.: A novel hybrid artificial neural network-parametric scheme for postprocessing medium-range precipitation forecasts. Adv. Water Resour. **151**, 103907 (2021)
8. Gneiting, T., Raftery, A.E., Westveld, A.H., Goldman, T.: Calibrated probabilistic forecasting using ensemble model output statistics and minimum CRPS estimation. Mon. Weather Rev. **133**(5), 1098–1118 (2005). https://doi.org/10.1175/MWR2904.1
9. Gneiting, T., Balabdaoui, F., Raftery, A.E.: Probabilistic forecasts, calibration and sharpness. J. R. Stat. Soc. B **69**, 243–268 (2007)
10. Gneiting, T., Schulz, B., Lerch, S.: Probabilistic solar forecasting: benchmarks, post-processing, verification. Sol. Energy **252**, 72–80 (2023)
11. Haupt, S.E., Chapman, W., Adams, S.V., Kirkwood, C., Hosking, J.S., Robinson, N.H., Lerch, S., Subramanian, A.C.: Towards implementing artificial intelligence post-processing in weather and climate: proposed actions from the Oxford 2019 workshop. Phil. Trans. R. Soc. A **379**(2194), 20200091 (2021). https://doi.org/10.1098/rsta.2020.0091
12. Hess, R.: Statistical postprocessing of ensemble forecasts for severe weather at Deutcher Wetterdienst. Nonlinar Process. Geophys. **27**, 473–487 (2020). https://doi.org/10.5194/npg-27-473-2020
13. Hirsch, T., Hess, R., Trepte, S., Primo, C., Glashoff, J., Reichert, B., Heizenreder, D.: ModelMIX—optimal combination of NWP model forecasts for AutoWARN. In: World Weather Open Science Conference, Montreal, Canada, pp. 16–21 (2014)
14. Horat, N., Lerch, S.: Deep learning for post-processing global probabilistic forecasts on subseasonal time scales. Working paper (2023). arXiv:2306.15956
15. Höhlein, K., Schulz, B., Westermann, R., Lerch, S.: Postprocessing of ensemble weather forecasts using permutation-invariant neural networks. Artif. Intell. Earth Syst. (2023) (In press)
16. Lang, M.N., Lerch, S., Mayr, G.J., Simon, T., Stauffer, R., Zeileis, A.: Remember the past: a comparison of time-adaptive training schemes for non-homogeneous regression. Nonlinear Process. Geophys. **27**(1), 23–34 (2020)
17. Lerch, S., Thorarinsdottir, T.L., Ravazzolo, F., Gneiting, T.: Forecaster's dilemma: extreme events and forecast evaluation. Stat. Sci. **32**, 106–127 (2017)
18. Messner, J.W., Mayr, G.J., Zeileis, A.: Nonhomogeneous boosting for predictor selection in ensemble postprocessing. Mon. Weather Rev. **145**(1), 137–147 (2017). https://doi.org/10.1175/MWR-D-16-0088.1

19. Murphy, A.H.: A new vector partition of the probability score. J. Appl. Meteorol. Climatol. **4**(12), 595–600 (1973). https://doi.org/10.1175/1520-0450(1973)012<0595:ANVPOT>2.0.CO;2
20. Pantillon, F., Lerch, S., Knippertz, P., Corsmeier, U.: Forecasting wind gusts in winter storms using a calibrated convection-permitting ensemble. Q. J. R. Meteorol. Soc. **144**(715), 1864–1881 (2018)
21. Peralta, C., Ben Bouallègue, Z., Theis, S.E., Gebhardt, C., Buchhold, M.: Accounting for initial condition uncertainties in COSMO-DE-EPS. J. Geophys. Res.: Atmos. **117**(D7) (2012). https://doi.org/10.1029/2011JD016581
22. Rasp, S., Lerch, S.: Neural networks for postprocessing ensemble weather forecasts. Mon. Weather. Rev. **146**(11), 3885–3900 (2018). 1805.09091. https://doi.org/10.1175/MWR-D-18-0187.1
23. Reinert, D., Prill, F., Frank, H., Denhard, M., Baldauf, M., Schraff, C., Gebhardt, C., Marsigli, C., Zängl, G.: DWD database reference for the global and regional ICON and ICON-EPS forecasting system. Technical report Version 222, Deutscher Wetterdienst (2023). https://www.dwd.de/DWD/forschung/nwv/fepub/icon_database_main.pdf
24. Sanders, F.: On subjective probability forecasting. J. Appl. Meteorol. **2**(2), 191–201 (1963). https://doi.org/10.1175/1520-0450(1963)002<0191:OSPF>2.0.CO;2
25. Scheuerer, M., Switanek, M.B., Worsnop, R.P., Hamill, T.M.: Using artificial neural networks for generating probabilistic subseasonal precipitation forecasts over California. Mon. Weather Rev. **148**(8), 3489–3506 (2020). https://doi.org/10.1175/MWR-D-20-0096.1
26. Schraff, C., Reich, H., Rhodin, A., Schomburg, A., Stephan, K., Periáñez, A., Potthast, R.: Kilometre-scale ensemble data assimilation for the COSMO model (KENDA). Q. J. R. Meteorol. Soc. **142**(696), 1453–1472 (2016). https://doi.org/10.1002/qj.2748
27. Schulz, B., Lerch, S.: Machine learning methods for postprocessing ensemble forecasts of wind gusts: a systematic comparison. Mon. Weather Rev. **150**(1), 235–257 (2022). 2106.09512. https://doi.org/10.1175/mwr-d-21-0150.1
28. Vannitsem, S., Wilks, D.S., Messner, J.W.: Statistical Postprocessing of Ensemble Forecasts. Elsevier (2018). https://doi.org/10.1016/c2016-0-03244-8
29. ...Vannitsem, S., Bremnes, J.B., Demaeyer, J., Evans, G.R., Flowerdew, J., Hemri, S., Lerch, S., Roberts, N., Theis, S., Atencia, A., Bouallègue, Z.B., Bhend, J., Dabernig, M., Cruz, L.D., Hieta, L., Mestre, O., Moret, L., Plenković, I.O., Schmeits, M., Taillardat, M., den Bergh, J.V., Schaeybroeck, B.V., Whan, K., Ylhaisi, J.: Statistical postprocessing for weather forecasts: review, challenges, and avenues in a big data world. Bull. Am. Meteor. Soc. **102**(3), E681–E699 (2021)
30. Veldkamp, S., Whan, K., Dirksen, S., Schmeits, M.: Statistical postprocessing of wind speed forecasts using convolutional neural networks. Mon. Weather Rev. **149**(4), 1141–1152 (2021). https://doi.org/10.1175/MWR-D-20-0219.1
31. Zhuang, F., Qi, Z., Duan, K., Xi, D., Zhu, Y., Zhu, H., Xiong, H., He, Q.: A comprehensive survey on transfer learning. Proc. IEEE **109**(1), 43–76 (2021). https://doi.org/10.1109/JPROC.2020.3004555

Chapter 9
AutoEncoding Cognitive Maps of Non-Trivial Topology

M. Beketov, K. Sorokin, M. Subbotin, V. Sotskov, K. Anokhin, and A. Ayzenberg

Abstract In present work we explore the applicability of autoencoders (AEs) as a tool to extract metric properties of a physical space from the neural activity signal of the hippocampal CA1 place cells, while the object (mouse) was exploring novel environments, flat arenas of different shapes. We show that an AE as shallow as 3+3 layers (3 layers both in the encoder and the decoder) can be trained to encode a cognitive map in its latent space—by introducing a simple additional loss of (supervised) metric reconstruction. The introduction of this loss term significantly changes the importance distribution of the AE encoder's last-layer neurons, making their statistical properties similar to those found in natural neural networks.

Keywords Neuroscience · Machine learning · Topology

9.1 Introduction and Problem Statement

Cognitive maps (introduced in [1]) is a term used to describe the inner mental representations of various spaces, including the physical space—perceived by an animal during its navigation in this space. These maps are encoded by the activity pattern of place cells—neurons mainly located in the field CA1 of the hippocampus, but also found in other brain structures [2]. These neurons are known to be principal for the formation of episodic and spatial memory [3]. At the same time, it is known that CA1 hippocampal place cells may also encode non-spatial representations [4, 5]. Studying the mechanisms standing behind the formation of ensembles of place cells coding novel environments is crucial for understanding how information is represented and

M. Beketov (✉) · K. Sorokin · A. Ayzenberg
HSE University, Moscow, Russia
e-mail: mbeketov@hse.ru; ksorokin@hse.ru

V. Sotskov · K. Anokhin
Institute for Advanced Brain Studies, Lomonosov Moscow State University, Moscow, Russia

M. Subbotin
Vozz.ai, London, UK

A. Ott et al. (eds.), *Applications of Mathematics in Sciences, Engineering, and Economics*, Springer Proceedings in Mathematics & Statistics 515,
https://doi.org/10.1007/978-3-032-01279-1_9

stored in the brain. In present work, in line with our previous approaches [6], we apply AutoEncoders (AEs, a family of artificial neural networks, ANNs) to the place cells' neural activity data recorded. The optical technique of calcium imaging [7] was used in our experiments. The mice were placed into arenas of various shape, or topology, flat arenas with 1, 2 and 3 "holes". The holes are modeled by impenetrable obstacles, where the mouse has no physical access. By the "topology" of space X we basically mean the number of these obstacles, which is the rank of the fundamental group $\pi_1(X)$ [8] (which is a free group) or the rank of first homology group $H_1(X; \mathbb{Z})$ (which is a free abelian group).

The general goal of our ongoing research is to find the algorithmical technique, which allows to predict some properties of environment by looking at the neural activity time series. In this paper the problem in each experiment is formulated as follows. We have true coordinates of mouse location at each time moment as well as its place cells' neural activity at each time moment. These data are synchronized. The problem is then to reconstruct the corresponding "true map" from the high-dimensional neural activity time series, in a way that arena's topology is preserved in this reconstruction. In previous works, we already tested the possibility of topology reconstruction from neural data alone [6] following the approach proposed in [9]. That approach was primarily based on construction of simplicial complexes from the time series of neural data and applying Nerve's theorem. In the same paper [6] we also tested another approach proposed in [10]—we sought the best way to project extremely high-dimensional point cloud of neural data to the 2-dimensional plane, and estimated how the projected data resemble the shape of arena. The main quantitative tool of statistical measurement of topology is persistent homology [11], we used 1-dimensional persistent homology in our work. In certain sense this invariant is metric-agnostic, but allows to compare topological features of data.

Compared to our previous approach in [6], in present work we go further and try to keep track of the metric structure of the arenas. It was previously shown that there might be approaches preserving the metric, like [12] but these studies were dedicated to open-field free exploration by the animals, which is a less controlled environment. Our approach also allows to measure the neurons' contribution into autoencoder's latent space representation, giving another elegant method not only for studying the neural cells involvement into encoding the information in general (and shed light on their dynamics [13], in particular, which we also studied using direct approaches [14]), but a new way of selecting informative place cells, without resorting to approaches based on mutual [15] or spatial [16] information. These methods could sometimes lead to selecting "false" place cells, an effect we were trying to overcome in our previous paper [6]. The approach we are testing in present paper does not require to manually filter out such uninformative cells—the AE automatically filters out such components of the signal—see the Results section for an illustration of the distribution of neurons w.r.t. importance of their contribution to the signal.

Another significant advantage of the proposed approach is its applicability to designing models of spiking neural networks, mimicking the hippocampus—one possible direction of our future research (we also presented a work dedicated to the dynamical aspects of the functional hippocampus' network architecture [17]). Resulting distributions of AE's latent layer neurons by their importance may be used for testing the "naturalness" of such ANNs as models of living neural networks.

9.2 Experimental Setup, Preprocessing and Data

For this study, we have taken the data first described in our previous paper [13]. In brief, mice with head-mounted miniscopes [7] were allowed to freely explore the novel environment with various number of obstacles in three consequent 10-min imaging sessions 24 h apart. During these sessions, calcium activity of hundreds of hippocampal CA1 neurons, as well as the trajectories of mice, were recorded. All manipulations with mice, including surgery, preparation, data acquiring, and processing were described in detail in our previous work [17].

The number of obstacles varied from 3 in the first imaging session, to 2 in the second session and then to 1 in the third imaging session Fig. 9.1

In each session, mice were exploring the environment for approximately 10 minutes while the calcium activity of hippocampal CA1 neurons was recorded with NVista HD miniscope at 20 fps. At the same time, the trajectories of mice were tracked using a camera fixed above the arena with the same frame rate. After that, the two video streams were handled. For calcium imaging data streams, this included pixel noise removal, cropping, image stabilization (both laterally and by height), and adjusting brightness and contrast parameters throughout the entire frame (polynomial background removal). To make the background as bright and the mouse as dark as possible in videos with moving rodents, we also adjusted the curve colour scheme. All the experiments made use of the same preset.

Then the processed stream with neural data was handled with tools of CaImAn [18], with CNMF (Constrained Nonnegative Matrix Factorization [19]) principle used for scanning each frame in order to tell the "true" neurons apart from the "fake" which are generally blood vessels or some kind of visual noise in the video. The ratio of accepted and rejected candidates to neurons was roughly equal 10 to 1 (numer-

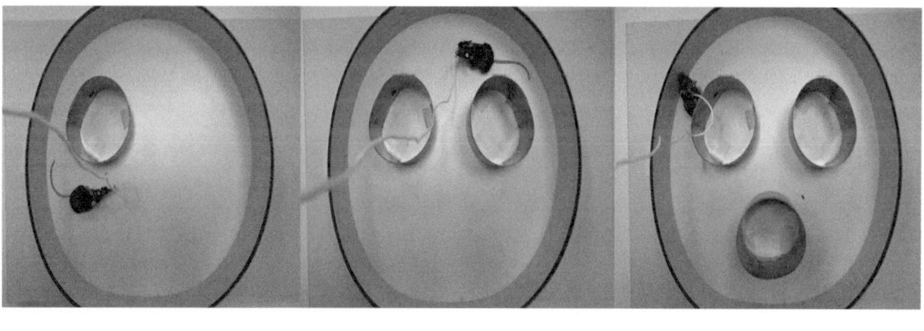

Fig. 9.1 Three arenas with different topologies ($\pi_1 = \mathbb{Z}, 2\mathbb{Z}, 3\mathbb{Z}$ correspondingly)

ically, it was typically about 500 total real and 50 fake neurons). The proportion remained constant, despite a few outlier cases with significantly fewer neurons.

The neural data we work as a result with is the activity of hundreds of place cells of several mice represented as the time series of dimension being the number of place cells recognised on the video frames.

The other dataframe, being mouse's trajectory was created by joining each centre of mass position. The mean of the coordinates between two good quality frames was used if any frames were missing or the distance between points was greater than a predetermined threshold (i.e., the animal's position varied by more than 25 pixels between two consecutive frames). The resultant trace is now a table that can be pointwise analysed to determine how much time the animal spent in each location and, if necessary, to calculate its pace and other parameters.

The prepared dataset will be used to:

1. Experimentally determine the best configuration of the main autoencoder;
2. Check if the autoencoder latent layer can be trained to simulate mouse place cells;
3. Experimental finding of lambda (regularization parameter of topological autoencoder).

9.3 Method: Autoencoders

9.3.1 Background

An AutoEncoder (AE) [20, 21] is an artificial neural network (ANN) that, given input data, transforms it in a nonlinear way, parameterized by the weights of its layers. One can depict an AE with what's called its computation graph—see Fig. 9.2.

Input data is a vector $\mathbf{x}_{i_0} \in \mathbb{R}^{D_{\text{input}}}$ that is "fed" to the input neurons, enumerated by i_0, with each element of $\mathbf{x}_{i_0}$ being "passed" to its corresponding neuron. After that the 1st layer transforms its input:

$$\mathbf{x}_{i_1} = \sigma\left(W_{i_1 i_0} x_{i_0} + \mathbf{b}_{i_1}\right) \tag{9.1}$$

first in a linear way: by multiplying x_{i_0} by a matrix of 1st (fully connected) layer's **weights** $W_{i_1 i_0}$ and then adding a vector of **biases** $\mathbf{b}_{i_1}$ (thus overall doing an affine transform); then applying some component-wise **nonlinear** activation function $\sigma(\cdot)$—turning it into a vector x_{i_1}—output of the 1st layer, and input to the 2nd layer on the next step. In this way, fed to layer by layer, data is transformed until one gets the output vector—that's why AEs are what's called *feedforward* NNs. As for the choice of $\sigma(\cdot)$, there are many possible options, in present work we're only working with what's called ReLU (rectified linear unit),—geometrically, a fully-connected layer followed by ReLU projects the input on some subspace of the affine image $W\mathbf{x} + \mathbf{b}$.

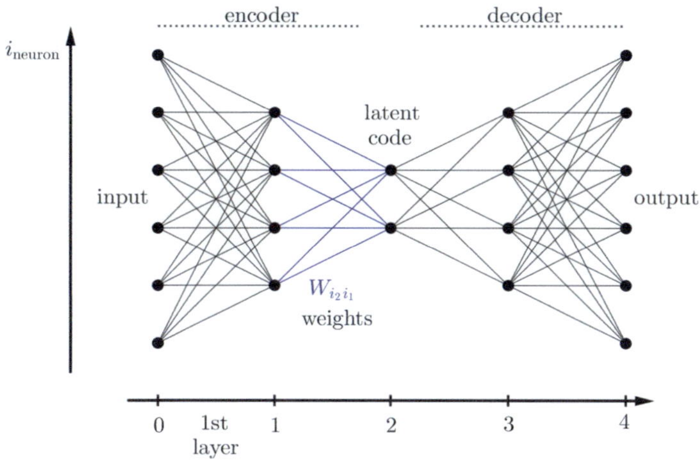

Fig. 9.2 Schematic of an AutoEncoder

$$\text{ReLU}(x) \equiv \begin{cases} x, & \text{if } x > 0 \\ 0 & \text{otherwise} \end{cases} \tag{9.2}$$

In an AE, layer sizes are typically decreasing from the input to some intermediary "bottleneck" layer, and then increasing back again, so $D_{\text{input}} = D_{\text{output}}$. The image of input vector $\mathbf{x}_{\text{input}}$ in this intermediary layer, denote it $\mathbf{z}$—is a lower-dimensional "nonlinear projection" of $\mathbf{x}_{\text{input}}$—called its "latent code" or "latent representation/embedding", hence the first half of the AE, $\mathbf{x}_{\text{input}} \to \mathbf{z}$ is the **encoder**:

$$\mathbf{z} \equiv \text{enc}(\mathbf{x}_{\text{input}}). \tag{9.3}$$

Thus all the (important) information contained in $\mathbf{x}_{\text{input}}$ should be "condensed" to this (typically) lower-dimensional latent code vector, $\mathbf{z}$. The second half of the AE, hence called the **decoder**, "decompresses" the data from this latent representation to produce output:

$$\mathbf{x}_{\text{output}} \equiv \text{dec}(\mathbf{z}). \tag{9.4}$$

An AE is "trained" on training data—a large enough (N_{train}) sample of input datapoints $\{\mathbf{x}^{(i)}_{\text{input}} \in \mathbb{R}^{D_{\text{input}}}\}_{i \in \{1,\ldots,N_{\text{train}}\}}$. Training means finding the parameters of the AE so that:

$$\mathcal{L}(\{\mathbf{x}_{\text{input}}\}) \equiv \frac{1}{N_{\text{train}}} \sum_{i=1}^{N_{\text{train}}} \left|\left| \mathbf{x}^{(i)}_{\text{input}} - \underbrace{\text{dec}(\overbrace{\text{enc}(\mathbf{x}^{(i)}_{\text{input}})}^{\equiv \mathbf{z}_{\text{input}} \text{ - latent enc.}})}_{\equiv \mathbf{x}^*_{\text{input}} \text{ - reconstruction}} \right|\right|^2 \to \min_{\text{AE parameters}} \tag{9.5}$$

where $||\cdot||$ denotes Euclidean 2-norm. This objective function (mean square reconstruction error, MSE) (9.5), referred to as the training **loss** function, makes sure that the AE reconstructs input data $\mathbf{x}_{\text{input}}$ well, even though $\mathbf{z}_{\text{input}}$ might be a times smaller-dimensional representation of the input signal. (So one can say that an AE is trained to approximate an identity function on input data—but one that compresses the input on its way.)

Minimizing (9.5) is done over the set of AE's parameters: the weight W_{ij} matrices and bias vectors $\mathbf{b}_i$ of each layer. It is done with either error back propagation [22] or stochastic gradient descent in this space of parameters—not on the whole training dataset, $\{\mathbf{x}_{\text{input}}\}$, but rather on small **batches** of it—tuning the parameters batch by batch (an iteration of optimization on one such batch is called an epoch).

9.3.2 *Metric Penalty*

A generic AE defined above is thus trained with one objective—to reconstruct the input signal (point cloud) as well as possible (minimizing reconstruction MSE (9.5)):

$$\mathcal{L}_{\text{recon.}}(\{\mathbf{x}\}_{\text{train}}) \propto \sum_{\mathbf{x}_{\text{train}}} \left|\left|\mathbf{x} - \mathbf{x}^*\right|\right|^2 \rightarrow \min_{\text{AE params}} \tag{9.6}$$

with the only **constraint** imposed on it being that it should "squeeze" the data into the low-dimensional latent space.

Our goal in present work was a little different—to see if an ANN can successfully distill, from the noisy input signal coming from living neurons, a very certain component—the **cognitive map**. A naïve understanding of this is that we just want $\mathbf{z}$, the latent representation of the signal—to just be (approximately) the **true coordinates** of the object in the physical space—-(x, y). The latter can be arbitrarily scaled, of course, so we can rescale them to vary from 0 to 1.

Denoting $\mathbf{z}_{\text{true}} = (x_{\text{true}}, y_{\text{true}})$—the true (Cartesian) coordinates of the object—a 2-dimensional vector, we thus modified the loss of our AE with what we call a **metric penalty** term:

$$\mathcal{L}_{\text{metric}}(\{\mathbf{x}\}_{\text{train}}) \equiv \left|\left|d_{ij}(\{\text{enc}(\mathbf{x})\}_{\text{train}}) - d_{ij}(\{\mathbf{z}_{\text{true}}\}_{\text{train}})\right|\right|_F^2 \tag{9.7}$$

where $d_{ij}(\cdot)$ are **pairwise distance matrices** of the (2-dimensional) point-clouds—the true coordinates of the object, $\{\mathbf{z}_{\text{true}}\}$; and the latent, "cognitive" coordinates (as output by the encoder)—being just encodings of $\mathbf{x}$ points—slices of the neural activity time series at corresponding (to that when the true coordinates are registered) moments of time (in our case, ~500-dimensional) of the training dataset. See Fig. 9.3 for an illustration of successfully doing such an encoding. Henceforth we'll refer to $\text{enc}(\mathbf{x})$ as **latent coordinates** (they may very well differ from the object's cognitive coordinates, but we aim to make them the best possible approximation of the latter).

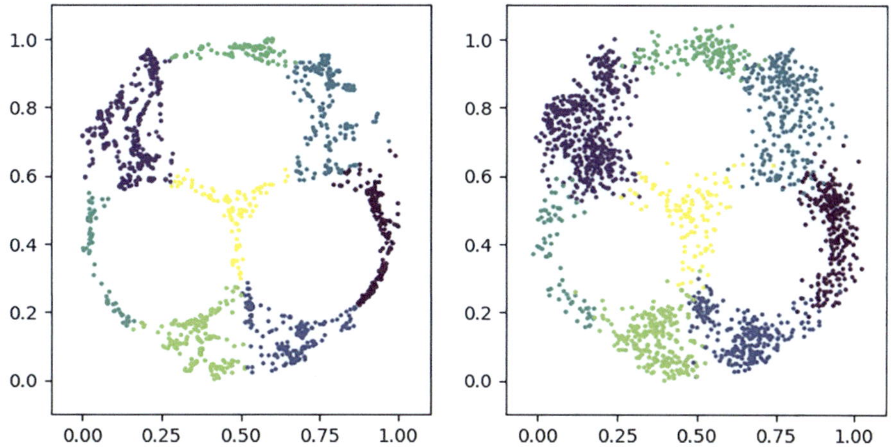

Fig. 9.3 *Left*: True (scaled) coordinates of the mouse (on a 3-hole arena), binned (k-NN algorithm) by proximity. *Right*: "Cognitive" coordinates (scaled) as encoded from the neural activity of the mouse's ~500 place cells with a 3-layer encoder (bins transferred from the left) = latent coordinates. *(Obtained on the test subsample of the dataset, "unseen" by the AE when training)*

With $||\cdot||_F$ being just the Frobenius matrix norm, $||A||_F^2 = \sum_{ij} |a_{ij}|^2$, and **z** vectors living in the 2-dimensional Euclidean space (where the distance between two points is the 2-norm of the difference of their position-vectors), this metric loss term (9.7) is **differentiable** w.r.t. the parameters (weights and biases) of the encoder enc$(\cdot)$, so, adding it to the total loss:

$$\mathcal{L}_{\text{total}} = \mathcal{L}_{\text{recon.}} + \beta\mathcal{L}_{\text{metric}} \tag{9.8}$$

one obtains a well-posed optimization problem $\mathcal{L}_{\text{total}} \to \min_{\text{AE params}}$, which is effectively solvable with gradient optimization techniques implemented in contemporary machine learning software packages (e.g. PyTorch [23] that we're using in the present work). This $\beta > 0$ parameter represents the strength of this metric penalty and allows to balance it with the reconstruction loss.

9.3.3 Architecture Search

Working on a specific problem with a specific dataset requires fixing an appropriate model. The schematic of an AE Fig. 9.2 is quite general: it only states that the layer sizes of the encoder are not increasing with depth (layer number), and that of the decoder are not decreasing—typically, in the same progression as the encoder, but backwards. In the middle, one has a "bottleneck"—the latent dimension is typically smaller than that of input, so the encoder is basically doing dimension-reduction. The given constants in our problem are:

- the **dimension** D **of input data**—given by the number of living neurons, the signal from which we're (auto-)encoding (of order $\sim$300–500),
- the **dimension** $d = 2$ **of the latent space**—equal to 2, the effective dimension of the physical space the object is moving in, the dimension of the Cartesian plane.

One is thus left free to choose any curve fixed at those two endpoints for the dependence of layer size on layer number. Practically, the less parameters specify this curve—the better, since finding optimal values of those parameters with some sort of cross-validation procedure requires a number of trials exponential in the number of parameters. With this in mind, we decided to encode the layer size dependence on depth with **one parameter**.

For the encoder, our layer size VS layer number curve should interpolate from D to d on an interval from 0 to N (N is the total number of encoder layers = the number of latent layer, if layers are enumerated from 0 for the input one, as on Fig. 9.2). So, if we scale both $[d, D]$ and $[0, N]$ to $[0, 1]$, we're left to choose a family of monotonic curves on the square $[0, 1] \times [0, 1]$—for that, we've simply chosen the family of power dependencies: $y = x^{\alpha}$. Adjusting for the need to have it decreasing, the final dependence (for the **decoder**) reads:

$$\text{layer\_size} = \left\lfloor d + (D-d) \cdot \left(\frac{\text{layer\_number}}{N} \right)^{\alpha} \right\rfloor \tag{9.9}$$

with layer number starting from 0 at the latent layer (this slightly differing from said above is legacy of our software implementation). We refer to this α parameter as "**bottlenecking slope**", see a schematic figure of AEs with different values of α at Fig. 9.4. (Although this is quite straightforward, we haven't seen any similar practical recipes for varying AE architectures with one parameter in the literature.)

With that, the **hyperparameters** to choose which specify a certain AE model are:

$$\text{AE hyperparameters} = (N, \alpha, \beta). \tag{9.10}$$

By the number_of_layers we count the number of layers in either the encoder or the decoder (they are the same), thus half of the total number of layers of the AE. α is the bottlenecking slope, β is the strength of the metric penalty term in the total loss.

We've tried **varying the number of layers** in the range $1-7$. Shallow AEs (1–2 layers) were not very effective (1-layer AE being just a slightly—ReLU—nonlinear version of PCA), while too deep ones (4+ layers) did not improve the reconstruction much, even having that much more parameters, plus there's an increasing risk of over-fitting. The optimal middle ground was found to have **3 layers** (in both encoder and decoder). Although it is known that feedforward ANNs this shallow can already be universal approximators [24–29]—that is, arbitrarily well (in some

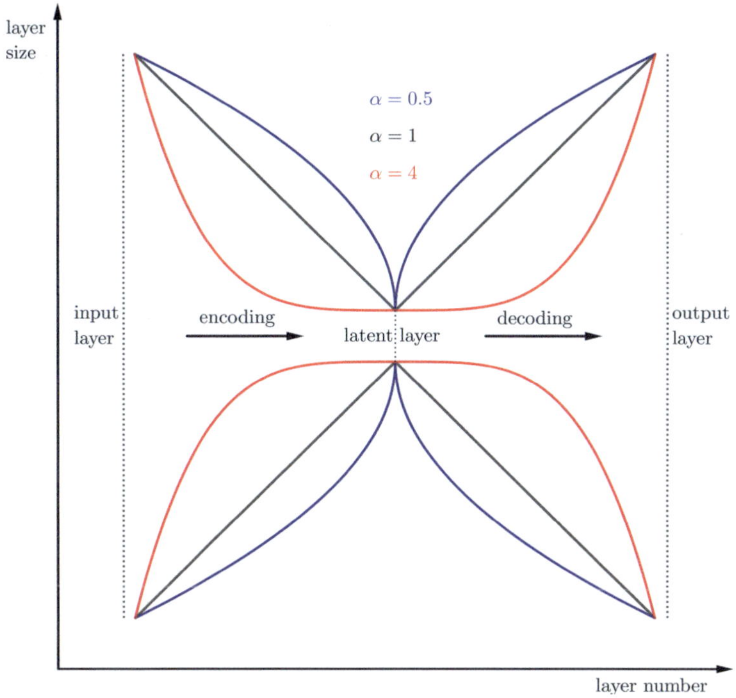

Fig. 9.4 Autoencoder architecrures of various bottlenecking slopes (α parameter)

norm) approximate continuous functions, this result seems to have some biological implications—see Results and discussion section.

With the total number of AE layers fixed to 3+3, we varied the bottlenecking slope coefficient. With the exponential nature of the dependence (9.9), it is natural to try α values **log-uniformly spread**—so we tried $\log_2 \alpha = [-2, -1, 0, 1, 2]$. With $\alpha = 2^{-2} = 0.25$ one has an AE shrinking steeply right near the bottleneck, very "thick" elsewhere, while another extreme, $\alpha = 2^2 = 4$ gives a quickly narrowing AE—see Fig. 9.4. Of course, on 4 discrete steps (3 layers) this curve is rough, but this gives some intuitive idea of how quickly the dimension is reduced by the encoder.

With that, the training speed of AEs of different α values was tracked during 250 epochs, see Fig. 9.5. It is commonly considered that sometime after the "elbow point" (inflection point at $\sim$100-150 epochs on the plots) the model has trained (the optimization converged). So, interestingly, "thicker" AEs were converging faster than their "thin" counterparts—even though the former have more parameters to tune than the latter. Also interestingly, taking more extreme $\alpha = 0.25$ (dashed line on Fig. 9.5) does not improve upon $\alpha = 0.5$ (the line is even above, but that may be due to randomness of the training procedure). With that, other **figures in this paper** were obtained an AE of $\alpha = 0.25$.

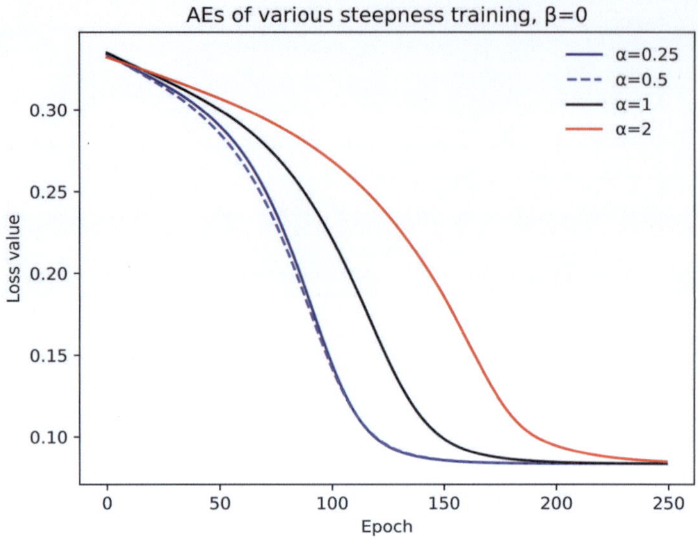

Fig. 9.5 Training dynamics of (3+3)-layer AEs of different "bottlenecking" slope, α

A typical (3+3)-layer AE with $\alpha = 0.25$, with input data dimension $D = 554$ had the following layer sizes progression:

$$\text{n\_neurons} = [\underbrace{554}_{\text{input}}, 515, 466, \overbrace{392, \underbrace{2}_{\text{latent}}}^{\text{last enc. layer}}, 392, ..., \underbrace{554}_{\text{output}}]. \tag{9.11}$$

9.4 Results and Discussion

We applied metric-preserving AutoEncoders (AEs) to the problem of cognitive map reconstruction in a supervised manner (by additionally penalising for metric dissimilarity of the inner, latent, representation of the time-series from the true coordinates). We found that mices' cognitive maps are well reconstructed by quite shallow (3+3 layers, 3 layers of the encoder) AEs. This fact provides certain empirical implication for neuroscience: it shows that autoencoder is a type of artificial architecture which models the correspondence between the actual spatial data and cognitive map in a nice manner.

Figure 9.6 demonstrates the dependence of latent maps' visual quality on the strength of metric reconstruction penalty. These latent maps were obtained on the test subsample of the dataset, "unseen" by the AE while training. Note the overlapping bins on the image corresponding to $\beta = 100$—the fact that bins (transferred from the binning of the true coordinates) are overlapping somewhere at their boundaries

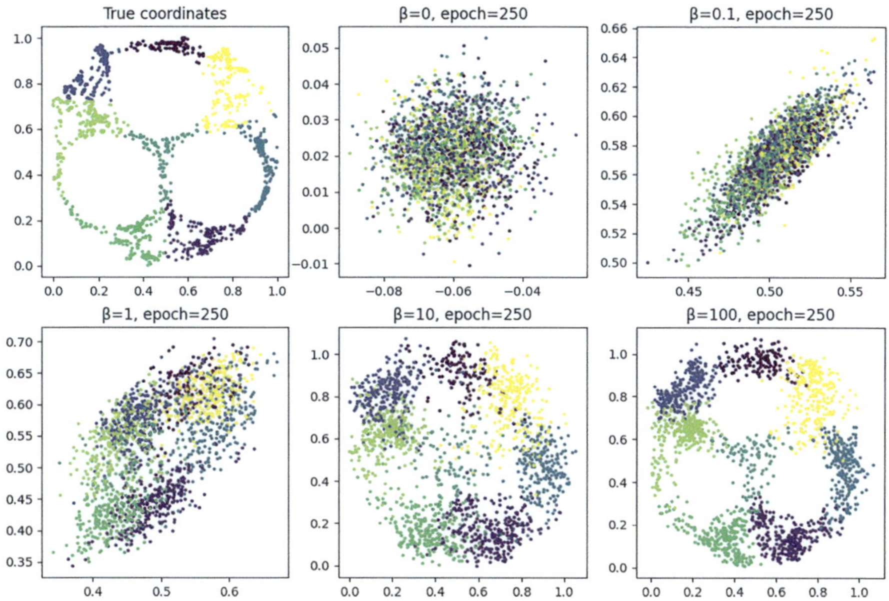

Fig. 9.6 True VS latent coordinates, obtained after 250 epochs of training AEs with various metric-regularization strength (β). *(Obtained on the test subsample of the dataset, "unseen" by the AE when training)*

indicates that the model reproduced the map overall, with minor inaccuracies in detail—which does not look like an overfit.

There is another observation supporting the similarity between autoencoders and natural neural networks. Figure 9.7 shows the distributions of encoders' last layer input neurons by their importance (simply defined as the sum of absolute values of the weight of their connections to the output neurons of that layer). With the metric penalty strength set to zero $\beta = 0$, this distribution is normal-like (shown in blue), while with a strong metric penalty $\beta = 100$ it is skewed towards more important neurons. This mimics the statistical properties of neurons found in living (natural) neural networks—there, typically, a few very important neurons (such are referred to as "rich club" [30]) dominate the whole network. This finding establishes a connection between such artificial models of intelligence and living ones, and is certainly worth further exploring.

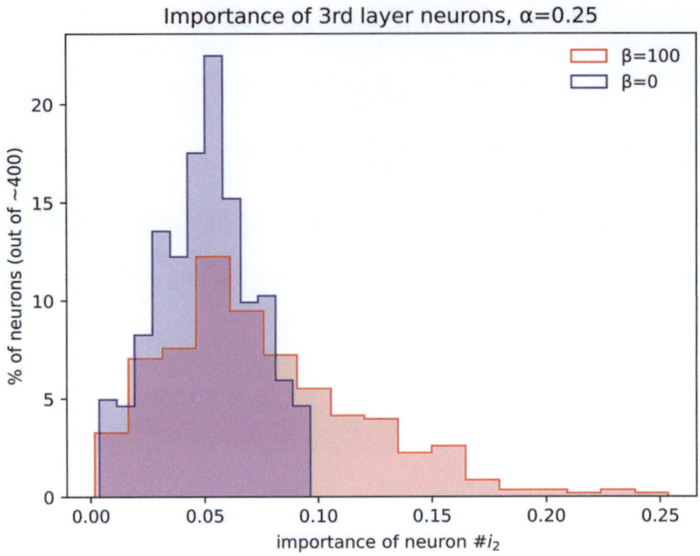

Fig. 9.7 Neurons of the 3rd (encoder's last) layer, distributed by importance (neuron # i_2) = $\Sigma_{i_3}|W_{i_3 i_2}|$ (= sum of its "contributions" to the latent x- and y- coordinates)

9.5 Future Prospects

We see several intriguing directions of developing upon present work.

9.5.1 Topological Autoencoders

Topological Autoencoders (TopoAEs), introduced in [31], are a modification of vanilla AEs that aims to preserve the topology of the dataset, meaning that if persistent homology groups of its latent representation (the encoder's output) differs from that of the origial data, TopoAE is penalized (with what's called topological loss term).

Critically, this is an unsupervised method (as opposed to our supervised approach)—TopoAE only preserves the topology that it finds in raw input data. As it is strongly expected that the topology of the stimulus space (physical space in particular) is present in the response signal of neurons specialized on it, TopoAEs look like a very promising method for our problem.

The only minor underdevelopment of the original TopoAE is that authors of [31] propose to equally penalize for the difference in homologies of all dimensions (with only 0- and 1-homologies available in authors' software implementation anyway)—while in our specific problem 1-homologies (loops) are certainly more important than 0-homologies (connected components). We're already working on such a generalization of TopoAEs applied to our problem.

9.5.2 *Sparse Autoencoders*

It might be interesting to explicitly enforce sparsity (in the AE's weights matrices) by adding a corresponding loss term—a technique known as sparse autoencoders [32, 33]. Findings in this direction might shed light on architectural nuances of ANNs successfully mimicking living NNs in the brain, and thus help understand the latter as well.

9.5.3 *Generative Models*

Another possible contribution is to develop generative models for such kind of neural data—one possible way of doing so being turning vanilla AEs into denoising autoencoders [34]. Having a generative model would have some ethical advantages—with enough training data collected, there is no need to further animal testing.

Acknowledgements The article was prepared within the framework of the project "Mirror Laboratories" HSE University, RF. This research was supported in part through computational resources of HPC facilities at HSE University [35].

Supporting Information

Competing Interests There are no competing interests.

Ethics Approval The animal study was reviewed and approved by All methods for animal care and all experimental protocols were approved by the National Research Center "Kurchatov Institute" Committee on Animal Care (NG-1/109PR of 13 February 2020) and were in accordance with the Russian Federation Order Requirements N 267 M3. Four C57BL/6 mice were used in this study, ages P0-P3 and 2–3 months. Mice were used without regard to gender.

References

1. Tolman, E.C.: Cognitive maps in rats and men. Psychol. Rev. **55**(4), 189–208 (1948). https://doi.org/10.1037/h0061626. PMID 18870876. S2CID 42496633
2. O'Keefe, J., Dostrovsky, J.: The hippocampus as a spatial map: preliminary evidence from unit activity in the freely-moving rat. Brain Res. (1971)
3. O'Keefe, J., Nadel, L.: The Hippocampus as a Cognitive Map. Oxford University Press, Clarendon Press (1978)

4. Mok, R.M., Love, B.C.: A non-spatial account of place and grid cells based on clustering models of concept learning. Nat. Commun. **10**(1), 5685 (2019). https://doi.org/10.1038/s41467-019-13760-8
5. Eichenbaum, H.: The hippocampus as a cognitive map of social space. Neuron **87**(1), 9–11 (2015). https://doi.org/10.1016/j.neuron.2015.06.013
6. Sorokin, K., Ayzenberg, A., Anokhin, K.V., Sotskov, V.P.., Beketov, M., Drynkin, R., Zaitsew, A.V.: Topology of cognitive maps. Cornell University. Series Computer Science (2022)
7. Ghosh, K.K., Burns, L.D., Cocker, E.D., Nimmerjahn, A., Ziv, Y., Gamal, A.E., Schnitzer, M.J.: Miniaturized integration of a fluorescence microscope. Nat. Methods **8**(10), 871–878 (2011). https://doi.org/10.1038/nmeth.1694
8. Hatcher, A.: Algebraic Topology. Cambridge University Press, Cambridge (2002)
9. Curto, C., Itskov, V.: Cell groups reveal structure of stimulus space. PLoS Comput. Biol. **4**(10), e1000205 (2008). https://doi.org/10.1371/journal.pcbi.1000205
10. Tenenbaum, J.B., de Silva, V., Langford, J.C.: A global geometric framework for nonlinear dimensionality reduction. Science. **290**(5500), 2319–2323 (2000). https://doi.org/10.1126/science.290.5500.2319
11. Carlsson, G.: Topology and data. AMS Bull. **46**(2), 255–308 (2009)
12. Giusti, C., Pastalkova, E., Curto, C., Itskov, V.: Clique topology reveals intrinsic geometric structure in neural correlations. Proc. Nat. Acad. Sci. **112**(44), 13455–13460 (2015). https://doi.org/10.1073/pnas.1506407112
13. Sotskov, V.P., Pospelov, N.A., Plusnin, V.V., Anokhin, K.V.: Calcium imaging reveals fast tuning dynamics of hippocampal place cells and CA1 population activity during free exploration task in mice. Int. J. Mol. Sci. **23**(2), 638 (2022). https://doi.org/10.3390/ijms23020638
14. Sotskov, V.P., Plusnin, V.V., Sorokin, K.S., Anokhin, K.V.: Rapid tuning dynamics of CA1 place codes in one-and two- dimensional free exploration tasks. Fourth Int. Conf. Neurotechnol. Neurointerfaces (CNN) **2022**, 168–171 (2022). https://doi.org/10.1109/CNN56452.2022.9912511
15. Li, Y., Deutsch C.V.: Mutual Information and Its Application In Spatial Statistics. Paper 122, CCG Annual Report 12 (2010)
16. Dr. Markus, E.J., Dr. Barnes, C.A., McNaughton, B.L., Gladden William, V.L., Skaggs, E.: Spatial information content and reliability of hippocampal CA1 neurons: effects of visual input. Hippocampus **4**(4) (1994)
17. Sorokin, K., Zaitsew, A.V., Levin, A.M., Magaj, G., Beketov, M., Sotskov, V.: Global cognitive graph properties dynamics of hippocampal formation/Cornell University. Series (2023). arxiv:q-bio.NC “arxiv Quantitative Biology”
18. Giovannucci, A., Friedrich, J., Gunn, P., Kalfon, J., Koay, S.A., Taxidis, J., Najafi, F., Gauthier, J.L., Zhou, P., Tank, D.W., Chklovskii, D.B., Pnevmatikakis, E.A.: CaImAn: an Open Source Tool for scalable Calcium Imaging data Analysis (2018)
19. Pnevmatikakis, E.A., Soudry, D., Gao, Y., Machado, T., Merel, J., Paninski, L.: Simultaneous denoising, deconvolution, and demixing of calcium imaging data. Neuron **89**(2), 285–299 (2016)
20. Kramer, M.A.: Nonlinear principal component analysis using autoassociative neural networks. AIChE J. **37**(2), 233–243 (1991)
21. Kramer, M.A.: Autoassociative neural networks. Comput. Chem. Eng. **16**(4), 313–328 (1992)
22. Rumelhart, D.E., Hinton, G.E., Williams, R.J.: Learning representations by back-propagating errors. Nature **323**(6088), 533–536 (1986)
23. Paszke, A., Gross, S., Massa, F., Lerer, A., Bradbury, J., Chanan, G., Chintala, S.: Pytorch: an imperative style, high-performance deep learning library. Adv. Neural Inf. Proc. Syst. **32** (2019)
24. Cybenko, G.: Approximation by superpositions of a sigmoidal function. Math. Control Signals Syst. **2**(4), 303–314 (1989)
25. Hornik, K.: Approximation capabilities of multilayer feedforward networks. Neural Netw. **4**(2), 251–257 (1991)

26. Funahashi, K.I.: On the approximate realization of continuous mappings by neural networks. Neural Netw. **2**(3), 183–192 (1989)
27. Hornik, K., Stinchcombe, M., White, H.: Multilayer feedforward networks are universal approximators. Neural Netw. **2**(5), 359–366 (1989)
28. Maiorov, V., Pinkus, A.: Lower bounds for approximation by MLP neural networks. Neurocomputing **25**(1–3), 81–91 (1999)
29. Guliyev, N.J., Ismailov, V.E.: Approximation capability of two hidden layer feedforward neural networks with fixed weights. Neurocomputing **316**, 262–269 (2018)
30. Griffa, A., Van den Heuvel, M.P.: Rich-Club neurocircuitry: function, evolution, and vulnerability. Dialog. Clin. Neurosci. **20**, (2), 121–32 (2018). https://doi.org/10.31887/DCNS.2018.20.2/agriffa. PMID: 30250389; PMCID: PMC6136122
31. Moor, M., Horn, M., Rieck, B., Borgwardt, K.: Topological autoencoders. In: Int. Conf. Mach. Learn. pp. 7045–7054. PMLR (2020)
32. Ng, A.: Sparse autoencoder. CS294A Lect. Notes **72**(2011), 1–19 (2011)
33. Makhzani, A., Frey, B.: K-sparse autoencoders (2013). arXiv:1312.5663
34. Bengio, Y., Yao, L., Alain, G., Vincent, P.: Generalized denoising auto-encoders as generative models. Adv. Neural Inf. Proc. Syst. **26** (2013)
35. Kostenetskiy, P.S., Chulkevich, R.A., Kozyrev, V.I.: HPC resources of the higher school of economics. J. Phys.: Conf. Ser. **1740**(1), 012050. IOP Publishing (2021)

Chapter 10
Residual Based Error Estimator for Chemical-Mechanically Coupled Battery Active Particles

Raphael Schoof, Lennart Flür, Florian Tuschner, and Willy Dörfler

Abstract Adaptive finite element methods are a powerful tool to obtain numerical simulation results in a reasonable time. Due to complex chemical and mechanical couplings in lithium-ion batteries, numerical simulations are very helpful to investigate promising new battery active materials such as amorphous silicon featuring a higher energy density than graphite. Based on a thermodynamically consistent continuum model with large deformation and chemo-mechanically coupled approach, we compare three different spatial adaptive refinement strategies: *Kelly-*, *gradient recovery-* and *residual based* error estimation. For the residual based case, the strong formulation of the residual is explicitly derived. With amorphous silicon as example material, we investigate two 3D representative host particle geometries, reduced with symmetry assumptions to a 1D unit interval and a 2D elliptical domain. Our numerical studies show that the Kelly estimator overestimates the error, whereas the gradient recovery estimator leads to lower refinement levels and a good capture of the change of the lithium flux. The residual based error estimator reveals a strong dependency on the cell error part which can be improved by a more suitable choice of constants to be more efficient. In a 2D domain, the concentration has a larger influence on the mesh distribution than the Cauchy stress.

Keywords Spatial adaptivity · Lithium-ion battery · Finite deformation · Finite elements · Numerical simulation

10.1 Introduction

Rechargeable batteries have become an integral part of our everyday lives [1, Sect. 1.2], especially lithium (Li)-ion batteries have convinced through high energy density and long life time and have become one of the most popular energy storage system [2–6]. Compared to the state of the art anode material graphite [7], amorphous

R. Schoof · L. Flür · F. Tuschner · W. Dörfler (✉)
Institute for Applied and Numerical Mathematics (IANM), Karlsruhe Institute of Technology (KIT), Karlsruhe, Germany
e-mail: willy.doerfler@kit.edu

A. Ott et al. (eds.), *Applications of Mathematics in Sciences, Engineering, and Economics*, Springer Proceedings in Mathematics & Statistics 515,
https://doi.org/10.1007/978-3-032-01279-1_10

silicon (aSi) has the advantage of a nearly tenfold theoretical capacity and is therefore a very promising candidate for next generation Li-ion batteries [8–11]. However, this benefit is accompanied by a volume change up to 300% compared to a volume change of approximately 10% in graphite anodes during lithium intake and extraction [7, 12]. To investigate the mechanical stresses resulting from the large volume changes, efficient numerical simulations are necessary to understand deterioration of the anode, the aging process and its implications on the battery's lifetime, [7, 12–15] and [16, Sect. 2.7].

In order to get numerical simulation results in reasonable time, adaptive refinement strategies in space and time are essential, e.g. for phase separation materials like lithium iron phosphate Li_xFePO_4 (LFP). This case is discussed in [17], where a gradient recovery estimator is used for the spatial refinement strategy as wells as a fully variable order—variable time step size for the time integration scheme. Since there is no moving phase front in aSi after the first half cycle of lithiation [18], no gradient motion of Li-concentration regarding the spatial coordinates can be observed as in the case of LFP. This makes it necessary to investigate whether this estimator is more efficient or not.

In this paper, we will compare three different types of spatial refinement strategies. Firstly, the *Kelly* error estimator, [19, 20] and [21, Sect. 4.2], approximates the error of the jump of the solution gradients along the faces of all cells. This estimator is ready for use as part of the open source finite element bibliography `deal.II` [20]. Since this bibliography is used as basis for our implementation, we will include this estimator in our comparison. Secondly, the *gradient recovery* error estimator [21, Chap. 4] is used as the spatial refinement criterion as in [17]. Thirdly, a *residual based* error estimator, see for more information [22, Sect. 9.2] and [23], is investigated, which leads to a reduction in the spatial error of the strong formulation of the respective differential equation to solve. The expressions for the residual based error indicator contains cell and face errors, whereby the inner face errors correspond in a similar way to the Kelly error estimator up to a chosen constant. See [19, 20, 24] for the choice of the constant and also for further information [25, Sect. 1.2], e.g. compare [26] for the p-laplace problem.

The aim of this paper is the numerical comparison of these three refinement strategies. As part of the cell errors of the residual error estimator, the strong form of the underlying differential system has to be derived in explicit form. Due to the strong coupling of chemical and mechanical effects, the derivation of specific parts are presented in detail. A numerical study in 1D shows the differences and similarities between the three refinement strategies, whereas a 2D example considers only the gradient recovery estimator and an updated version of the residual based error estimator.

The rest of this work is organized as follows: the theory of our used model approach is introduced in Sect. 10.2 as well as the derivation of the terms of the strong formulation needed for the residual based error estimation. In Sect. 10.3, the numerical solution algorithm is briefly presented, followed by the numerical results and discussions in Sect. 10.4. The last Sect. 10.5 summarizes our main findings and provides an outlook.

10.2 Theory

Firstly, this section reviews and summarizes the theory from [27, 28] for the formulation of our chemo-mechanically coupled model for battery active particles. Our theory is based on the thermodynamically consistent approach [29]. In the end, the derivation of all needed terms for the residual based error estimator is given.

Finite Deformation. Due to the large volume changes of aSi, we introduce the total deformation gradient tensor $\mathsf{F} \in \mathbb{R}^{d,d}$ being the derivative of the mapping from the referential Lagrangian domain $\Omega_0 \subset \mathbb{R}^d$ to the current Eulerian domain $\Omega \subset \mathbb{R}^d$ with the dimension $d \in \{1, 2, 3\}$, see Fig. 10.1 (for more information about the mapping, see [30, Sect. 2], [31, Chap. VI] and [27–29]). In our case, the total deformation gradient F is completely reversible and can be multiplicatively split up into an elastic part $\mathsf{F}_{\text{el}}(\bar{c}, \nabla_0 \mathbf{u}) \in \mathbb{R}^{d,d}$, caused by mechanical stress, and a chemical part $\mathsf{F}_{\text{ch}}(\bar{c}) \in \mathbb{R}^{d,d}_{\text{sym}}$, caused by changes in the lithium concentration:

$$\mathsf{Id} + \nabla_0 \mathbf{u} = \mathsf{F}(\nabla_0 \mathbf{u}) = \mathsf{F}_{\text{ch}}(\bar{c})\mathsf{F}_{\text{el}}(\bar{c}, \nabla_0 \mathbf{u}).$$

Both, the normalized concentration $\bar{c}(t, \mathbf{X}_0) = c(t, \mathbf{X}_0)/c_{\max} \in [0, 1]$ as well as the displacement $\mathbf{u}(t, \mathbf{X}_0) \in \mathbb{R}^d$ depend on the position vector $\mathbf{X}_0 \in \Omega_0$ in space and the time $t \in [0, t_{\text{end}}]$. The chemical part of the deformation gradient is defined as $\mathsf{F}_{\text{ch}}(\bar{c}) = \lambda_{\text{ch}}(\bar{c})\mathsf{Id}$ with $\lambda_{\text{ch}}(\bar{c}) = \sqrt[3]{1 + v_{\text{pmv}} c_{\max} \bar{c}}$ and the elastic part $\mathsf{F}_{\text{el}}(\bar{c}, \nabla_0 \mathbf{u}) = \lambda_{\text{ch}}^{-1}(\bar{c})\mathsf{F}(\nabla_0 \mathbf{u})$. All values for the material parameters, such as the partial molar volume v_{pmv}, can be found in Table 10.3 in Sect. B.

Free Energy. Guaranteeing a strictly positive entropy production [29, 32–34] for a thermodynamically consistent model, we base our model approach on a positive free energy $\psi(\bar{c}, \nabla_0 \mathbf{u})$, considering chemical and mechanical effects:

$$\psi(\bar{c}, \nabla_0 \mathbf{u}) = \psi_{\text{ch}}(\bar{c}) + \psi_{\text{el}}(\bar{c}, \nabla_0 \mathbf{u}).$$

The chemical part of the free energy density is formulated using an experimentally obtained open-circuit voltage (OCV) curve U_{OCV} in V [29, 32, 35–37]:

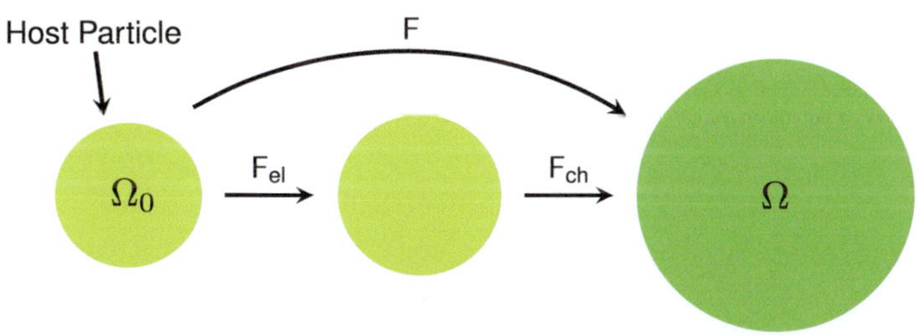

Fig. 10.1 Multiplicative decomposition of the total deformation gradient F into an elastic part F_{el} and a chemical part F_{ch}, compare [27, Fig. 1]

$$\rho\psi_{\mathrm{ch}}(\bar{c}) = -c_{\max}\int_0^{\bar{c}} \mathrm{Fa}\, U_{\mathrm{OCV}}(z)\,\mathrm{d}z \tag{10.1}$$

with the density ρ of aSi and the Faraday constant Fa. We base the model for the elastic part on the linear elastic approach (Saint Venant–Kirchhoff model) as in [30, Sect. 6.5], [31, Sect. VI §3] and [17, 29]:

$$\rho\psi_{\mathrm{el}} = \frac{1}{2}\mathsf{E}_{\mathrm{el}} : \mathbb{C}\left[\mathsf{E}_{\mathrm{el}}\right] \quad \text{and} \quad \mathbb{C}\left[\mathsf{E}_{\mathrm{el}}\right] = \lambda \operatorname{tr}(\mathsf{E}_{\mathrm{el}})\mathsf{Id} + 2G\mathsf{E}_{\mathrm{el}}, \tag{10.2}$$

the constant, isotropic stiffness tensor $\mathbb{C}$ of aSi, first and second Lamé constants $\lambda = 2G\nu/(1-2\nu)$ and $G = E/[2(1+2\nu)]$, Young's modulus E and Poisson's ratio ν. The elastic strain tensor $\mathsf{E}_{\mathrm{el}}(\bar{c}, \boldsymbol{\nabla}_0\mathbf{u}) \in \mathbb{R}^{d,d}_{\mathrm{sym}}$, which is known in literature as the Green–Saint Venant strain tensor or simply *the* Lagrangian strain tensor [38, Sect. 8.1], is defined as:

$$\mathsf{E}_{\mathrm{el}} = \frac{1}{2}\left(\mathsf{F}_{\mathrm{el}}^{\mathsf{T}}\mathsf{F}_{\mathrm{el}} - \mathsf{Id}\right) = \frac{1}{2}\left(\lambda_{\mathrm{ch}}^{-2}\mathsf{F}^{\mathsf{T}}\mathsf{F} - \mathsf{Id}\right) = \frac{1}{2}\left(\lambda_{\mathrm{ch}}^{-2}\mathsf{C} - \mathsf{Id}\right)$$

with the right Cauchy–Green tensor $\mathsf{C} := \mathsf{F}^{\mathsf{T}}\mathsf{F} \in \mathbb{R}^{d,d}_{\mathrm{sym}}$ [30, Sect. 2.5].

Chemistry. The change in lithium concentration inside the host material in the Lagrangian domain Ω_0 can be described by a continuity equation [27–29]:

$$\partial_t \bar{c} = -\boldsymbol{\nabla}_0 \cdot \mathbf{N}(\bar{c}, \boldsymbol{\nabla}_0\mathbf{u}) \qquad \text{in } (0, t_{\mathrm{end}}) \times \Omega_0.$$

The lithium flux $\mathbf{N}$ is given as $\mathbf{N}(\bar{c}, \boldsymbol{\nabla}_0\mathbf{u}) = -m(\bar{c}, \boldsymbol{\nabla}_0\mathbf{u})\boldsymbol{\nabla}_0\mu(\bar{c}, \boldsymbol{\nabla}_0\mathbf{u}) \in \mathbb{R}^d$ with the mobility $m(\bar{c}, \boldsymbol{\nabla}_0\mathbf{u}) := D(\partial_c\mu(\bar{c}, \boldsymbol{\nabla}_0\mathbf{u}))^{-1} > 0$, the constant diffusion $D > 0$ and the chemical potential $\mu(\bar{c}, \boldsymbol{\nabla}_0\mathbf{u}) \in \mathbb{R}$, derived as the derivative of the free energy density with respect to c [29] as

$$\mu(\bar{c}, \boldsymbol{\nabla}_0\mathbf{u}) = \partial_c(\rho\psi)(\bar{c}, \boldsymbol{\nabla}_0\mathbf{u}) = -\mathrm{Fa}\, U_{\mathrm{OCV}}(\bar{c}) + \partial_c\mathsf{E}_{\mathrm{el}}(\bar{c}, \boldsymbol{\nabla}_0\mathbf{u}) : \mathbb{C}[\mathsf{E}_{\mathrm{el}}(\bar{c}, \boldsymbol{\nabla}_0\mathbf{u})], \tag{10.3}$$

where the last term is additionally given as

$$\begin{aligned}\partial_c\mathsf{E}_{\mathrm{el}} : \mathbb{C}[\mathsf{E}_{\mathrm{el}}] &= \frac{1}{2}(-2)\lambda_{\mathrm{ch}}^{-3}\partial_c\lambda_{\mathrm{ch}}\mathsf{C} : \mathbb{C}[\mathsf{E}_{\mathrm{el}}] = -\lambda_{\mathrm{ch}}^{-3}\frac{v_{\mathrm{pmv}}}{3}\lambda_{\mathrm{ch}}^{-2}\mathsf{C} : \mathbb{C}[\mathsf{E}_{\mathrm{el}}] \\ &= -\frac{v_{\mathrm{pmv}}}{3}\lambda_{\mathrm{ch}}^{-5}(\mathsf{F}^{\mathsf{T}}\mathsf{F}) : \mathbb{C}[\mathsf{E}_{\mathrm{el}}] = -\frac{v_{\mathrm{pmv}}}{3\lambda_{\mathrm{ch}}^{3}}(\lambda_{\mathrm{ch}}^{-2}\mathsf{F}\mathbb{C}[\mathsf{E}_{\mathrm{el}}]) : \mathsf{F}\end{aligned}$$

due to symmetry of $\mathbb{C}[\mathsf{E}_{\mathrm{el}}]$ and calculation rules for $\square : \square = \operatorname{tr}(\square^{\mathsf{T}}\square)$. The representative host particle is cycled with a uniform and constant external flux N_{ext} with either positive or negative sign. This external flux is applied at the boundary of Ω_0 and measured in terms of the charging-rate (C-rate), for which we refer to [17, 39]. With this definition, the simulation time t and the state of charge (SOC) regarding the maximal capacity [40] can be related by

$$\mathrm{SOC} = \frac{c_0}{c_{\max}} + N_{\mathrm{ext}}\,[\mathrm{C}] \cdot t[\mathrm{h}].$$

Mechanics. The deformation of the active material is characterized by a static balance of linear momentum [27–29], which yields the equation:

$$\mathbf{0} = -\nabla_0 \cdot \mathsf{P}(\bar{c}, \nabla_0 \mathbf{u}) \qquad \text{in } (0, t_{\mathrm{end}}) \times \Omega_0$$

with the first Piola–Kirchhoff stress tensor thermodynamical consistent derived as $\mathsf{P} = \partial_{\mathsf{F}}(\rho\psi) = 2\mathsf{F}\partial_{\mathsf{C}}(\rho\psi) = \lambda_{\mathrm{ch}}^{-2}\mathsf{F}\mathbb{C}\left[\mathsf{E}_{\mathrm{el}}\right] \in \mathbb{R}^{d,d}$ and the Cauchy stress $\boldsymbol{\sigma} \in \mathbb{R}^{d,d}_{\mathrm{sym}}$ in the Eulerian configuration Ω coupled via $\boldsymbol{\sigma} = \mathsf{P}\mathsf{F}^{\mathsf{T}} / \det(\mathsf{F})$, compare [30, Sects. 3.1 and 6.1].

Derivation of the strong formulation. For the implementation of the residual based error estimator, both the divergence of the lithium flux $\mathbf{N}$ and the divergence of the first Piola–Kirchhoff tensor P are needed in explicit form. These can be obtained via calculation rules from [30, Sect. 1.8] for the derivatives. In a first step, $\nabla_0 \cdot \mathbf{N}(\bar{c}, \nabla_0 \mathbf{u})$ is derived. The derivation of the explicit form of $\nabla_0 \cdot \mathsf{P}(\bar{c}, \nabla_0 \mathbf{u})$ follows subsequently.

Computation of $\nabla_0 \cdot \mathbf{N}$. With the definition of the lithium flux $\mathbf{N}(\bar{c}, \nabla_0 \mathbf{u})$, it follows

$$\begin{aligned}
\nabla_0 \cdot \mathbf{N} &= \nabla_0 \cdot (-m(\bar{c}, \nabla_0 \mathbf{u}) \nabla_0 \mu(\bar{c}, \nabla_0 \mathbf{u})) \\
&= -\nabla_0 m(\bar{c}, \nabla_0 \mathbf{u}) \cdot \nabla_0 \mu(\bar{c}, \nabla_0 \mathbf{u}) - m(\bar{c}, \nabla_0 \mathbf{u}) \nabla_0 \cdot \nabla_0 \mu(\bar{c}, \nabla_0 \mathbf{u})) \\
&= -\nabla_0 m(\bar{c}, \nabla_0 \mathbf{u}) \cdot \nabla_0 \mu(\bar{c}, \nabla_0 \mathbf{u}) - m(\bar{c}, \nabla_0 \mathbf{u}) \Delta_0 \mu(\bar{c}, \nabla_0 \mathbf{u})) \\
&= -\big(\partial_c m(\bar{c}, \nabla_0 \mathbf{u}) \nabla_0 c + \nabla_0^2 \mathbf{u}[\partial_{\nabla_0 \mathbf{u}} m(\bar{c}, \nabla_0 \mathbf{u})]\big) \cdot \nabla_0 \mu(\bar{c}, \nabla_0 \mathbf{u}) \\
&\quad - m(\bar{c}, \nabla_0 \mathbf{u}) \Delta_0 \mu(\bar{c}, \nabla_0 \mathbf{u})).
\end{aligned} \tag{10.4}$$

Equation 10.4 results from an application of the chain rule to the mobility with the Hessian $\nabla_0^2 \mathbf{u}(t, \mathbf{X}_0) \in \mathbb{R}^{d,d,d}$ of the displacement $\mathbf{u}$, and an index reduction of the last two indices in $\nabla_0^2 \mathbf{u}\left[\partial_{\nabla_0 \mathbf{u}} m(\bar{c}, \nabla_0 \mathbf{u})\right] \in \mathbb{R}^d$. The first term in Eq. 10.4 can be formulated by using the definition of the mobility $m(\bar{c}, \nabla_0 \mathbf{u}) = D(\partial_c \mu(\bar{c}, \nabla_0 \mathbf{u}))^{-1}$ and another application of the chain rule as

$$\partial_c m(\bar{c}, \nabla_0 \mathbf{u}) = D\partial_c(\partial_c \mu(\bar{c}, \nabla_0 \mathbf{u})^{-1}) = -D(\partial_c \mu(\bar{c}, \nabla_0 \mathbf{u}))^{-2} \partial_c^2 \mu(\bar{c}, \nabla_0 \mathbf{u}).$$

Using the definition for the chemical potential $\mu(\bar{c}, \nabla_0 \mathbf{u}) = \mu_{\mathrm{ch}}(\bar{c}) + \mu_{\mathrm{el}}(\bar{c}, \nabla_0 \mathbf{u})$, compare Eqs. 10.3, 10.1 and 10.2, we have

$$\begin{aligned}
\partial_c \mu(\bar{c}, \nabla_0 \mathbf{u}) &= \partial_c \mu_{\mathrm{ch}}(\bar{c}) + \partial_c \mu_{\mathrm{el}}(\bar{c}, \nabla_0 \mathbf{u}) = \partial_c^2 (\rho\psi)_{\mathrm{ch}}(\bar{c}) + \partial_c^2 (\rho\psi)_{\mathrm{el}}(\bar{c}, \nabla_0 \mathbf{u}) \\
&= -\frac{1}{c_{\max}} \mathrm{Fa}\, U'_{\mathrm{OCV}} + \partial_c^2 \mathsf{E}_{\mathrm{el}} : \mathbb{C}[\mathsf{E}_{\mathrm{el}}] + \partial_c \mathsf{E}_{\mathrm{el}} : \mathbb{C}[\partial_c \mathsf{E}_{\mathrm{el}}]
\end{aligned}$$

and

$$\partial_c^2 \mu(\bar{c}, \nabla_0 \mathbf{u}) = \partial_c^2 \mu_{\text{ch}}(\bar{c}) + \partial_c^2 \mu_{\text{el}}(\bar{c}, \nabla_0 \mathbf{u}) = \partial_c^3 (\rho\psi)_{\text{ch}}(\bar{c}) + \partial_c^3 (\rho\psi)_{\text{el}}(\bar{c}, \nabla_0 \mathbf{u})$$
$$= -\frac{1}{c_{\max}^2} \text{Fa}\, U_{\text{OCV}}'' + \partial_c^3 \mathsf{E}_{\text{el}} : \mathbb{C}[\mathsf{E}_{\text{el}}] + 3\partial_c^2 \mathsf{E}_{\text{el}} : \mathbb{C}[\partial_c \mathsf{E}_{\text{el}}].$$

Furthermore, we need

$$\partial_{\nabla_0 \mathbf{u}} m(\bar{c}, \nabla_0 \mathbf{u}) = D \partial_{\nabla_0 \mathbf{u}} (\partial_c \mu(\bar{c}, \nabla_0 \mathbf{u}))^{-1}$$
$$= -D(\partial_c \mu(\bar{c}, \nabla_0 \mathbf{u}))^{-2} \partial_{\nabla_0 \mathbf{u}} \partial_c \mu(\bar{c}, \nabla_0 \mathbf{u})$$

with the last term

$$\partial_{\nabla_0 \mathbf{u}} \partial_c \mu(\bar{c}, \nabla_0 \mathbf{u}) = \partial_{\nabla_0 \mathbf{u}} \partial_c \big(\mu_{\text{ch}}(\bar{c}) + \mu_{\text{el}}(\bar{c}, \nabla_0 \mathbf{u})\big) = \partial_{\nabla_0 \mathbf{u}} \partial_c \mu_{\text{el}}(\bar{c}, \nabla_0 \mathbf{u})$$
$$= \partial_{\nabla_0 \mathbf{u}} \big(\partial_c^2 \mathsf{E}_{\text{el}} : \mathbb{C}[\mathsf{E}_{\text{el}}] + \partial_c \mathsf{E}_{\text{el}} : \mathbb{C}[\partial_c \mathsf{E}_{\text{el}}]\big)$$
$$= \partial_{\nabla_0 \mathbf{u}} \partial_c^2 \mathsf{E}_{\text{el}}[\mathbb{C}[\mathsf{E}_{\text{el}}]] + \partial_{\nabla_0 \mathbf{u}} (\mathbb{C}[\mathsf{E}_{\text{el}}])[\partial_c^2 \mathsf{E}_{\text{el}}] \tag{10.5}$$
$$+ 2\partial_{\nabla_0 \mathbf{u}} \partial_c \mathsf{E}_{\text{el}}[\mathbb{C}[\partial_c \mathsf{E}_{\text{el}}]]. \tag{10.6}$$

The second term in Eq. 10.5 can be computed with a symmetric tensor $\mathsf{S} \in \mathbb{R}^{d,d}_{\text{sym}}$ as

$$\partial_{\nabla_0 \mathbf{u}} (\mathbb{C}[\mathsf{E}_{\text{el}}])[\mathsf{S}] = 2G \partial_{\nabla_0 \mathbf{u}} \mathsf{E}_{\text{el}}[\mathsf{S}] + \lambda \mathsf{Id} \partial_{\nabla_0 \mathbf{u}} \mathsf{E}_{\text{el}}[\mathsf{S}]$$
$$= (2G + \lambda) \partial_{\nabla_0 \mathbf{u}} \mathsf{E}_{\text{el}}[\mathsf{S}]$$
$$= (2G + \lambda) \lambda_{\text{ch}}^{-2} \mathsf{F} \mathsf{S}$$

and the term in Eq. 10.6 as

$$\partial_{\nabla_0 \mathbf{u}} (\partial_c \mathsf{E}_{\text{el}} : \mathbb{C}[\partial_c \mathsf{E}_{\text{el}}]) = 2\partial_{\nabla_0 \mathbf{u}} \partial_c \mathsf{E}_{\text{el}}[\mathbb{C}[\partial_c \mathsf{E}_{\text{el}}]]$$
$$= 2((-2)\frac{v_{\text{pmv}}}{3} \lambda_{\text{ch}}^{-5} \mathsf{F})\, (2G \partial_c \mathsf{E}_{\text{el}} + \lambda \operatorname{tr}(\partial_c \mathsf{E}_{\text{el}}) \mathsf{Id}).$$

Computation of $\nabla_0 \cdot \mathsf{P}$. The divergence of the first Piola–Kirchhoff tensor $\mathsf{P}(\bar{c}, \nabla_0 \mathbf{u})$ is given as

$$\nabla_0 \cdot \mathsf{P}(\bar{c}, \nabla_0 \mathbf{u}) = \nabla_0 \cdot \big(\lambda_{\text{ch}}^{-2}(\bar{c}) \mathsf{F}(\nabla_0 \mathbf{u}) \mathbb{C}[\mathsf{E}_{\text{el}}(\bar{c}, \nabla_0 \mathbf{u})]\big)$$
$$= \mathsf{F}\mathbb{C}[\mathsf{E}_{\text{el}}] \nabla_0 \lambda_{\text{ch}}^{-2} + \lambda_{\text{ch}}^{-2} \nabla_0 \cdot (\mathsf{F}\mathbb{C}[\mathsf{E}_{\text{el}}])$$
$$= (-2)\frac{v_{\text{pmv}}}{3} \lambda_{\text{ch}}^{-5} \mathsf{F}\mathbb{C}[\mathsf{E}_{\text{el}}] \nabla_0 c + \lambda_{\text{ch}}^{-2} \nabla_0 \cdot (\mathsf{F}\mathbb{C}[\mathsf{E}_{\text{el}}])$$
$$= \partial_{\nabla_0 \mathbf{u}} \partial_c \mathsf{E}_{\text{el}}[\mathbb{C}[\mathsf{E}_{\text{el}}]] \nabla_0 c + \lambda_{\text{ch}}^{-2} \nabla_0 \cdot (\mathsf{F}\mathbb{C}[\mathsf{E}_{\text{el}}]).$$

The final term computes to

$$\nabla_0 \cdot \big(\mathsf{F}(\nabla_0\mathbf{u})\mathbb{C}[\mathsf{E}_{\text{el}}(c, \nabla_0\mathbf{u})]\big) = \nabla_0\mathsf{F}(\nabla_0\mathbf{u})[\mathbb{C}[\mathsf{E}_{\text{el}}(c, \nabla_0\mathbf{u})]] \\ + \mathsf{F}(\nabla_0\mathbf{u})\nabla_0 \cdot (\mathbb{C}[\mathsf{E}_{\text{el}}(c, \nabla_0\mathbf{u})])$$

with $\nabla_0\mathsf{F}(\nabla_0\mathbf{u})[\mathsf{S}] = \nabla_0^2\mathbf{u}[\mathsf{S}]$ with a symmetric tensor $\mathsf{S} \in \mathbb{R}^{d,d}_{\text{sym}}$. The last term is turned into an explicit form using $\nabla_0 \cdot \mathsf{A} = \operatorname{tr}(\nabla_0\mathsf{A}) = \nabla_0\mathsf{A}[\mathsf{Id}]$:

$$\begin{aligned}
\nabla_0 \cdot (\mathbb{C}[\mathsf{E}_{\text{el}}]) &= 2G\nabla_0 \cdot \mathsf{E}_{\text{el}}(\bar{c}, \nabla_0\mathbf{u}) + \lambda\nabla_0 \cdot \big(\operatorname{tr}(\mathsf{E}_{\text{el}}(\bar{c}, \nabla_0\mathbf{u}))\mathsf{Id}\big) \\
&= 2G\nabla_0 \cdot \Big(\frac{1}{2}(\lambda_{\text{ch}}^{-2}(\bar{c})\mathsf{C}(\nabla_0\mathbf{u}) - \mathsf{Id})\Big) \\
&\quad + \lambda\nabla_0 \cdot \Big(\operatorname{tr}(\frac{1}{2}(\lambda_{\text{ch}}^{-2}(\bar{c})\mathsf{C}(\nabla_0\mathbf{u}) - \mathsf{Id}))\mathsf{Id}\Big) \\
&= 2G\frac{1}{2}\big(\mathsf{C}\nabla_0\lambda_{\text{ch}}^{-2} + \lambda_{\text{ch}}^{-2}\nabla_0 \cdot \mathsf{C}\big) + \lambda\frac{1}{2}\nabla_0 \cdot \big(\lambda_{\text{ch}}^{-2}\operatorname{tr}(\mathsf{C})\mathsf{Id}\big) \\
&= G\big(\mathsf{C}(-2)\lambda_{\text{ch}}^{-3}\partial_c\lambda_{\text{ch}}\nabla_0 c + \lambda_{\text{ch}}^{-2}\nabla_0\mathsf{C}[\mathsf{Id}]\big) \\
&\quad + \lambda\frac{1}{2}\mathsf{Id}\nabla_0\big(\lambda_{\text{ch}}^{-2}\operatorname{tr}(\mathsf{C})\big) + \mathbf{0} \\
&= G\big(\mathsf{C}(-2)\lambda_{\text{ch}}^{-3}\frac{v_{\text{pmv}}}{3}\lambda_{\text{ch}}^{-2}\nabla_0 c + \lambda_{\text{ch}}^{-2}\nabla_0^2\mathbf{u}[\partial_{\nabla\mathbf{u}}\mathsf{F}[\partial_{\mathsf{F}}\mathsf{C}[\mathsf{Id}]]]\big) \\
&\quad + \lambda\frac{1}{2}\big(\operatorname{tr}(\mathsf{C})\nabla_0\lambda_{\text{ch}}^{-2} + \lambda_{\text{ch}}^{-2}\nabla_0\operatorname{tr}(\mathsf{C})\big) \\
&= G\big(2\partial_c\mathsf{E}_{\text{el}}\nabla_0 c + \lambda_{\text{ch}}^{-2}\nabla_0^2\mathbf{u}[2\mathsf{F}]\big) \\
&\quad + \frac{1}{2}\lambda\big(\operatorname{tr}(\mathsf{C})(-2)\lambda_{\text{ch}}^{-3}\frac{v_{\text{pmv}}}{3}\lambda_{\text{ch}}^{-2}\nabla_0 c + \lambda_{\text{ch}}^{-2}\nabla_0^2\mathbf{u}[\partial_{\nabla\mathbf{u}}\mathsf{F}[\partial_{\mathsf{F}}\mathsf{C}[\mathsf{Id}]]]\big) \\
&= 2G\big(\partial_c\mathsf{E}_{\text{el}}\nabla_0 c + \lambda_{\text{ch}}^{-2}\nabla_0^2\mathbf{u}[\mathsf{F}]\big) + \lambda\big(\operatorname{tr}(\partial_c\mathsf{E}_{\text{el}})\nabla_0 c + \lambda_{\text{ch}}^{-2}\nabla_0^2\mathbf{u}[\mathsf{F}]\big) \\
&= \mathbb{C}[\partial_c\mathsf{E}_{\text{el}}]\nabla_0 c + (2G + \lambda)\lambda_{\text{ch}}^{-2}\nabla_0^2\mathbf{u}[\mathsf{F}].
\end{aligned}$$

Now, we have all terms to apply the residual based error estimator.

10.3 Numerical Solution Procedure

Now, all important aspects for the numerical treatment and the three adaptive refinement possibilities, being numerically compared, are shortly introduced.

Problem Formulation. Firstly, we refer to Table 10.2 for the normalization of our used model parameters. From now on, we consider only the dimensionless variables without any accentuation. Secondly, we obtain our problem formulation of the dimensionless initial boundary value problem from Sect. 10.2: let $t_{\text{end}} > 0$ the final simulation time and Ω_0 a bounded electrode particle as reference configuration with dimension $d = 3$. Find the normalized concentration $c\colon [0, t_{\text{end}}] \times \overline{\Omega_0} \to [0, 1]$, the chemical potential $\mu\colon [0, t_{\text{end}}] \times \overline{\Omega_0} \to \mathbb{R}$ and the displacement $\mathbf{u}\colon [0, t_{\text{end}}] \times \overline{\Omega_0} \to \mathbb{R}^d$ satisfying

$$\begin{cases} \partial_t c = -\nabla_0 \cdot \mathbf{N}(c, \nabla_0 \mathbf{u}) & \text{in } (0, t_{\text{end}}) \times \Omega_0, & (10.7a)\\ \mu(c, \nabla_0 \mathbf{u}) = \partial_c(\rho\psi)(c, \nabla_0 \mathbf{u}) & \text{in } (0, t_{\text{end}}) \times \Omega_0, & (10.7b)\\ \mathbf{0} = -\nabla_0 \cdot \mathsf{P}(c, \nabla_0 \mathbf{u}) & \text{in } (0, t_{\text{end}}) \times \Omega_0, & (10.7c)\\ \mathbf{N}(c, \nabla_0 \mathbf{u}) \cdot \mathbf{n}_0 = N_{\text{ext}} & \text{on } (0, t_{\text{end}}) \times \partial\Omega_0, & (10.7d)\\ \mathsf{P}(c, \nabla_0 \mathbf{u}) \cdot \mathbf{n}_0 = \mathbf{0} & \text{on } (0, t_{\text{end}}) \times \partial\Omega_0, & (10.7e)\\ c(0, \cdot) = c_0 & \text{in } \Omega_0 & (10.7f) \end{cases}$$

with an initial condition c_0 being consistent with the boundary conditions. Our quantities of interest are the concentration c and the Cauchy stress σ, computed in a postprocessing stress with the help of the three solution variables. In the case of lithiation of the host particle, the sign of the external lithium flux N_{ext} is positive. Contrarily, in the case of delithiation, the sign of the external lithium flux is negative. Appropriate boundary conditions for the displacement are used to exclude rigid body motions. The original formulation of the chemical deformation gradient F_{ch} is written for the three-dimensional case, however, it is applicable and mathematically valid also for all variables and equations in lower dimensions as well.

Numerical Solution Procedure. We apply the finite element method on the model equations of Eqs. 10.7a–10.7f for the spatial discretization similar to [17, 28]. Furthermore, we use the family of numerical differentiation formulas (NDFs) in a variable-step, variable-order algorithm [41–44], because the properties of the resulting differential algebraic equation (DAE) are similar to stiff ordinary differential equations [17]. The change in time step sizes τ_n and order k_n is handled by an error control, see Algorithm 1 in [17]. The adaptivity of Algorithm 1 is then achieved with a mixed error control using thresholds RelTol_t, AbsTol_t, RelTol_x and AbsTol_x, compare [17] and [44, Sect. 1.4]. Finally, the resulting nonlinear system is linearized with the Newton–Raphson method and is solved for the updates with a direct LU-decomposition. In this work, we compare three error estimators for the local criterion of the adaptive mesh refinement and coarsening: the *Kelly-*, the *gradient recovery-* and the *residual based* error estimator, as described in the introduction. For later use, we state the definition of the residual:

$$\mathbf{R}_h^{n+1} = \begin{pmatrix} \frac{1}{\hat{\tau}_n}\mathsf{M}(\mathbf{y}^{n+1} - \Phi^n) + \nabla_0 \cdot \mathbf{N}_h^{n+1} \\ \mu_h^{n+1} - \partial_c(\rho\psi)_h^{n+1} \\ \nabla_0 \cdot \mathsf{P}_h^{n+1} \end{pmatrix}$$

with the mass matrix M of the finite element space of dimension N regarding the concentration c, the discrete solution vector $\mathbf{y}\colon [0, t_{\text{end}}] \to \mathbb{R}^{(2+d)N}$ with $\mathbf{y}^{n+1} \approx \mathbf{y}(t_{n+1})$, containing all three discrete components c_h, μ_h and $\mathbf{u}_h$. $\hat{\tau}_n$ is a modified time step size with coefficients of the selected order k_n at time t_n and Φ^n consists of the solutions of some former time steps $\mathbf{y}^n, \ldots, \mathbf{y}^{n-k_n}$ [42, Sect. 2.3].

In total, the residual based error estimation est_x can be split up into a cell fraction and a face fraction, whereby the latter one includes also the boundary parts. For the first one, we sum up the L^2-norm of the residual over all cells K of our triangulation, weighted with the largest diagonal of the cell h_K in Eq. 10.8. For the latter one, we add up all jumps $[\![\cdot]\!]$ of the function in square brackets at the inner faces $\mathcal{F}^{\circ}$ of all cells in Eq. 10.9 and additionally as well as the boundary conditions for all faces $\mathcal{F}^{\partial\Omega_0}$ at all boundary cells in Eq. 10.10. Both face parts are weighted with $h_K/24$ as proposed in [21, Sect. 4.2]. In summary, we have the overall error estimator with constants $\gamma_1 > 0$ and $\gamma_2 > 0$ on one cell K computed as:

$$\mathrm{est}_K^2 := \gamma_1 \eta_{\mathrm{cell,K}}^2 + \gamma_2 \eta_{\mathrm{face,K}}^2 \tag{10.8}$$

$$= \gamma_1 h_K^2 \|\mathbf{R}_h^{n+1}\|_{L^2(K)}^2$$

$$+ \gamma_2 \frac{h_K}{24} \sum_{F \in \mathcal{F}^{\circ} \cap \partial K} \left(\|[\![\mathbf{N}_h^{n+1} \cdot \mathbf{n}_0]\!]\|_{L^2(F)}^2 + \|[\![\mathsf{P}_h^{n+1} \cdot \mathbf{n}_0]\!]\|_{L^2(F)}^2 \right) \tag{10.9}$$

$$+ \gamma_2 \frac{h_K}{24} \sum_{F \in \mathcal{F}^{\partial\Omega_0} \cap \partial K} \left(\|\mathbf{N}_h^{n+1} \cdot \mathbf{n}_0 - N_{\mathrm{ext}}\|_{L^2(F)}^2 + \|\mathsf{P}_h^{n+1} \cdot \mathbf{n}_0 - \mathbf{0}\|_{L^2(F)}^2 \right). \tag{10.10}$$

The total error est_x is then given as $\mathrm{est}_x^2 = \sum_K \mathrm{est}_K^2$. Keep in mind that we have to add further boundary conditions for some additional artificial boundaries for the specified computational domain as explained in Sect. 10.4.1.

10.4 Numerical Studies

In this section, we specify our simulation setup, present and discuss our simulation results of the comparison of the three spatial refinement strategies for the presented model of Sect. 10.2.

10.4.1 Simulation Setup

We choose aSi as host material for our derived theory with material parameters and dimensionless values given in Table 10.3. The used experimental OCV-curve is taken from [29, Equation (SI-51)] and displayed in [27, Fig. 2]. For lithiation, we apply an external constant lithium flux $N_{\mathrm{ext}} = 1\mathrm{C}$ and for delithiation, the value $N_{\mathrm{ext}} = -1\mathrm{C}$ is given. Following [28, 29, 45], we start with a constant concentration $c_0 = 0.02$ and determine one cycle period of lithiation or delithiation with a duration 0.90 h, so we are in the range of SOC $\in [0.02, 0.92]$ and time $t \in [0.0\,\mathrm{h}, 0.90\,\mathrm{h}]$ for lithiation, followed by a delithiation with the same duration of 0.90 h and decreasing SOC. After the discharging, another charging cycle is added and so on. For a total simulation time of $t = 2.7$ h, we have following cycles: charging–discharging–charging.

Geometrical Setup. Similar to [45], we choose a representative 3D spherical particle with domain Ω_0, reduced to a 1D computational domain Ω_{com} as shown in Fig. 10.2 in terms of the radial variable r, and a quarter ellipsoid as computational domain Ω_{com} resulting from a 3D nanowire domain Ω with no changes in z-direction and symmetry around the x- and y-axis as in Fig. 10.3. For the adaption of the quadrature weight and the initial conditions, we refer to the literature cited.

However, we have to mention the additional artificial inner boundary conditions, indicated with the subscript $\square_{\text{in}}$. For the 1D computational domain, we have

$$\mathbf{N} \cdot \mathbf{n}_0 = 0, \quad u = 0, \quad \text{on } (0, t_{\text{end}} \times \Gamma_{\text{in}})$$

and for the 2D computational domain

$$\begin{aligned} \mathbf{N} \cdot \mathbf{n}_0 &= 0, \quad u_y = 0, \quad \text{on } (0, t_{\text{end}} \times \Gamma_{\text{in},x}), \\ \mathbf{N} \cdot \mathbf{n}_0 &= 0, \quad u_x = 0, \quad \text{on } (0, t_{\text{end}} \times \Gamma_{\text{in},y}). \end{aligned}$$

These new terms also appear in the computation of the boundary face terms of the residual base error estimator of Eq. 10.10. For the 1D computational domain, we have to add

$$\gamma_2 \frac{h_K}{24} \sum_{F \in \mathcal{F}^{\Gamma_{\text{in}}} \cap K} \left(\|\mathbf{N}_h^{n+1} \cdot \mathbf{n}_0 - 0\|^2_{L^2(F)} + \|u^{n+1} - 0\|^2_{L^2(F)} \right)$$

to the already mentioned face terms and for the 2D computational domain

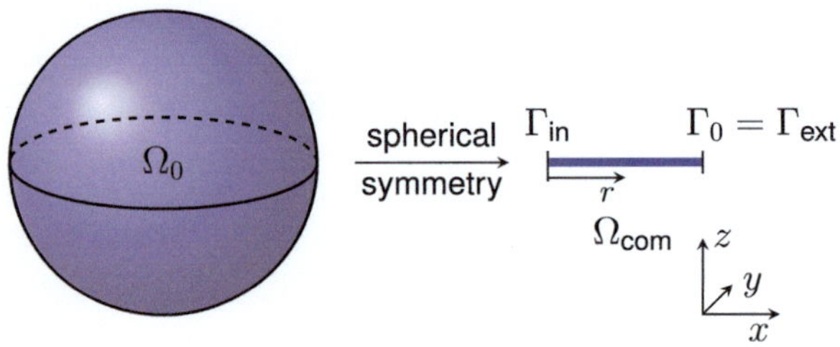

Fig. 10.2 Dimension reduction of a 3D unit sphere to the 1D unit interval with spherical symmetry, based on [46, Fig. B.1]

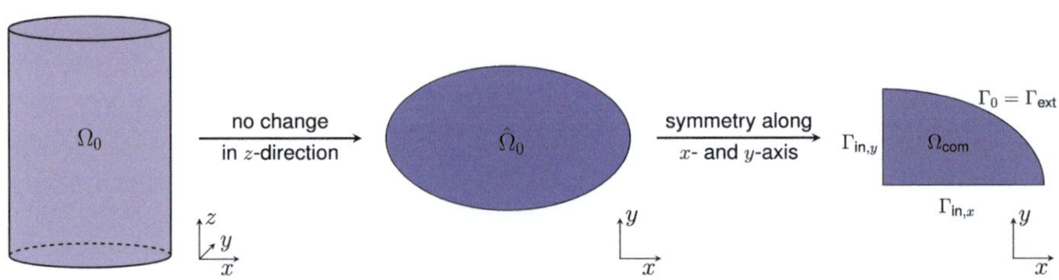

Fig. 10.3 Dimension reduction of a 3D elliptical nanowire to the 2D quarter ellipsoid, based on [28, Fig. 5]

$$\gamma_2 \frac{h_K}{24} \sum_{F \in \mathcal{F}^{\Gamma_{\text{in},x}} \cap K} \left(\|\mathbf{N}_h^{n+1} \cdot \mathbf{n}_0 - 0\|^2_{L^2(F)} + \|u_y^{n+1} - 0\|^2_{L^2(F)} \right)$$
$$+ \gamma_2 \frac{h_K}{24} \sum_{F \in \mathcal{F}^{\Gamma_{\text{in},y}} \cap K} \left(\|\mathbf{N}_h^{n+1} \cdot \mathbf{n}_0 - 0\|^2_{L^2(F)} + \|u_x^{n+1} - 0\|^2_{L^2(F)} \right).$$

Implementation Details. We use an isoparametric fourth-order Lagrangian finite element method with isoparametric representation of the curved boundary for our numerical simulations. The `C++` finite element library `deal.II` [20] is taken as basis as well as the interface to the Trilinos library [47, Version 12.8.1] and the UMFPACK package [48, Version 5.7.8] for the LU-decomposition. All simulations for the 1D computational domain are performed on a desktop computer with 64 GB RAM, Intel i5-9500 CPU, GCC compiler version 10.5 and operating system Ubuntu 20.04.6 LTS, whereas the simulations for the 2D computational domain are executed with GCC compiler version 12.1 with a single node of the BwUniCluster 2.0 with 40 Intel Xeon Gold 6230 with 2.1 GHz and 96 GB RAM, on [49]. The highly resolved solution, which is used to compute the error in Table 10.1, is also computed on BwUniCluster 2.0. Unless otherwise stated, we set for the space and time adaptive algorithm following parameters: tolerances $\text{RelTol}_t = \text{RelTol}_x = 1 \times 10^{-5}$, $\text{AbsTol}_t = \text{AbsTol}_x = 1 \times 10^{-8}$, initial time step size $\tau_0 = 1 \times 10^{-6}$, final simulation time $t_{\text{end}} = 2.7$, maximal time step size $\tau_{\max} = 0.1$, number of initial refinements 7, minimal refinement level 3 for the 1D computational domain and level 1 for the 2D computational domain, maximal refinement level of 20 and marking parameters for the local mesh coarsening and refinement, $\theta_c = 0.05$ and $\theta_r = 0.3$. Furthermore, we use automatic differentiation (AD) for the computation of the Newton matrix [45] and OpenMP Version 4.5 for shared memory parallelization for assembling the system matrices, residuals and the Kelly- and gradient recovery estimator.

10.4.2 Numerical Results

Example 1: 1D Spherical Symmetry. In a first step, we compare the three adaptive refinement strategies at four different time steps during our simulation with $t_{\text{end}} = 2.7$ h. Figure 10.4 shows the refinement levels for the Kelly error estimator, the gradient recovery estimator and the residual based error estimator with $\gamma_1 = 1.0$ and $\gamma_2 = 1.0$ over the particle radius at $t = 0.48$ h corresponding to SOC $= 0.50$ in Subfigure (a), at $t = 0.9001$ h corresponding to SOC $= 0.9199$ in Subfigure (b), at $t = 1.8001$ h corresponding to SOC $= 0.0201$ in Subfigure (c) and the final simulation time at $t_{\text{end}} = 2.7$ h corresponding to SOC $= 0.92$ in Subfigure (d). For the simulation with the Kelly error estimator, we have increased the maximal refinement level to 25 and also updated the tolerances to $\text{RelTol}_t = \text{RelTol}_x = 1 \times 10^{-4}$, $\text{AbsTol}_t = \text{AbsTol}_x = 1 \times 10^{-7}$ to get a stable simulation. Is is clearly evident that the Kelly error estimator over-refines the boundary parts, since the estimator has no information about the exact boundary terms. The gradient recovery estimator has only at some range around $r = 0.8$ a higher refinement level in Fig. 10.4a, d, whereas shortly after the sign change of the external lithium flux also a higher refinement level is reached at the particle surface around $r = 1.0$ in the other two subfigures. The residual based error estimator features some stair-like behavior with increasing refinement level from smaller to larger radius values in Fig. 10.4a, d. Shortly after the change of the sign of the lithium flux in Fig. 10.4b, c, there is a drop of the refinement level at the particle surface. This could be explained by the large influence of the cell error fraction on the residual based refinement strategy, compare the two cell and face fractions, which are additionally added in Fig. 10.4d. Due to the change of the lithium flux, the cell fraction error is smaller (since the lithium flux direction close to particle surface around $r = 1.0$ has also changed) and therefore, a smaller refinement level is sufficient. The behavior of the face error of the residual based strategy in Fig. 10.4d leads to a similar refinement as the gradient recovery strategy.

It is known in literature that the edge fraction is the dominating one and the more important one for the error estimation, see for example [50] and [51, Sect. 1.6]. There-

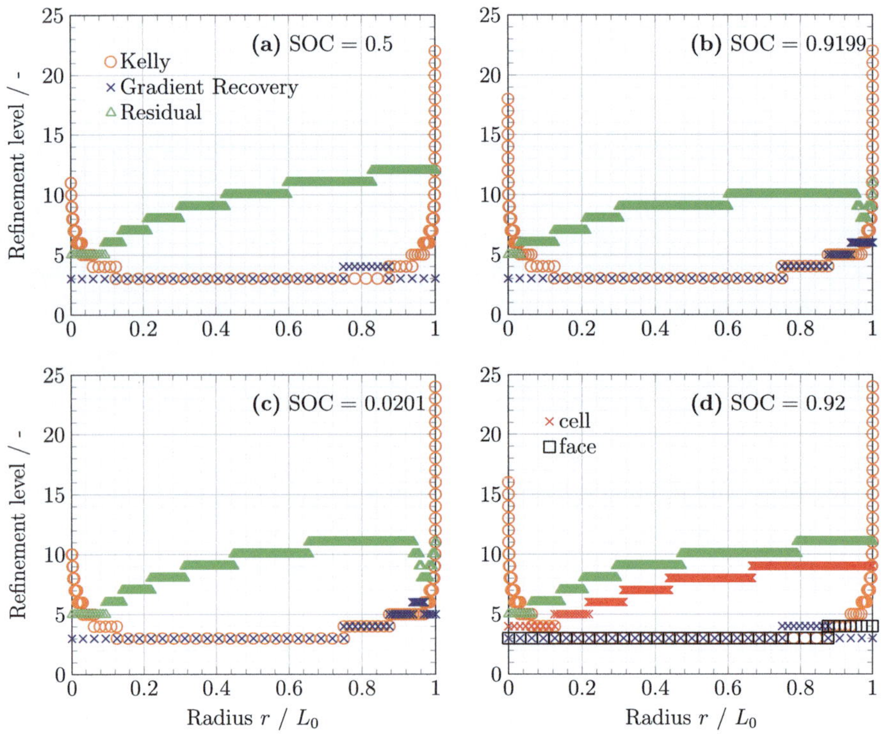

Fig. 10.4 Comparison of three different refinement strategies (Kelly error estimator, gradient recovery error estimator and residual based error estimator with $\gamma_1 = 1.0$ and $\gamma_2 = 1.0$) over the particle radius at different SOC: SOC = 0.5 in (**a**), SOC = 0.9199 in (**b**), SOC = 0.0201 in (**c**) and SOC = 0.92 in (**d**), whereas the last one contains also the two fractions of the cell and the face error of the residual based error estimator

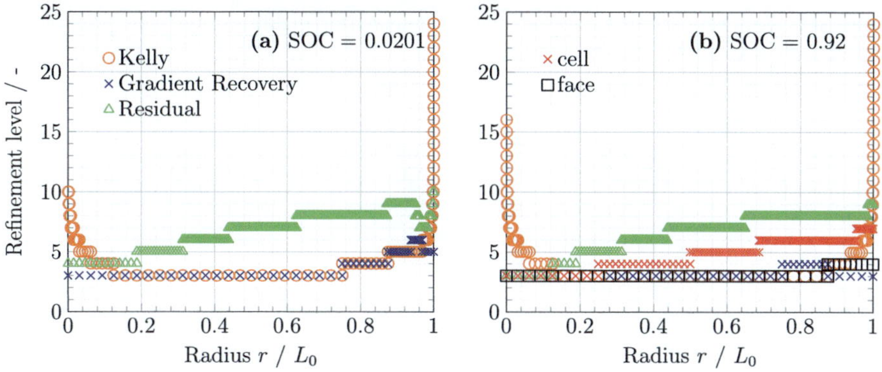

Fig. 10.5 Same situation as in Fig. 10.4c, d, but with updated version of the residual based error estimator with $\gamma_1 = 0.001$ and $\gamma_2 = 1.0$

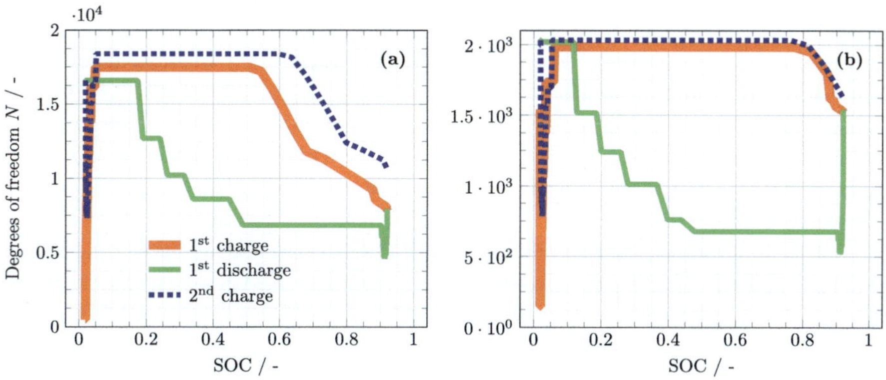

Fig. 10.6 Number of DOFs over SOC for the residual based error estimator with $\gamma_1 = 1.0$ and $\gamma_2 = 1.0$ in (**a**) and updated version with $\gamma_1 = 0.001$ and $\gamma_2 = 1.0$ in (**b**)

fore, we have additionally performed some simulations with an updated version of the residual based error estimator with $\gamma_1 = 0.001$ and $\gamma_2 = 1.0$. Our choice of γ_1 is suitable for reducing the influence of the cell proportion, but requires further efforts, e.g. possible influence of the chosen finite element order or a possible comparison in the order of magnitude to the computed numerical error. However, these considerations will be postponed to future work. The smaller fraction of the cell error clearly reduces the refinement level of the residual based refinement strategy, comparing Figs. 10.5a, b–10.4c, d.

This influence of the cell error fraction is further investigated in Fig. 10.6, where the number of degrees of freedom (DOFs) $(2 + d)N$ is plotted for the residual based error estimator with $\gamma_1 = 1.0$ and $\gamma_2 = 1.0$. in Subfigure (a) and the updated version with $\gamma_1 = 0.001$ and $\gamma_2 = 1.0$ in Subfigure (b). Both plots show some qualitative similarities, which are the steep increase of the number of DOFs after the start of

Table 10.1 Error between the numerical solution compared to a highly resolved numerical solution regarding the respective norm: $\|\mathbf{y} - \mathbf{y}_{\text{highly\_resolved}}\|_{\square}$

Error estimator	DOFs	L^2-Norm	H^1-Norm
Gradient recovery	111	1.20×10^{-6}	3.02×10^{-6}
Residual	15 939	3.08×10^{-7}	1.07×10^{-6}
Residual updated	1575	5.44×10^{-7}	1.65×10^{-6}

the first lithiation followed by some plateau and a decrease in the end of the first charge. The first discharge looks also similar, since the number of DOFs gets smaller but reaches a higher level at the end of the first discharge. For the second charge, there is the same behavior in both plots visible featuring a large drop in the number of DOFs and following then the same way as the first charge. However, there is one difference: the absolute number of DOFs is reduced by a factor of 10 in the updated version. This shows again the significant influence of the cell error fraction.

In Table 10.1, we display the L^2-Norm and H^1-Norm of the numerical solution $\mathbf{y}$ compared with a highly resolved solution $\mathbf{y}_{\text{highly\_resolved}}$ at time $t = 0.2$ h. For the numerical results of the table, we use the exact computation of the Newton matrix and no AD. The high resolved solution is computed with $N = 1.5$ million DOFs and updated $\text{RelTol}_t = 1 \times 10^{-11}$ and $\text{AbsTol}_t = 1 \times 10^{-14}$. The numerical solution $\mathbf{y}$ is computed with $\text{RelTol}_t = 1 \times 10^{-7}$, $\text{AbsTol}_t = 1 \times 10^{-10}$, initial time step size $\tau_0 = 1 \times 10^{-10}$ and initial refinement level with 17. We compare the results of the gradient recovery-, the residual based- and the updated residual based error estimator. On the one hand, it can be seen that the gradient recovery displays the lowest number of DOFs but also the largest errors. On the other hand, residual based methods feature smaller norm errors but have also significant larger number of DOFs. Interestingly, the updated case leads to a factor of ten fewer number of DOFs, whereas the error norms are at a comparable level. From this, we can conclude that it is more efficient to use the updated version of the residual based error estimator due to the smaller number of DOFs. Further, we are accurate enough with the updated version and do not need to solve the numerical simulations for the solutions with the higher number of DOFs.

Figure 10.7 displays the development of the absolute value of the maximal hydrostatic stress σ_{h} over the particle radius during the three half cycles. The hydrostatic Cauchy stress is defined as $\sigma_{\text{h}} = (\sigma_{\text{r}} + 2\sigma_{\phi})/3$ with the stress in radial direction σ_{r} and in tangential direction σ_{ϕ}. The evolution of the stress reveals high stresses at low SOC values and lower stresses at high SOC values. The increase of the first charge at low SOC results from the constant initial condition of the concentration. The small drop of the first discharge at high SOC and the large drop of the second charge at low SOC emerge from the change of stress curvature. For more information about this curvature change, see [28, Fig. 7]. In total, it can be concluded from the SOC ranges and the behavior of the number of DOFs of Figs. 10.6 and 10.7 that for

Fig. 10.7 Absolute value of the maximal hydrostatic Cauchy stress $\sigma_{\text{h, max}}$ over SOC for charging–discharging–charging cycle for the residual based error estimator with updated constants $\gamma_1 = 0.001$ and $\gamma_2 = 1.0$

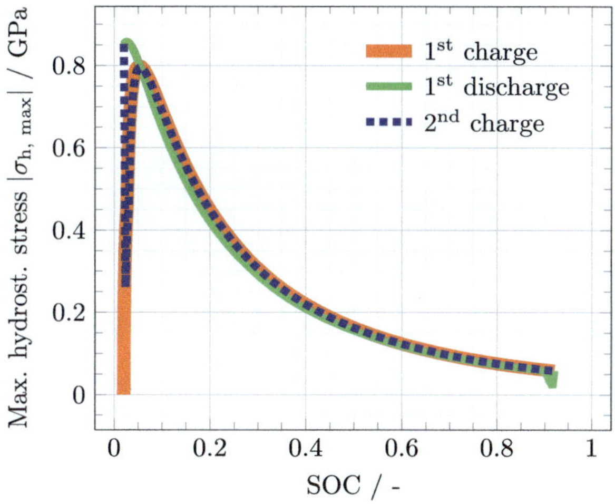

the 1D spherical symmetric case larger maximal Cauchy stress values are related with a larger number of DOFs.

Example 2: 2D Quarter Ellipse. The 2D computational domain is a proof of concept to show the feasibility of the gradient recovery estimator and updated version of the residual based error estimator as adaptive refinement strategies. For the simulations, we have modified some parameters in the following way: the number of initial refinements to 6, initial time step size to 1×10^{-8} and $t_{\text{end}} = 0.9$. In Fig. 10.8, the numerical solution of the concentration c in Subfigure (a) and of the von Mises stress $\sigma_{\text{vM}} = \sqrt{\sigma_{11}^2 + \sigma_{22}^2 - \sigma_{11}\sigma_{22} + 3\sigma_{12}^2}$ in Subfigure (b) are shown with the underlying mesh of the updated version of the residual based error estimator with $\gamma_1 = 0.001$ and $\gamma_2 = 1.0$. Due to the elliptical domain, more lithium gets inside or out of the host particle at the area with the highest surface to volume ratio which is at the particle surface of the larger half axis [52, Sect. 2.2.3]. We point out that the mesh has higher refinements at areas of higher concentration values although the highest stresses are at the particle surface of the smaller half axis. The large von Mises stress σ_{vM} at the smaller half axis occurs due to the forced zero stress in normal direction and the large tangential stresses in this area. This indicates that the concentration c has a larger influence than the von Mises stress σ_{vM} on the spatial mesh strategy in the two-dimensional distribution at one time step.

The numerical solution of the concentration c with the underlying mesh of the gradient recovery refinement strategy is displayed in Fig. 10.9. Here, also a less refined mesh fulfills the tolerances of the adaptive solution algorithm compared to the updated version of the residual based strategy. This is similar to the findings of the 1D spherical symmetric case.

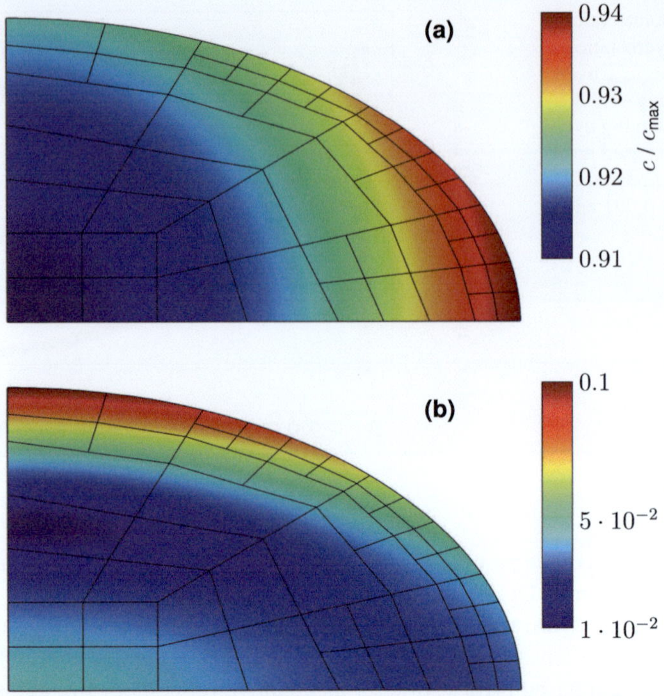

Fig. 10.8 Numerical solution of the concentration c in (a) and von Mises stress σ_{vM} in (b) with the used mesh at the final simulation time $t_{\mathrm{end}} = 0.9$ h for the updated version of the residual based error estimator with $\gamma_1 = 0.001$ and $\gamma_2 = 1.0$

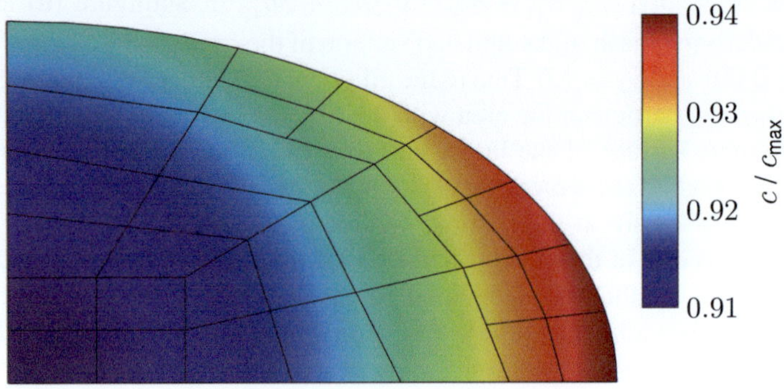

Fig. 10.9 Numerical solution of the concentration c with the used mesh at the final simulation time $t_{\mathrm{end}} = 0.9$ h with the gradient recovery estimator

10.5 Conclusion and Outlook

Conclusion. In this work, three different spatial refinement strategies are investigated which have been applied to a chemo-elastic model for Li-diffusion within aSi particles. The ready to use Kelly error estimator [20] of `deal.II`, the gradient recovery estimator of [17] and a residual based error estimator are compared. For the usage of the residual based error estimator, the strong formulation of the residual is derived explicitly in Sect. 10.2 including many dependencies of our chemo-mechanical coupled model approach. In Sect. 10.4, we considered the numerical results for a 3D spherical domain, reduced to 1D computational domain, and a 3D elliptical nanowire, reduced to 2D computational domain. For the 1D computational domain, we found out that the Kelly error estimator over-estimates the error at the boundary and therefore cannot be recommended for our application case. The gradient recovery estimator features higher refinement levels during the change of the lithium flux and rather lower refinement levels during solely the Li-diffusion periods. The residual based error estimator shows a stair-like behavior from the particle center to the particle surface. However, the cell error part leads to significant larger refinement levels compared to the face error part. Therefore, we have introduced an updated residual based error estimator with $\gamma_1 = 0.001$ (suitably chosen) and $\gamma_2 = 1.0$, which reduces the influence of the cell error part. The updated version leads to a more efficient error estimator, since the number of DOFs is smaller by a factor of 10 and the updated error estimator results in a comparable L^2- and H^1-error compared to the residual case with $\gamma_1 = 1.0$ and $\gamma_2 = 1.0$.

During a charging and discharging process, higher Cauchy stress values are related with a higher number of DOFs, whereas in the 2D computational domain, the distribution of Li-concentration has a larger influence on the mesh distribution compared to the Cauchy stress distribution at one specific time step.

Outlook. Since we pointed out the significant influence of the cell error fraction of the residual based error estimator, further work is required for a closer examination and detailed calibration of the ratio of the cell and face errors of the residual based error estimator or for a potential influence of the finite element order. However, as charging and discharging is a highly dynamic process, a suitable calibration should be carefully checked. Another application of great interest is the consideration of another definition of the elastic strain tensor E_{el} as in [45] leading to an easy usage for further elasto-plastic couplings. In addition, the investigation of other battery active materials like graphite-silicon composites or materials for sodium-ion batteries is possible. Other 2D geometries and 3D geometries could also be taken into account.

CrediT authorship contribution statement

R. Schoof: Conceptualization, Data curation, Formal Analysis, Investigation, Methodology, Software, Validation, Visualization, Writing—original draft **L. Flür:** Formal Analysis, Methodology, Software, Visualization, Writing—review & editing **F. Tuschner:** Data curation, Writing—review & editing **W. Dörfler:** Conceptualization, Funding acquisition, Project administration, Resources, Supervision, Writing—review & editing

Acknowledgements The authors thank G. F. Castelli for the software basis, L. von Kolzenberg and L. Köbbing for intensive and constructive discussions about modeling silicon particles and T. Laufer for proofreading. R. S. and F. T. acknowledge financial support by the German Research Foundation (DFG) through the Research Training Group 2218 SiMET—Simulation of Mechano-Electro-Thermal processes in Lithium-ion Batteries, project number 281041241. The authors acknowledge support by the state of Baden-Württemberg through bwHPC.

Appendices

A Abbreviations and Symbols

Abbreviations

AD	Automatic differentiation
aSi	Amorphous silicon
C-rate	Charging rate
DAE	Differential algebraic equation
DOF	Degree of freedom
NDF	Numerical differentiation formula
OCV	Open-circuit voltage
pmv	Partial molar volume
SOC	State of charge

Symbol	**Description**
Latin symbols	
c	Concentration
C_{el}	Right Cauchy-Green tensor
$\mathbb{C}$	Fourth-order stiffness tensor
E	Young's modulus
E_{el}	Elastic strain tensor
F	Deformation gradient
$\mathsf{F} = \mathsf{F}_{\mathrm{ch}}\mathsf{F}_{\mathrm{el}}$	Multiplicative decomposition of F
F_{ch}	Chemical deformation gradient
F_{el}	Elastic deformation gradient
Fa	Faraday constant
G	Shear modulus/second Lamé constant
Id	Identity tensor
$J = J_{\mathrm{ch}}J_{\mathrm{el}}$	Multiplicative decomposition of volume change
k_n	Order of time integration algorithm at time t_n
m	Scalar valued mobility
$\mathbf{n}, \mathbf{n}_0$	Normal vector on Ω, Ω_0
$\mathbf{N}$	Lithium flux
N_{ext}	External lithium flux
P	First Piola–Kirchhoff stress tensor
t	Time
t_n	Time step
t_{cycle}	Cycle time
U_{OCV}	OCV curve
$\mathbf{u} = \mathbf{x} - \mathbf{X}_0$	Time dependent displacement vector
v_{pmv}	Partial molar volume of lithium
$\mathbf{x}$	Position in Eulerian domain
$\mathbf{X}_0$	Initial placement
Greek symbols	
γ_1	Factor for the cell error of the residual based error estimator
γ_2	Factor for the face error of the residual based error estimator
Γ_{in}	Inner artificial boundary part
λ	First Lamé constant
λ_{ch}	Factor of concentration induced deformation gradient
μ	Chemical potential
ν	Poisson's ratio
Ω	Eulerian domain
Ω_0	Lagrangian domain
$\psi = \psi_{\mathrm{ch}} + \psi_{\mathrm{el}}$	Total energy
ψ_{ch}	Chemical part of total energy
ψ_{el}	Elastic part of total energy
ρ	Density
σ	Cauchy stress tensor
τ_n	Time step size at time t_n
Mathematical symbols	
$\partial\Omega_0$	Boundary of Ω_0
∇_0	Gradient vector in Lagrangian domain
$\Box : \tilde{\Box}$	Reduction of two dimensions of two tensors $\Box$ and $\tilde{\Box}$
$\mathcal{A}[\mathsf{B}]$	Reduction of the last two dimensions of the third order tensor $\mathcal{A}$ and the second order tensor B
$\mathbb{A}[\mathsf{B}]$	Reduction of the last two dimensions of the fourth order tensor $\mathbb{A}$ and the second order tensor B
$\partial_{\Box}$	Partial derivative with respect to $\Box$
Indices	
$\Box_0$	Considering variable in Lagrangian domain or initial condition
$\Box_{\mathrm{ch}}$	Chemical part of $\Box$
$\Box_{\mathrm{com}}$	Computational part of $\Box$
$\Box_{\mathrm{el}}$	Elastic part of $\Box$
$\Box_{\mathrm{max}}$	Maximal part of $\Box$
$\Box_{\mathrm{min}}$	Minimal part of $\Box$

Table 10.2 Dimensionless variables of the used model equations

$\tilde{\mathbf{X}}_0 = \mathbf{X}_0/L_0$	$\tilde{t} = t/t_{\text{cycle}}$	$\tilde{\mathbf{u}} = \mathbf{u}/L_0$	$\tilde{c} = c/c_{\max}$	$\tilde{v}_{\text{pmv}} = v_{\text{pmv}} c_{\max}$
$\widetilde{\rho\psi} = \rho\psi/R_{\text{gas}} T c_{\max}$		$\tilde{\mu} = \mu/R_{\text{gas}} T$	$\tilde{E} = E/R_{\text{gas}} T c_{\max}$	
$\text{Fo} = D t_{\text{cycle}}/L_0^2$	$\tilde{N}_{\text{ext}} = N_{\text{ext}} t_{\text{cycle}}/L_0 c_{\max}$		$\tilde{U}_{\text{OCV}} = \text{Fa}\, U_{\text{OCV}}/R_{\text{gas}} T$	

Table 10.3 Model parameters for numerical experiments [29]

Description	Symbol	Value	Unit	Dimensionless
Universal gas constant	R_{gas}	8.314	$\text{J mol}^{-1}\ \text{K}^{-1}$	1
Faraday constant	Fa	96485	$\text{J V}^{-1}\ \text{mol}^{-1}$	1
Operation temperature	T	298.15	K	1
Silicon				
Particle length scale	L_0	50×10^{-9}	m	1
Diffusion coefficient	D	1×10^{-17}	$\text{m}^2\ \text{s}^{-1}$	14.4
OCV curve	U_{OCV}	[29]	V	[29]
Young's modulus	E	90.13×10^9	Pa	116.74
Partial molar volume	v_{pmv}	10.96×10^{-6}	$\text{m}^3\ \text{mol}^{-1}$	3.41
Maximal concentration	$c_{\max}$	311.47×10^3	mol m^{-3}	1
Initial concentration	c_0	6.23×10^3	mol m^{-3}	2×10^{-2}
Poisson's ratio	ν	0.22	–	0.22

B Simulation Parameters

The dimensionless variables of the considered model equations are given in Table 10.2 and the used model parameters for the numerical simulations are listed in Table 10.3.

References

1. Writer, B.: Lithium-Ion Batteries: A Machine-Generated Summary of Current Research. Springer, Cham (2019). https://doi.org/10.1007/978-3-030-16800-1
2. Armand, M., Axmann, P., Bresser, D., Copley, M., Edström, K., Ekberg, C., Guyomard, D., Lestriez, B., Novák, P., Petranikova, M., Porcher, W., Trabesinger, S., Wohlfahrt-Mehrens, M., Zhang, H.: Lithium-ion batteries - current state of the art and anticipated developments. J. Power Sources **479**, 228708 (2020). https://doi.org/10.1016/j.jpowsour.2020.228708

3. Xu, J., Cai, X., Cai, S., Shao, Y., Hu, C., Lu, S., Ding, S.: High-energy lithium-ion batteries: recent progress and a promising future in applications. Energy Environ. Mater. **6**(5) (2023). https://doi.org/10.1002/eem2.12450
4. Zhan, R., Wang, X., Chen, Z., Seh, Z.W., Wang, L., Sun, Y.: Promises and challenges of the practical implementation of prelithiation in lithium-ion batteries. Adv. Energy Mater. **11**(35) (2021). https://doi.org/10.1002/aenm.202101565
5. Martí-Florences, M., Cecilia, A., Costa-Castelló, R.: Modelling and estimation in lithium-ion batteries: a literature review. Energies **16**(19), 6846 (2023). https://doi.org/10.3390/en16196846
6. Nzereogu, P.U., Omah, A.D., Ezema, F.I., Iwuoha, E.I., Nwanya, A.C.: Anode materials for lithium-ion batteries: a review. Appl. Surf. Sci. Adv. **9**, 100233 (2022). https://doi.org/10.1016/j.apsadv.2022.100233
7. Zhao, Y., Stein, P., Bai, Y., Al-Siraj, M., Yang, Y., Xu, B.-X.: A review on modeling of electro-chemo-mechanics in lithium-ion batteries. J. Power Sources **413**, 259–283 (2019). https://doi.org/10.1016/j.jpowsour.2018.12.011
8. Zuo, X., Zhu, J., Müller-Buschbaum, P., Cheng, Y.: Silicon based lithium-ion battery anodes: a chronicle perspective review. Nano Energy **31**, 113–143 (2017). https://doi.org/10.1016/j.nanoen.2016.11.013
9. Tomaszewska, A., Chu, Z., Feng, X., O'Kane, S., Liu, X., Chen, J., Ji, C., Endler, E., Li, R., Liu, L., Li, Y., Zheng, S., Vetterlein, S., Gao, M., Du, J., Parkes, M., Ouyang, M., Marinescu, M., Offer, G., Wu, B.: Lithium-ion battery fast charging: a review. eTransportation **1**, 100011 (2019). https://doi.org/10.1016/j.etran.2019.100011
10. Li, P., Kim, H., Myung, S.-T., Sun, Y.-K.: Diverting exploration of silicon anode into practical way: a review focused on silicon-graphite composite for lithium ion batteries. Energy Stor. Mater. **35**, 550–576 (2021). https://doi.org/10.1016/j.ensm.2020.11.028
11. Tian, H., Xin, F., Wang, X., He, W., Han, W.: High capacity group-IV elements (Si, Ge, Sn) based anodes for lithium-ion batteries. J. Materiomics **1**(3), 153–169 (2015). https://doi.org/10.1016/j.jmat.2015.06.002
12. Zhang, W.-J.: A review of the electrochemical performance of alloy anodes for lithium-ion batteries. J. Power Sources **196**(1), 13–24 (2011). https://doi.org/10.1016/j.jpowsour.2010.07.020
13. Miranda, D., Costa, C.M., Almeida, A.M., Lanceros-Méndez, S.S.: Computer simulations of the influence of geometry in the performance of conventional and unconventional lithium-ion batteries. Appl. Energy **165**, 318–328 (2016). https://doi.org/10.1016/j.apenergy.2015.12.068
14. Xu, R., Zhao, K.: Electrochemomechanics of electrodes in Li-Ion batteries: a review. J. Electrochem. Energy Convers. Storage **13**(3), 030803 (2016). https://doi.org/10.1115/1.4035310
15. Abu, S.M., Hannan, M.A., Hossain Lipu, M.S., Mannan, M., Ker, P.J., Hossain, M.J., Indra Mahlia, T.M.: State of the art of lithium-ion battery material potentials: an analytical evaluations, issues and future research directions. J. Clean. Prod. **394**, 136246 (2023). https://doi.org/10.1016/j.jclepro.2023.136246
16. Korthauer, R.: Lithium-Ion Batteries: Basics and Applications. Springer, Berlin, Heidelberg (2018). https://doi.org/10.1007/978-3-662-53071-9
17. Castelli, G.F., von Kolzenberg, L., Horstmann, B., Latz, A., Dörfler, W.: Efficient simulation of chemical-mechanical coupling in battery active particles. Energy Technol. **9**(6), 2000835 (2021). https://doi.org/10.1002/ente.202000835
18. Di Leo, C.V., Rejovitzky, E., Anand, L.: A Cahn-Hilliard-type phase-field theory for species diffusion coupled with large elastic deformations: application to phase-separating li-ion electrode materials. J. Mech. Phys. Solids **70**, 1–29 (2014). https://doi.org/10.1016/j.jmps.2014.05.001
19. Kelly, D.W., de S. R. Gago, J.P., Zienkiewicz, O.C., Babuška, I.: A posteriori error analysis and adaptive processes in the finite element method. I. Error analysis. Internat. J. Numer. Methods Engrg. **19**(11), 1593–1619 (1983). https://doi.org/10.1002/nme.1620191103
20. Arndt, D., Bangerth, W., Bergbauer, M., Feder, M., Fehling, M., Heinz, J., Heister, T., Heltai, L., Kronbichler, M., Maier, M., Munch, P., Pelteret, J-.P., Turcksin, B., Wells, D., Zampini, S.:

The deal. II library, version 9.5. J. Numer. Math. **31**(3), 231–246 (2023). https://doi.org/10.1515/jnma-2023-0089
21. Ainsworth, M., Oden, J.T.: A Posteriori Error Estimation in Finite Element Analysis. Pure and Applied Mathematics. Wiley, New York (2000). https://doi.org/10.1002/9781118032824
22. Brenner, S.C., Scott, L.R.: The Mathematical Theory of Finite Element Methods. Springer, New York (2008). https://doi.org/10.1007/978-0-387-75934-0
23. Babuška, I., Gatica, G.N.: A residual-based a posteriori error estimator for the Stokes-Darcy coupled problem. SIAM J. Numer. Anal. **48**(2), 498–523 (2010). https://doi.org/10.1137/080727646
24. Ainsworth, M., Oden, J.T.: A posteriori error estimation in finite element analysis. Comput. Methods Appl. Mech. Engrg. **142**(1–2), 1–88 (1997). https://doi.org/10.1016/S0045-7825(96)01107-3
25. Verfürth, R.: A review of a posteriori error estimation and adaptive mesh-refinement techniques. Wiley-Teubner series in advances in numerical mathematics. Wiley [u.a.], Chichester (1996)
26. Carstensen, C., Klose, R.: A posteriori finite element error control for the p-Laplace problem. SIAM J. Sci. Comput. **25**(3), 792–814 (2003). https://doi.org/10.1137/S1064827502416617
27. Schoof, R., Castelli, G.F., Dörfler, W.: Parallelization of a finite element solver for chemo-mechanical coupled anode and cathode particles in lithium-ion batteries. In: Kvamsdal, T., Mathisen, K.M., Lie, K.-A., Larson, M.G. (eds.) 8th European Congress on Computational Methods in Applied Sciences and Engineering (ECCOMAS Congress 2022), Barcelona, 2022. CIMNE. https://doi.org/10.23967/eccomas.2022.106
28. Schoof, R., Castelli, G.F., Dörfler, W.: Simulation of the deformation for cycling chemo-mechanically coupled battery active particles with mechanical constraints. Comput. Math. Appl. **149**, 135–149 (2023). https://doi.org/10.1016/j.camwa.2023.08.027
29. von Kolzenberg, L., Latz, A., Horstmann, B.: Chemo-mechanical model of sei growth on silicon electrode particles. Batter. Supercaps **5**(2), e202100216 (2022). https://doi.org/10.1002/batt.202100216
30. Holzapfel, G.A.: Nonlinear Solid Mechanics. Wiley, Chichester (2010)
31. Braess, D.: Finite Elements, 3rd edn. Cambridge University Press, Cambridge (2007). https://doi.org/10.1007/978-3-540-72450-6
32. Latz, A., Zausch, J.: Multiscale modeling of lithium ion batteries: thermal aspects. Beilstein J. Nanotechnol. **6**, 987–1007 (2015). https://doi.org/10.3762/bjnano.6.102
33. Latz, A., Zausch, J.: Thermodynamic consistent transport theory of Li-ion batteries. J. Power Sources **196**(6), 3296–3302 (2011). https://doi.org/10.1016/j.jpowsour.2010.11.088
34. Schammer, M., Horstmann, B., Latz, A.: Theory of transport in highly concentrated electrolytes. J. Electrochem. Soc. **168**(2), 026511 (2021). https://doi.org/10.1149/1945-7111/abdddf
35. Chan, C.K., Peng, H., Liu, G., McIlwrath, K., Zhang, X.F., Huggins, R.A., Cui, Y.: High-performance lithium battery anodes using silicon nanowires. Nat. Nanotechnol. **3**(1), 31–35 (2007). https://doi.org/10.1038/nnano.2007.411
36. Keil, P., Schuster, S.F., Wilhelm, J., Travi, J., Hauser, A., Karl, R.C., Jossen, A.: Calendar aging of lithium-ion batteries. J. Electrochem. Soc. **163**(9), A1872–A1880 (2016). https://doi.org/10.1149/2.0411609jes
37. Latz, A., Zausch, J.: Thermodynamic derivation of a Butler-Volmer model for intercalation in Li-ion batteries. Electrochim. Acta **110**, 358–362 (2013). https://doi.org/10.1016/j.electacta.2013.06.043
38. Lubliner, J.: Plasticity Theory. Pearson Education, Inc., New York (2006). https://doi.org/10.1115/1.2899459
39. Deng, D.: Li-ion batteries: basics, progress, and challenges. Energy Sci. Eng. **3**(5), 385–418 (2015). https://doi.org/10.1002/ese3.95
40. Piller, S., Perrin, M., Jossen, A.: Methods for state-of-charge determination and their applications. J. Power Sources **96**(1), 113–120 (2001). https://doi.org/10.1016/s0378-7753(01)00560-2
41. Reichelt, M.W., Shampine, L.F., Kierzenka, J.: Matlab `ode15s` (1997). http://www.mathworks.com. Copyright 1984–2020 The MathWorks, Inc

42. Shampine, L.F., Reichelt, M.W.: The MATLAB ODE suite. SIAM J. Sci. Comput. **18**(1), 1–22 (1997). https://doi.org/10.1137/S1064827594276424
43. Shampine, L.F., Reichelt, M.W., Kierzenka, J.A.: Solving index-1 DAEs in MATLAB and Simulink. SIAM Rev. **41**(3), 538–552 (1999). https://doi.org/10.1137/S003614459933425X
44. Shampine, L.F., Gladwell, I., Thompson, S.: Solving ODEs with MATLAB. Cambridge University Press, Cambridge (2003). https://doi.org/10.1017/CBO9780511615542
45. Schoof, R., Niermann, J., Dyck, A., Böhlke, T., Dörfler, W.: Efficient modeling and simulation of chemo-elasto-plastically coupled battery active particles. Comput. Mech. (2023). https://doi.org/10.48550/arXiv.2310.05440
46. Castelli, G.F.: Numerical Investigation of Cahn–Hilliard-Type Phase-Field Models for Battery Active Particles. PhD thesis, Karlsruhe Institute of Technology (KIT) (2021). https://doi.org/10.5445/IR/1000141249
47. The Trilinos Project Team. The Trilinos Project Website (2023). https://trilinos.github.io
48. Davis, T.A.: Algorithm 832: UMFPACK V4.3—an unsymmetric-pattern multifrontal method. ACM Trans. Math. Software **30**(2), 196–199 (2004). https://doi.org/10.1145/992200.992206
49. WIKI bw HPC Team. BwUniCluster 2.0 Hardware and Architecture (2024). https://wiki.bwhpc.de/e/BwUniCluster_2.0_Hardware_and_Architecture
50. Carstensen, C., Verfürth, R.: Edge residuals dominate a posteriori error estimates for low order finite element methods. SIAM J. Numer. Anal. **36**(5), 1571–1587 (1999). https://doi.org/10.1137/S003614299732334X
51. Verfürth, R.: A posteriori error estimation techniques for finite element methods. Numerical mathematics and scientific computation. Oxford University Press, Oxford (2013)
52. Huttin, M.: Phase-field modeling of the influence of mechanical stresses on charging and discharging processes in lithium ion batteries. PhD thesis, Karlsruher Institut für Technologie (KIT), Karlsruhe (2014)

Chapter 11
Analysis of Kerr Frequency Combs in Microresonators with Two-Photon Absorption and Free-Carriers

Lukas Bengel and Wolfgang Reichel

Abstract We study optical Kerr frequency combs in nonlinear microresonators in the presence of two-photon absorption (TPA) and free-carrier absorption (FCA) described by an extended Lugiato-Lefever system of equations. We give a criterion for modulation instability of constant stationary solutions and establish existence of non-constant solutions. Both results are accompanied by numerical simulations, which show that the abstract conditions both for modulation instability and existence are satisfied in large parts of the parameter regions.

Keywords Frequency combs · Lugiato-Lefever equation · Two-photon absorption · Free-carrier absorption · Modulation instability · Continuation

11.1 Introduction

Kerr frequency combs in nonlinear microresonators are versatile tools for a wide field of applications ranging from optical communications [18], spectroscopy [25], ultrafast optical measurements [25, 27] to high precision clocks [4, 28]. The typical design of a Kerr nonlinear microresonator is based on specifically composed silica and silicon nitrade waveguides. Frequency combs, i.e., stable signals which are highly localized in time-domain and have a broad and equally spaced spectral signature in frequency domain, emerge under suitable operating conditions of the microresonator as the outcome of nonlinear interaction of the resonantly enhanced pump field leading to modulation instability, and, ultimately to the formation of a stable stationary pattern. The mathematical model [1, 2, 15] for the dynamics of frequency combs is known as the Lugiato-Lefever equation,

---

L. Bengel (✉) · W. Reichel
Institute for Analysis (IANA), Karlsruhe Institute of Technology, Karlsruhe, Germany
e-mail: lukas.bengel@kit.edu

W. Reichel
e-mail: wolfgang.reichel@kit.edu

A. Ott et al. (eds.), *Applications of Mathematics in Sciences, Engineering, and Economics*, Springer Proceedings in Mathematics & Statistics 515,
https://doi.org/10.1007/978-3-032-01279-1_11

$$\mathrm{i}a_t(x,t) = -da_{xx}(x,t) + \big(\zeta - \mathrm{i} - |a(x,t)|^2\big)a(x,t) + \mathrm{i}f \tag{11.1}$$

which is a damped, driven and detuned nonlinear Schrödinger equation. Here $a = a(x,t)$ with $(x,t) \in [0, L] \times \mathbb{R}$ represents the optical field inside a ringresonator of length L which is coupled to a continuous strong input laser and subject to damping losses and material dispersion. The material dispersion is given by d ($d < 0$ stands for normal dispersion, $d > 0$ for anomalous dispersion), the forcing by $f > 0$ (arising form the continuous strong laser source), and the damping term by $-\mathrm{i}a$ (describing losses). Additionally, there is a detuning variable $\zeta \in \mathbb{R}$, which describes the offset between the frequency of the input laser and the nearest resonance frequency of the linear dispersion relation of the ringresonator.

There has been a lot of results in the physics, engineering, and mathematical literature about the properties of (11.1). Although not at the center of our interest in this paper, let us briefly indicate some results which are in the same spirit as the ones we shall present. The most elementary states of (11.1) are those, where only the pumped mode is excited and no other mode of the ringresonator. These states are given as spatially constant steady states of (11.1). Of interest are of course spatially localized states, and they can be investigated, e.g., via bifurcation analysis from spatially constant states as in [7, 17, 19–22] or via their dynamical stability/instability properties as in [3, 8, 12, 14, 24].

Recently, new designs of Kerr microresonators based on silicon have entered scientific discussions. Their advantage is the well-established material platform and the much higher Kerr effect. However, there are drawbacks: two-photon absorption (TPA) and free-carrier absorption (FCA), cf. [10, 11, 23, 26] . TPA describes the effect that two photons simultaneously excite an electron from the valence band into the conduction band. Within the conduction band additional photons can excite the electrons into even higher excitation states (FCA). Altogether, TPA and FCA act against Kerr comb formation due to their cumulative damping effect. This effect leads to a coupled model, where next to the optical field $a = a(x,t)$ also the free-carrier density $N = N(x,t)$ needs to be considered. In (11.1) the reference frame has already been chosen such that stationarity on (11.1) amounts to movement with group-velocity v. Since free carriers do not move with the optical field, the reference frame of the free carrier density N is retarded relative to a. As a result, cf. [13, 16, 26], the following model has been derived

$$\begin{aligned} \mathrm{i}a_t(x,t) =& -da_{xx}(x,t) + \big(\zeta - \mathrm{i} - (1+\mathrm{i}r)|a(x,t)|^2\big)a(x,t) + \mathrm{i}f \\ &- \sigma(\mathrm{i} - \mu)N(x - tv, t)a(x,t), \\ N_t(x,t) =& r|a(x+tv,t)|^4 - \frac{N(x,t)}{\tau}. \end{aligned} \tag{11.2}$$

Here, $\tau > 0$ is the effective life-time (dwell time) of the free carriers, $r > 0$ quantifies the TPA effect, $\sigma > 0$ is the FCA cross section, and $\mu \in \mathbb{R}$ the FCA dispersion. Notice the relative retardation of N in the first equation and the relative advancement of a

in the second equation. The values of all physical quantities (taken from [26]) will be specified in Sect. 11.2.2, where we perform numerical simulations.

A mathematically simple way to remove the relative retardation is to define $n(x,t) = N(x - tv, t)$ as the moving coordinate version of the free carrier density. The resulting equation is then given by

$$\begin{aligned} \mathrm{i}a_t &= -da_{xx} + \big(\zeta - \mathrm{i} - \sigma(\mathrm{i} - \mu)n - (1 + \mathrm{i}r)|a|^2\big)a + \mathrm{i}f, \\ n_t &= r|a|^4 - vn_x - \frac{n}{\tau}, \end{aligned} \tag{11.3}$$

and it is this equation on which we base our analysis. It is easy to see that homogeneous steady states $n_0 \in \mathbb{R}$ and $a_0 \in \mathbb{C}$ of (11.3) exist and satisfy

$$\big(\zeta - \mathrm{i} - \sigma(\mathrm{i} - \mu)n_0 - (1 + \mathrm{i}r)|a_0|^2\big)a_0 + \mathrm{i}f = 0, \quad r|a_0|^4 - \frac{n_0}{\tau} = 0. \tag{11.4}$$

It is reasonable to conjecture that the additional effects caused by TPA and FCA may prevent the existence of other, non-constant steady states – provided $r, \sigma > 0$ are too large. The only way to counterbalance this is the value of the dwell time τ. Indeed, on a technical level, the free carriers can be removed from the microresonator by a reverse biased p-i-n-junction so that their dwell time within the device can be substantially reduced.

The goal of this paper is two-fold.

(i) We will provide conditions on the parameters $\zeta, \sigma, \tau, d, \mu$ such that a steady state (a_0, n_0) of (11.3) becomes unstable. This instability can be seen as the onset of the formation of non-trivial, patterned states which ultimately lead to frequency combs. This is done in Theorem 11.1 in Sect. 11.2, and it provides a rigorous back-up to a very similar result obtained in [26] based on a substantial number of physically well-justified but mathematically non-rigorous approximations.
(ii) We will give rigorous existence results for particular non-homogeneous solutions of (11.3). This is done in Theorem 11.3 in Sect. 11.3. We are not aware of any other existence results for (11.3) except for [5], where a much simplified version with TPA but without FCA has been studied. Our results build upon [5] and show existence for values of σ and μ near zero by a bifurcation argument based on the Lyapunov-Schmidt reduction.

Our results are illustrated with numerical simulations. The code (based on Matlab and `pde2path`) is available at Repository KITopen.[1] We finish the paper with a discussion and conclusion in Sect. 11.4.

[1] https://doi.org/10.5445/IR/1000168160.

11.2 Modulation Instability of Constant Steady States

Constant steady states of (11.3) are solutions $(a_0, n_0) \in \mathbb{C} \times \mathbb{R}$ of the algebraic equations (11.4). Note that $|a_0|^2$ is related to the forcing f by

$$\left((\zeta + \sigma\mu r\tau|a_0|^4 - |a_0|^2)^2 + (1 + \sigma r\tau|a_0|^4 + r|a_0|^2)^2\right)|a_0|^2 = f^2.$$

Linearization of (11.3) about a constant steady state $(a_0, n_0) \in \mathbb{C} \times \mathbb{R}$ yields

$$\begin{aligned} \mathrm{i}a_t &= -da_{xx} + \big(\zeta - \mathrm{i} - \sigma(\mathrm{i} - \mu)n_0\big)a - \sigma(\mathrm{i} - \mu)a_0 n - (1 + \mathrm{i}r)(2|a_0|^2 a + a_0^2\bar{a}), \\ n_t &= 2r|a_0|^2(a_0\bar{a} + \bar{a}_0 a) - vn_x - \frac{n}{\tau}. \end{aligned}$$

In the following we will look for conditions which lead to exponentially growing solutions of the linearized equations. This phenomenon is called modulation instability. Our investigation of modulation instability consists of analytical results in Sect. 11.2.1 and numerical illustrations in Sect. 11.2.2.

11.2.1 Analysis of the Linearized Equations

In the equation for a there is a coupling between a and its complex conjugate $\bar{a}$. For n the situation is different since $n = \bar{n}$, provided $n(\cdot, 0)$ is a real-valued and hence physical initial condition. Nevertheless, it is helpful to adjoin the following equations both for $\bar{a}$ and $\bar{n}$ to the previous two equations

$$\begin{aligned} -\mathrm{i}\bar{a}_t &= -d\bar{a}_{xx} + (\zeta + \mathrm{i} + \sigma(\mathrm{i} + \mu)n_0)\bar{a} + \sigma(\mathrm{i} + \mu)\bar{a}_0\bar{n} - (1 - \mathrm{i}r)(2|a_0|^2\bar{a} + \bar{a}_0^2 a), \\ \bar{n}_t &= 2r|a_0|^2(\bar{a}_0 a + a_0\bar{a}) - v\bar{n}_x - \frac{\bar{n}}{\tau} \end{aligned}$$

so that altogether we obtain a system of four equations for $(a, \bar{a}, n, \bar{n})$. Using the notation $\mathbb{T} = \mathbb{R}/2\pi\mathbb{Z}$ for the one dimensional torus of length 2π we define next the following differential operators

$$\begin{aligned} A &: H^2(\mathbb{T}) \to L^2(\mathbb{T}) \text{ with } A := -d\partial_x^2 + \zeta - \mathrm{i} - \sigma(\mathrm{i} - \mu)n_0 - (1 + \mathrm{i}r)2|a_0|^2, \\ B &: (H^2(\mathbb{T}))^2 \times (H^1(\mathbb{T}))^2 \to (L^2(\mathbb{T}))^4 \text{ with} \end{aligned}$$

$$B := \begin{bmatrix} A & -(1 + \mathrm{i}r)a_0^2 & -\sigma(\mathrm{i} - \mu)a_0 & 0 \\ -(1 - \mathrm{i}r)\bar{a}_0^2 & A^* & 0 & \sigma(\mathrm{i} + \mu)\bar{a}_0 \\ 2r|a_0|^2\bar{a}_0 & 2r|a_0|^2 a_0 & -v\partial_x - \frac{1}{\tau} & 0 \\ 2r|a_0|^2\bar{a}_0 & 2r|a_0|^2 a_0 & 0 & -v\partial_x - \frac{1}{\tau} \end{bmatrix}.$$

Then the operator

$$L := JB : (H^2(\mathbb{T}))^2 \times (H^1(\mathbb{T}))^2 \to (L^2(\mathbb{T}))^4 \text{ with } J := \mathrm{diag}(-\mathrm{i}, \mathrm{i}, 1, 1) \in \mathbb{C}^{4\times 4} \tag{11.5}$$

allows us to write the linearized equation as

$$\mathbf{v}_t = L\mathbf{v}, \quad \mathbf{v} = (a, \bar{a}, n, \bar{n})^\top.$$

We will show that under certain conditions on the parameters, specified in the theorem below, there exist (spectrally) unstable constant steady states, i.e., $\sigma(L) \cap \{z \in \mathbb{C} : \operatorname{Re} z > 0\} \neq \emptyset$. For the statement of the following theorem we use the notation $[x]$ to denote the closest integer to $x \in \mathbb{R}$ (if x has distance $1/2$ from the integers one may pick either of the neighbouring integers).

Theorem 11.1 *Let $d \in \mathbb{R} \setminus \{0\}$, $r, \sigma, \tau > 0$, $\mu, \upsilon, \zeta, f \in \mathbb{R}$, and $\varepsilon > 0$ be given parameters and assume that $(a_0, r\tau|a_0|^4) \in \mathbb{C} \times \mathbb{R}$ is a steady state of* (11.3). *Define*

$$k^* := \left[\sqrt{\frac{2|a_0|^2 - \sigma\mu r\tau|a_0|^4 - \zeta}{d}}\right], \quad \Delta := \frac{d(k^*)^2 + \zeta}{|a_0|^2} + \sigma\mu r\tau|a_0|^2 - 2.$$

Then the steady state $(a_0, r\tau|a_0|^4) \in \mathbb{C} \times \mathbb{R}$ is spectrally unstable provided the following three conditions hold:

$$2r < \sqrt{1 + r^2 - \Delta^2}, \quad \sigma\tau < \frac{1}{(1+\varepsilon)4r}\left(\sqrt{1 + r^2 - \Delta^2} - 2r\right)^2 \tag{11.6}$$

and

$$|a_0|^2 \in (\xi_1(\varepsilon), \xi_2(\varepsilon))$$

where $0 < \xi_1(\varepsilon) < \xi_2(\varepsilon)$ denote the roots of the polynomial

$$q_\varepsilon(\xi) = -1 - \varepsilon + (\sqrt{1 + r^2 - \Delta^2} - 2r)\xi - \sigma r\tau\xi^2$$

and

$$\left(\varepsilon + \frac{|\omega_1| + |\omega_4|}{\varepsilon}\right)^2 < \left(-1 + (\sqrt{1 + r^2 - \Delta^2} - 2r)|a_0|^2 - \sigma r\tau|a_0|^4 + \frac{1}{\tau}\right)^2 + \frac{\upsilon^2|a_0|^2}{|d|} \tag{11.7}$$

where

$$\omega_1 := \frac{2\sigma r|a_0|^4}{(1+r^2-\Delta^2)}\left((r-\mu(1+\Delta))\sqrt{1+r^2-\Delta^2}-(1+r^2-\Delta^2)\right),$$
$$\omega_4 := \frac{2\sigma r|a_0|^4}{(1+r^2-\Delta^2)}\left((\mu(1+\Delta)-r)\sqrt{1+r^2-\Delta^2}-(1+r^2-\Delta^2)\right).$$

Moreover, the instability appears for modes with wave number k^*.

Remark 11.1 We have implicitly assumed that the expressions in the two square roots of the above theorem are non-negative.

Proof Using the notation from (11.5), let L denote the linearization about the steady state $(a_0, r\tau|a_0|^4)$. Exploiting the Fourier transform of periodic L^2-functions we decompose the spectrum of the linearization

$$\sigma(L) = \bigcup_{k\in\mathbb{Z}} \sigma(L_k)$$

where the matrizes L_k are given by

$$L_k = \begin{bmatrix} \ell_1(k) & \ell_2 & \ell_3 & 0 \\ \bar{\ell}_2 & \bar{\ell}_1(k) & 0 & \bar{\ell}_3 \\ \bar{\ell}_4 & \ell_4 & \ell_5(k) & 0 \\ \bar{\ell}_4 & \ell_4 & 0 & \ell_5(k) \end{bmatrix} \in \mathbb{C}^{4\times 4},$$

with coefficients

$$\ell_1(k) = -\mathrm{i}dk^2 - (\mathrm{i}\zeta+1) - \sigma r\tau(1+\mathrm{i}\mu)|a_0|^4 + 2(\mathrm{i}-r)|a_0|^2, \quad \ell_2 = (\mathrm{i}-r)a_0^2,$$
$$\ell_3 = -\sigma(1+\mathrm{i}\mu)a_0, \quad \ell_4 = 2r|a_0|^2 a_0, \quad \ell_5(k) = -\mathrm{i}vk - \frac{1}{\tau}.$$

In particular, it is enough to show that there exists $k \in \mathbb{Z}$ such that $\sigma(L_k) \cap \{z : \operatorname{Re} z > 0\} \neq \emptyset$. Let us fix some $k \in \mathbb{Z}$ and consider the matrix L_k. To get a better understanding of the spectrum of L_k we perform several similarity transformations, eventually allowing us to locate eigenvalues of the matrix. If we define $T_1 = \operatorname{diag}(1, 1, \ell_3, \bar{\ell}_3) \in \mathbb{C}^{4\times 4}$ then one finds

$$L_{1,k} := T_1 L_k T_1^{-1} = \begin{bmatrix} \ell_1(k) & \ell_2 & 1 & 0 \\ \bar{\ell}_2 & \bar{\ell}_1(k) & 0 & 1 \\ \ell_3\bar{\ell}_4 & \ell_3\ell_4 & \ell_5(k) & 0 \\ \bar{\ell}_3\bar{\ell}_4 & \bar{\ell}_3\ell_4 & 0 & \ell_5(k) \end{bmatrix}.$$

The matrix

$$A_k := \begin{bmatrix} \ell_1(k) & \ell_2 \\ \bar{\ell}_2 & \bar{\ell}_1(k) \end{bmatrix}$$

has the eigenvalues

$$\begin{aligned}\lambda_\pm(k) &= \operatorname{Re}\ell_1(k) \pm \sqrt{(\operatorname{Re}\ell_1(k))^2 - |\ell_1(k)|^2 + |\ell_2|^2} \\ &= -1 - \sigma r\tau|a_0|^4 - 2r|a_0|^2 \\ &\quad \pm \sqrt{|a_0|^4(1+r^2) - \left(dk^2 + \zeta + \sigma\mu r\tau|a_0|^4 - 2|a_0|^2\right)^2}.\end{aligned}$$

Observe that

$$\max\{\operatorname{Re}\lambda_\pm(k) : k \in \mathbb{Z}\} = \lambda_+(\pm k^*) = -1 + |a_0|^2\left(\sqrt{1+r^2-\Delta^2} - 2r\right) - \sigma r\tau|a_0|^4,$$

where $k^* = \left[\sqrt{(2|a_0|^2 - \sigma\mu r\tau|a_0|^4 - \zeta)/d}\right]$ is the integer which is closest to $\sqrt{(2|a_0|^2 - \sigma\mu r\tau|a_0|^4 - \zeta)/d}$. Thus anticipating that L_{k^*} has an unstable eigenvalue close to $\lambda_+(\pm k^*)$ we will from now on set $k = k^*$.

The matrix A_{k^*} is diagonalizable through conjugation with the matrix

$$S := \begin{bmatrix} a_0^2(\mathrm{i}-r) & a_0^2(\mathrm{i}-r) \\ |a_0|^2\left(\sqrt{1+r^2-\Delta^2} + \mathrm{i}\Delta\right) & |a_0|^2\left(-\sqrt{1+r^2-\Delta^2} + \mathrm{i}\Delta\right)\end{bmatrix}.$$

Consequently, after introducing the block diagonal matrix $T_2 = \operatorname{diag}(S, S) \in \mathbb{C}^{4\times 4}$ we obtain

$$L_{2,k^*} = T_2^{-1} L_{1,k^*} T_2 = \begin{bmatrix} \lambda_+(k^*) & 0 & 1 & 0 \\ 0 & \lambda_-(k^*) & 0 & 1 \\ \omega_1 & \omega_2 & \ell_5(k^*) & 0 \\ \omega_3 & \omega_4 & 0 & \ell_5(k^*) \end{bmatrix}$$

where the coefficients ω_i are computed from

$$\begin{bmatrix}\omega_1 & \omega_2 \\ \omega_3 & \omega_4\end{bmatrix} = S^{-1}\begin{bmatrix}\ell_3\bar{\ell}_4 & \ell_3\ell_4 \\ \bar{\ell}_3\bar{\ell}_4 & \bar{\ell}_3\ell_4\end{bmatrix} S.$$

The eigenvalues of L_{2,k^*} are precisely the roots of the characteristic polynomial

$$\begin{aligned} p(\lambda) &= \det(L_{2,k^*} - \lambda) \\ &= (\lambda_+(k^*) - \lambda)(\lambda_-(k^*) - \lambda)(\ell_5(k^*) - \lambda)^2 - \omega_1(\lambda_-(k^*) - \lambda)(\ell_5(k^*) - \lambda) \\ &\quad - \omega_4(\lambda_+(k^*) - \lambda)(\ell_5(k^*) - \lambda).\end{aligned}$$

Clearly, $p(\ell_5(k^*)) = 0$ which implies that $\ell_5(k^*)$ is a stable eigenvalue of L_{2,k^*}. The remaining roots of p are determined from the equation $\tilde{p}(\lambda) = 0$ where

$$\tilde{p}(\lambda) = (\lambda_+(k^*) - \lambda)(\lambda_-(k^*) - \lambda)(\ell_5(k^*) - \lambda) - \omega_1(\lambda_-(k^*) - \lambda) - \omega_4(\lambda_+(k^*) - \lambda).$$

To locate the roots of $\tilde{p}$, we observe that they coincide with the eigenvalues of the matrix

$$L_{3,k^*,\varepsilon} := \begin{bmatrix} \lambda_+(k^*) & 0 & \varepsilon \\ 0 & \lambda_-(k^*) & \varepsilon \\ \frac{\omega_1}{\varepsilon} & \frac{\omega_4}{\varepsilon} & \ell_5(k^*) \end{bmatrix} \quad \text{for arbitrary } \varepsilon > 0,$$

and where ω_1, ω_4 are the values given in the statement of the theorem.

The following result allows us to derive bounds on the location of the eigenvalues of $L_{3,k^*,\varepsilon}$.

Theorem 11.2 (Gerschgorin's Circle Theorem, cf. *[6]***)** *Let $N \in \mathbb{N}$, $A = (a_{ij})_{i,j=1,\dots,N} \in \mathbb{C}^{N\times N}$ and $G_i := \{z \in \mathbb{C}: \ |z - a_{ii}| \le \sum_{j:j\neq i} |a_{ij}|\}$. Then we have*

$$\sigma(A) \subset \bigcup_{i=1}^{N} G_i.$$

Moreover, if $G_i \cap G_j = \emptyset$ for fixed i and all $j \neq i$, then $\sigma(A) \cap G_i \neq \emptyset$.

The Gerschgorin circles of $L_{3,k^*,\varepsilon}$ are given by

$$\begin{aligned} G_1 &= \{z \in \mathbb{C}: \ |z - \lambda_+(k^*)| \le \varepsilon\}, \\ G_2 &= \{z \in \mathbb{C}: \ |z - \lambda_-(k^*)| \le \varepsilon\}, \\ G_3 &= \left\{z \in \mathbb{C}: \ |z - \ell_5(k^*)| \le \frac{|\omega_1| + |\omega_4|}{\varepsilon}\right\}. \end{aligned}$$

We aim to show that

$$G_1 \subset \{z \in \mathbb{C}: \ \operatorname{Re} z > 0\}, \tag{11.8}$$
$$G_1 \cap G_2 = \emptyset, \tag{11.9}$$
$$G_1 \cap G_3 = \emptyset, \tag{11.10}$$

which, using Theorem 11.2, implies that $G_1 \cap \sigma(L_{3,k^*,\varepsilon}) \neq \emptyset$, and the instability of $(a_0, r\tau|a_0|^4)$ then follows from $\sigma(L_{3,k^*,\varepsilon}) \subset \sigma(L_{k^*}) \subset \sigma(L)$. To prove (11.8) let us consider the polynomial

$$q_\varepsilon(\xi) = -1 - \varepsilon + (\sqrt{1 + r^2 - \Delta^2} - 2r)\xi - \sigma r\tau\xi^2.$$

A short calculation shows that q_ε has two positive roots $0 < \xi_1(\varepsilon) < \xi_2(\varepsilon)$ if and only if

$$2r < \sqrt{1 + r^2 - \Delta^2}, \quad 0 < \sigma\tau < \frac{1}{(1+\varepsilon)4r}\left(\sqrt{1 + r^2 - \Delta^2} - 2r\right)^2,$$

which is precisely (11.6). Consequently, we find $\lambda_+(k^*) > \varepsilon$ provided $|a_0|^2 \in (\xi_1(\varepsilon), \xi_2(\varepsilon))$, which yields the inclusion (11.8). The explicit formula for $\lambda_-(k^*)$ also shows that $\lambda_-(k^*) \le -\varepsilon$, thus implying $|\lambda_+(k^*) - \lambda_-(k^*)| > 2\varepsilon$ so that we have established (11.9). Finally, (11.10) follows from $\varepsilon + \frac{|\omega_1|+|\omega_4|}{\varepsilon} < |\lambda_+(k^*) - \ell_5(k^*)|$ which is equivalent to our assumption (11.7). Thus, the proof of Theorem 11.1 is complete. □

Remark 11.2 Assumption (11.6) in Theorem 11.1 yields an upper bound for the TPA parameter since

$$2r < \sqrt{1+r^2-\Delta^2} \leq \sqrt{1+r^2} \quad \Longrightarrow \quad r < \frac{1}{\sqrt{3}}.$$

In the context of instabilities of steady-states for (11.2) this bound was already derived in [5, 26].

In the case of anomalous dispersion $d > 0$, a simplified version of Theorem 11.1 can be obtained as follows. First, by optimizing the detuning parameter $\zeta \in \mathbb{R}$ in such a way that $\zeta = A - \sigma\mu r\tau A^2$, $A = |a_0|^2$, we obtain the maximal power transfer between the forcing f and the power A of the $k = 0$-mode. The above choice of ζ implies that the constant steady state a_0 is real-valued and $A = a_0^2$ satisfies

$$f = \big(1 + \sigma r\tau A^2 + rA\big)\sqrt{A}.$$

Thus, $(\sqrt{A}, r\tau A^2) \in \mathbb{R} \times \mathbb{R}$ is a real-valued steady state of (11.2). Second, by slightly adjusting the forcing f, we may assume that $\sqrt{A/d} \in \mathbb{N}$. As a consequence, we obtain that $\Delta = 0$ in Theorem 11.1. These observations lead to the following corollary.

Corollary 11.1 *Let* $d, \sigma, \tau > 0$, $\mu, \upsilon \in \mathbb{R}$, $r \in (0, 1/\sqrt{3})$. *Assume that* $(\sqrt{A}, r\tau A^2) \in \mathbb{R} \times \mathbb{R}$ *is a real-valued steady-state of* (11.3) *with* $f = \big(1 + \sigma r\tau A^2 + rA\big)\sqrt{A}$, $\zeta = A - \sigma\mu r\tau A^2$ *such that* $\sqrt{A/d} \in \mathbb{N}$. *Then the steady state* $(\sqrt{A}, r\tau A^2) \in \mathbb{R} \times \mathbb{R}$ *is spectrally unstable provided there exists* $\varepsilon > 0$ *such that the following three conditions hold:*

$$\sigma\tau < \frac{1}{(1+\varepsilon)4r}\Big(\sqrt{1+r^2} - 2r\Big)^2 \tag{11.11}$$

and

$$A \in (\tilde{\xi}_1(\varepsilon), \tilde{\xi}_2(\varepsilon)) \tag{11.12}$$

where $0 < \tilde{\xi}_1(\varepsilon) < \tilde{\xi}_2(\varepsilon)$ *are the positive roots of the polynomial*

$$\tilde{q}(\xi) := -1 - \varepsilon + (\sqrt{1+r^2} - 2r)\xi - \sigma r\tau\xi^2,$$

and

$$\left(\varepsilon + \frac{|\omega_1| + |\omega_4|}{\varepsilon}\right)^2 < \left(-1 + (\sqrt{1+r^2} - 2r)A - \sigma r\tau A^2 + \frac{1}{\tau}\right)^2 + \frac{\upsilon^2 A}{d} \tag{11.13}$$

where

$$\omega_1 := \frac{2\sigma r A^2}{(1+r^2)}\Big((r-\mu)\sqrt{1+r^2}-(1+r^2)\Big),$$
$$\omega_4 := \frac{2\sigma r A^2}{(1+r^2)}\Big((\mu-r)\sqrt{1+r^2}-(1+r^2)\Big).$$

Moreover the instability is found for modes with wave number $k^* = \sqrt{A/d}$.

11.2.2 Numercial Illustration

Here we present numerical computations of the instability region as established in Corollary 11.1. Motivated by [26], and under the normalization $L = 2\pi$, we fix the values of the physical quantities as $d = 0.025$, $\sigma = 0.5$, $\mu = 1$, $v = 190$. The values ζ for the detuning and f for the forcing will be specified below, and the TPA strength r and dwell time τ will be allowed to vary within physically reasonable ranges. The parameter $\varepsilon > 0$ is a free parameter (in principle). However, small values of ε typically lead to larger regions where instability is guaranteed. Numerical experiments showed that our choice of $\varepsilon = 0.05$ led to good results (see discussion below).

The numerical algorithm is as follows. For a given $0 < r < 1/\sqrt{3}$ and $0 < \tau$ we check the condition (11.11). If it is satisfied, we compute the smallest number $A \in (\tilde{\xi}_1(\varepsilon), \xi_2(\varepsilon))$ such that $\sqrt{A/d} \in \mathbb{N}$ and set $f := \big(1+\sigma r\tau A^2 + rA\big)\sqrt{A}$, $\zeta := A - \sigma\mu r\tau A^2$. Then $(\sqrt{A}, r\tau A^2) \in \mathbb{R}\times\mathbb{R}$ is a constant steady-state. If the remaining conditions (11.11), (11.13) of Corollary 11.1 are met, we indicate this by a colored dot in the $r - \sigma\tau$ plane[2], where the color encodes the value of the forcing f^2. Note that our choice of A yields the minimal value of f such that all conditions of Corollary 11.1 are satisfied, and we therefore refer to it as the threshold power, denoted by f_{thres}. The result is depicted in the left panel of Fig. 11.1.

Corollary 11.1 provides only a sufficient condition for instability so that the threshold power f may be overestimated while the instability region may be underestimated. We therefore wanted to see how well Corollary 11.1 predicts the full instability region. To illustrate this we computed the eigenvalues of L_{2,k^*} (see the proof of Theorem 11.1) for k^* as in Corollary 11.1 and where the forcing is treated as a free parameter (which in turn determines a_0 and ζ). The minimal necessary forcing value which leads to an eigenvalue with positive real parts is called f_{num}. The result is shown in the right panel of the Fig. 11.1. The difference $f^2_{\text{thres}} - f^2_{\text{num}} \geq 0$ for each tuple $(r, \sigma\tau)$ of the instability region is shown in the bottom panel of Fig. 11.1. Our conclusions are the following:

---

[2] We scale the y-axis by σ to improve readability of the plots.

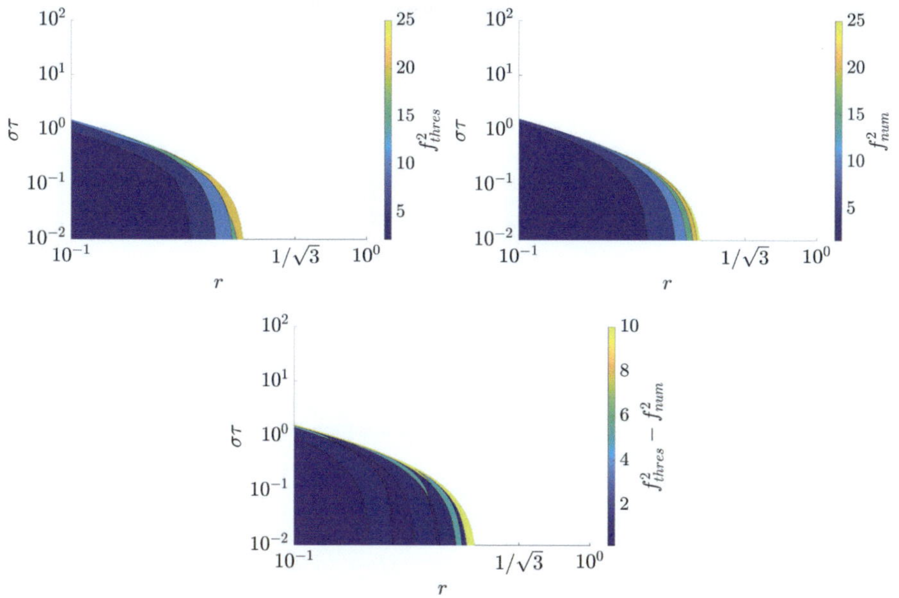

Fig. 11.1 Illustration of the instability region in the $r - \sigma\tau$-plane with the threshold power encoded by the different colors. Parameters are set to $d = 0.025$, $\sigma = 0.5$, $\mu = 1$, $\upsilon = 190$, $\varepsilon = 0.05$. The left panel shows the instability region predicted by Corollary 11.1. The right panel displays the instability region obtained by numerical approximations of unstable eigenvalues of L_{2,k^*}. The bottom panel shows the difference between f_{thres} and f_{num}

(i) the shape of the instability region as predicted by the analytical results of Corollary 11.1 (left panel) is in very good agreement with the numerically computed instability region (right panel),
(ii) both results are also in good agreement with [26],
(iii) quantitatively, the values of f_{thres} and f_{num} agree well in large parts of the instability region but differ more significantly near its border (which is to be expected since there f_{num} and f_{thres} both tend to infinity).

Similarly, in Fig. 11.2, computations are presented for $\mu = -1$ while leaving all other parameters unchanged.

11.3 Existence of Non-homogeneous Solutions

This section is devoted to the existence of non-constant 2π-periodic solutions of

$$\begin{aligned} -da'' + (\zeta - \mathrm{i} - (\mathrm{i}\sigma - \mu)n - (1 + \mathrm{i}r)|a|^2)a + \mathrm{i}f &= 0, \\ \upsilon n' + \frac{n}{\tau} - r|a|^4 &= 0, \end{aligned} \tag{11.14}$$

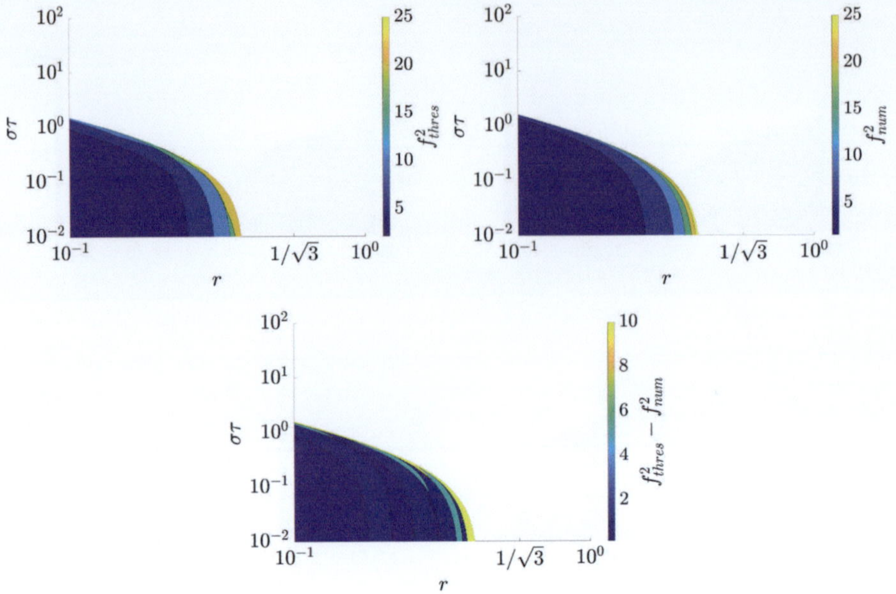

Fig. 11.2 This figure shows computations of the instability region as in Fig. 11.1 for the TPA parameter $\mu = -1$, while leaving all other parameters unchanged

which yield space-periodic and, in the case of the free carriers via $N(x, t) := n(x + vt)$, also time-periodic solutions of (11.3).

Remark 11.3 In (11.14) we write the coupling constant as $\mathrm{i}\sigma - \mu$ instead of $\sigma(\mathrm{i} - \mu)$ which is more suitable for our purposes. Clearly, both choices of couplings are equivalent provided $\sigma \neq 0$.

In the following, our considerations are divided into three parts. In the first part, we exploit continuation methods based on the Implicit Function Theorem to establish Theorem 11.3, which is our main result on the existence of non-constant solutions of (11.14). In the second part, we present numerical simulations of non-constant solutions illustrating the findings of Theorem 11.3. Both, the analytical and the numerical existence analysis require that a continuability condition is satisfied. In the last part, we give a detailed discussion of this condition in the limit where damping, forcing, and the two-photon absorption are small.

11.3.1 Continuation Analysis for Non-homogeneous Solutions

We continue to use the notation of Sect. 11.2, in particular $\mathbb{T}$ is the one dimensional torus of length 2π. We consider all function spaces as vector spaces over the field $\mathbb{R}$ and we use on $L^2(\mathbb{T})$ the scalar product $\langle f, g\rangle_{L^2} = \operatorname{Re}\int_{\mathbb{T}} f\bar{g}\,dx$.

Since the equation for n is a linear first order inhomogeneous ODE with periodicity condition in x we can solve it by the Duhamel formula. More precisely, for fixed $a \in H^2(\mathbb{T})$ the solution of the ODE for n is given by

$$\mathcal{N}(a) = r\left(v\partial_x + \frac{1}{\tau}\right)^{-1} |a|^4.$$

The solution operator $\mathcal{N}$ can be computed explicitly as a non-local integral operator, which is given by the following expression:

$$\mathcal{N}: \begin{cases} H^1(\mathbb{T}) \to H^2(\mathbb{T}), \\ \mathcal{N}(a)(x) := \frac{r}{v}\left(\frac{1}{e^{\frac{2\pi}{\tau v}}-1}\int_0^{2\pi} e^{\frac{y-x}{\tau v}}|a(y)|^4 dy + \int_0^x e^{\frac{y-x}{\tau v}}|a(y)|^4 dy\right), \end{cases} \tag{11.15}$$

and allows us to write (11.14) in the compact form

$$-da'' + \left(\zeta - \mathrm{i} - (\mathrm{i}\sigma - \mu)\mathcal{N}(a) - (1+\mathrm{i}r)|a|^2\right)a + \mathrm{i}f = 0. \tag{11.16}$$

For $\sigma = \mu = 0$, (11.16) becomes

$$-da'' + \left(\zeta - \mathrm{i} - (1+\mathrm{i}r)|a|^2\right)a + \mathrm{i}f = 0 \tag{11.17}$$

and the existence of non-constant periodic solutions of (11.17) is proven in [5] provided that the nonlinear damping parameter satisfies $0 \le r \le r_*$. The value of r_* is available explicitly and satisfies $r_* \in (0, 1/\sqrt{3})$ provided that $f^2 \ge 1$. We start with such a solution under the assumption of its non-degeneracy according to the following definition.

Definition 11.1 A non-constant solution $a_0 \in H^2(\mathbb{T})$ of (11.17) is called non-degenerate if the kernel of the linearized operator

$$\mathcal{L}_0\phi = -d\phi'' + (\zeta - \mathrm{i} - 2(1+\mathrm{i}r)|a_0|^2)\phi - (1+\mathrm{i}r)a_0^2\bar{\phi}, \quad \phi \in H^2(\mathbb{T}) \tag{11.18}$$

satisfies $\operatorname{Ker}(\mathcal{L}_0) = \operatorname{Span}\{a_0'\}$.

Remark 11.4 Note that $\mathcal{L}_0 : H^2(\mathbb{T}) \to L^2(\mathbb{T})$ is a compact perturbation of the isomorphism $-d\partial_x^2 + \operatorname{sign}(d) : H^2(\mathbb{T}) \to L^2(\mathbb{T})$ and hence an index-zero Fredholm operator. Notice also that $\operatorname{Span}\{a_0'\}$ always belongs to the kernel of $\mathcal{L}_0$ due to translation invariance in x of (11.17). Non-degeneracy means that except for the obvious candidate a_0' (and its real multiples) there is no other element in the kernel of $\mathcal{L}_0$.

Our strategy is to show that non-constant, non-degenerate solutions of (11.17) as in Definition 11.1 can be continued into the regime of $(\sigma, \mu) \neq (0, 0)$ provided that a continuability condition is satisfied. Before we state our existence result, we note that the non-degeneracy assumption implies that the adjoint operator $\mathcal{L}_0^* : H^2(\mathbb{T}) \to L^2(\mathbb{T})$ also has a one-dimensional kernel and thus there exists $a_0^* \in H^2(\mathbb{T})$ such that $\mathrm{Ker}(\mathcal{L}_0^*) = \mathrm{Span}\{a_0^*\}$.

Theorem 11.3 *Let $d, \tau, v \neq 0, \zeta, f \in \mathbb{R}, r > 0$ be fixed and $a_0 \in H^2(\mathbb{T})$ be a non-degenerate solution of* (11.17) *as in Definition 11.1. Assume that*

$$\alpha := \int_0^{2\pi} \mathcal{N}(a_0) a_0 \bar{a}_0^* dx \neq 0,$$

where $\mathcal{N}$ is the integral operator defined in (11.15)*. Then the following is true:*

(i) If $\mathrm{Re}\,\alpha \neq 0$, then there exist $\sigma^* > 0$ and a continuously differentiable curve $(-\sigma^*, \sigma^*) \ni \sigma \mapsto (\mu(\sigma), a(\sigma)) \in \mathbb{R} \times H^2(\mathbb{T})$ such that $a(\sigma)$ is a non-trivial solution of (11.16) *for the coupling parameters $\sigma, \mu(\sigma)$ and*

$$\mu(0) = 0, \qquad \|a(\sigma) - a_0\|_{H^2} = O(|\sigma|) \quad \text{as } \sigma \to 0.$$

Moreover, $\mu'(0) = -\mathrm{Im}\,\alpha / \mathrm{Re}\,\alpha$.

(ii) If $\mathrm{Im}\,\alpha \neq 0$, then there exist $\mu^* > 0$ and a continuously differentiable curve $(-\mu^*, \mu^*) \ni \mu \mapsto (\sigma(\mu), a(\mu)) \in \mathbb{R} \times H^2(\mathbb{T})$, such that $a(\mu)$ is a non-trivial solution of (11.16) *for the coupling parameters $\sigma(\mu), \mu$ and*

$$\sigma(0) = 0, \qquad \|a(\mu) - a_0\|_{H^2} = O(|\mu|) \quad \text{as } \mu \to 0.$$

Moreover, $\sigma'(0) = -\mathrm{Re}\,\alpha / \mathrm{Im}\,\alpha$.

Proof We seek solutions of (11.16) of the form

$$a = a_0 + b, \quad \text{with } b \in H^2(\mathbb{T}) : \ b \perp_{L^2} a_0'.$$

After defining the space $X_0 := \mathrm{Span}\{a_0'\}^{\perp_{L^2}} \cap H^2(\mathbb{T})$ this means that $b \in X_0$. Plugging the ansatz into (11.16) yields

$$\mathcal{F}(b, \sigma, \mu) := \mathcal{L}_0 b - (1 + \mathrm{i}r)\mathcal{R}(b) - (\mathrm{i}\sigma - \mu)\mathcal{N}(a_0 + b)(a_0 + b) = 0,$$

where $\mathcal{L}_0$ is the linearization about a_0 such that $\mathrm{Ker}(\mathcal{L}_0) = \mathrm{Span}\{a_0'\}$ by our non-degeneracy assumption and

$$\mathcal{R}(b) := \bar{a}_0 b^2 + 2a_0|b|^2 + |b|^2 b.$$

Note that $\mathcal{F} \in C^1(X_0 \times \mathbb{R} \times \mathbb{R}, L^2(\mathbb{T}))$. Next, we use the Lyapunov-Schmidt reduction to find non-trivial solutions of the equation $\mathcal{F}(b, \sigma, \mu) = 0$. Let us define the orthogonal projections

$$P : L^2(\mathbb{T}) \to \text{Span}\{a_0^*\} \quad \text{and} \quad P^\perp := I - P : L^2(\mathbb{T}) \to \text{Span}\{a_0^*\}^{\perp_{L^2}} = \text{Ran}(\mathcal{L}_0).$$

Then, $\mathcal{F}(b, \sigma, \mu) = 0$ is equivalent to the system

$$P^\perp \mathcal{F}(b, \sigma, \mu) = 0, \tag{11.19}$$

$$P\mathcal{F}(b, \sigma, \mu) = 0, \tag{11.20}$$

which we solve subsequently. Consider the auxiliary function

$$\mathcal{G} : X_0 \times \mathbb{R} \times \mathbb{R} \to \text{Ran}(\mathcal{L}_0) = \text{Span}\{a_0^*\}^{\perp_{L^2}}, \quad \mathcal{G}(b, \sigma, \mu) := P^\perp \mathcal{F}(b, \sigma, \mu).$$

Then, $\mathcal{G}$ is continuously differentiable, $\mathcal{G}(0, 0, 0) = 0$, and

$$\partial_b \mathcal{G}(0, 0, 0) = P^\perp \mathcal{L}_0 : X_0 \to \text{Span}\{a_0^*\}^{\perp_{L^2}}$$

is an isomorphism by construction. The Implicit Function Theorem yields that there exist open neighborhoods $U \subset X_0$ of 0, $V \subset \mathbb{R}^2$ of 0 and a continuous differentiable map $V \ni (\sigma, \mu) \to b(\sigma, \mu) \in U$ such that $b(0, 0) = 0$ and

$$(b, (\sigma, \mu)) \in U \times V \text{ solves } \mathcal{G}(b, \sigma, \mu) = 0$$
$$\Leftrightarrow \; (\sigma, \mu) \in V, \; (b, \sigma, \mu) = (b(\sigma, \mu), \sigma, \mu).$$

This means that for $(\sigma, \mu) \in V$ the triple $(b(\sigma, \mu), \sigma, \mu)$ solves (11.19). We substitute the triple $b(\sigma, \mu), \sigma, \mu)$ into the bifurcation equation (11.20) to find

$$P\mathcal{F}(b(\sigma, \mu), \sigma, \mu) = 0.$$

The equation is equivalent to the 2-dimensional problem $g(\sigma, \mu) = 0$ where

$$g(\sigma, \mu) := \big\langle (1 + \mathrm{i}r)\mathcal{R}\big(b(\sigma, \mu)\big) + (\mathrm{i}\sigma - \mu)\mathcal{N}\big(a_0 + b(\sigma, \mu)\big)\big(a_0 + b(\sigma, \mu)\big), a_0^* \big\rangle_{L^2}.$$

It is straight forward to check that $g(0, 0) = 0$ and

$$\nabla g(0, 0) = -\begin{bmatrix} \operatorname{Im} \alpha \\ \operatorname{Re} \alpha \end{bmatrix}.$$

Now, the assertions of Theorem 11.3 follow directly by another application of the Implicit Function Theorem to the problem $g(\sigma, \mu) = 0$ which finishes the proof. □

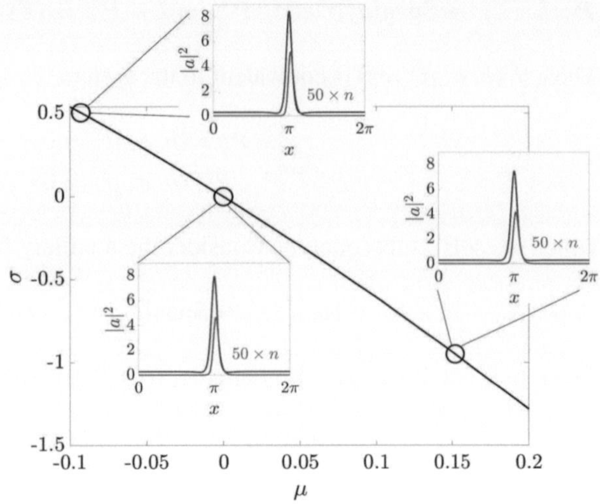

Fig. 11.3 Bifurcation diagram with inserts of solution profiles of (a, n). The free-carrier density is scaled by a factor of 50 to improve the visibility. Parameters: $d = 0.1, \zeta = 3.9, f = 2, r = 0.02, \tau = 0.1, v = 1$

11.3.2 Numerical Illustration of the Existence Result

In the following, we describe an algorithm to numerically approximate the solution curve obtained in Theorem 11.3. Our method relies upon the continuation and bifurcation algorithm of the Matlab package `pde2path`, cf. [29]. We recall that the continuation procedure of Theorem 11.3 requires a non-trivial 2π-periodic starting solution a_0 of (11.17). Such solutions can be found numerically from bifurcations of the curve of constant solutions of (11.17) parameterized by ζ and we refer to [5] for a detailed numerical bifurcation analysis of (11.17).

For the first step of our algorithm, we fix the parameters d, ζ, f, r, τ, v and compute a non-trivial starting solution a_0 of (11.17). In the second step, we implement the orthogonality condition $\langle b, a_0' \rangle_{L^2} = 0$ and perform a two parameter continuation in the parameters σ, μ, which yields a curve of solutions to (11.14). Projecting this curve onto the σ-μ-plane yields the result depicted in Fig. 11.3, where each point on this curve now corresponds to an approximation of a non-homogeneous solution of (11.14). The top and right inserts of the figure display profiles of the approximated solutions for $\sigma\mu \neq 0$, and the bottom insert corresponds to the solution a_0, n_0 at the bifurcation point. Note that the free carrier densities n can be computed by applying a discretized version of the solution operator (11.15) to the functions a. In this example the solution curve only exists if $\sigma\mu \leq 0$. The questions whether non-constant solutions also exist if $\sigma\mu > 0$ remains open, even after performing numerous simulations for different configurations of parameters and starting points a_0.

11.3.3 Continuability Condition in the Limit of Small Damping and Forcing

The continuability condition $\alpha \neq 0$ of Theorem 11.3 becomes more explicit in the limit of small damping and forcing and small two-photon absorption. This means that instead of (11.14) we consider now

$$
\begin{aligned}
-da'' + (\zeta - \mathrm{i}\varepsilon - (\mathrm{i}\sigma - \mu)n - (1 + \mathrm{i}\varepsilon r)|a|^2)a + \mathrm{i}\varepsilon f &= 0, \\
vn' + \frac{n}{\tau} - \varepsilon r|a|^4 &= 0,
\end{aligned}
\tag{11.21}
$$

with $|\varepsilon| \ll 1$. By resolving the second equation in (11.21) we obtain the compact form

$$
-da'' + \big(\zeta - \mathrm{i}\varepsilon - (\mathrm{i}\sigma - \mu)\mathcal{N}_\varepsilon(a) - (1 + \mathrm{i}\varepsilon r)|a|^2\big)a + \mathrm{i}\varepsilon f = 0
$$

where we have set $\mathcal{N}_\varepsilon = \varepsilon\mathcal{N}$ while we have kept the definition (11.15) for $\mathcal{N}$. Let $a_0 \in H^2(\mathbb{T})$ be a non-degenerate solution of

$$
-da'' + (\zeta - \mathrm{i}\varepsilon - (1 + \mathrm{i}\varepsilon r)|a|^2)a + \mathrm{i}\varepsilon f = 0, \quad |\varepsilon| \ll 1. \tag{11.22}
$$

Using regular perturbation theory, we expand a_0 in powers of ε:

$$
a_0 = a_{00} + \varepsilon a_{01} + O(\varepsilon^2), \tag{11.23}
$$

where a_{00} is a solution of the nonlinear Schrödinger equation (NLS)

$$
-da'' + \zeta a - |a|^2 a = 0
$$

and a_{01} is found from the linear inhomogeneous equation

$$
-da_{01}'' + \zeta a_{01} - 2|a_{00}|^2 a_{01} - a_{00}^2 \bar{a}_{01} = \mathrm{i}(1 + r|a_{00}|^2)a_{00} - \mathrm{i}f.
$$

In particular, (11.23) yields an expansion of a_0^* of the form

$$
a_0^* = a_{00}' + O(|\varepsilon|)
$$

allowing us to obtain an expansion of α in ε:

$$
\begin{aligned}
\alpha &= \int_0^{2\pi} N_\varepsilon(a_0) a_0 \bar{a}_0^* dx \\
&= \int_0^{2\pi} \varepsilon N(a_{00}) a_{00} \bar{a}_{00}' dx + O(\varepsilon^2) \\
&= -\frac{1}{2} \int_0^{2\pi} \varepsilon \partial_x N(a_{00}) |a_{00}|^2 dx + O(\varepsilon^2) \\
&= \frac{1}{2} \left(\frac{1}{\tau v} \int_0^{2\pi} \varepsilon N(a_{00}) |a_{00}|^2 dx - \frac{\varepsilon r}{v} \|a_{00}\|_{L^6}^6 \right) + O(\varepsilon^2).
\end{aligned}
$$

A sufficient condition for $\alpha \neq 0$ is that

$$
\begin{aligned}
&\frac{\frac{1}{\tau v}}{e^{\frac{2\pi}{\tau v}} - 1} \int_0^{2\pi} \int_0^{2\pi} e^{\frac{y-x}{\tau v}} |a_{00}(y)|^4 dy |a_{00}(x)|^2 dx \\
&\qquad + \frac{1}{\tau v} \int_0^{2\pi} \int_0^{x} e^{\frac{y-x}{\tau v}} |a_{00}(y)|^4 dy |a_{00}(x)|^2 dx \neq \|a_{00}\|_{L^6}^6,
\end{aligned}
$$

provided that $\varepsilon \neq 0$ is sufficiently small. Since the function $\phi : (0, \infty) \to \mathbb{R}$ defined by

$$
\begin{aligned}
\phi(\xi) := &\frac{\xi}{e^{2\pi\xi} - 1} \int_0^{2\pi} \int_0^{2\pi} e^{(y-x)\xi} |a_{00}(y)|^4 dy |a_{00}(x)|^2 dx \\
&+ \xi \int_0^{2\pi} \int_0^{x} e^{(y-x)\xi} |a_{00}(y)|^4 dy |a_{00}(x)|^2 dx
\end{aligned}
$$

is in general not constant, $\alpha \neq 0$ can be achieved for suitable choices of the parameters τ, v.

In the limit $\tau v \to +\infty$ (corresponding to $\xi \to 0$), we find that $\alpha \neq 0$ further simplifies to

$$
\|a_{00}\|_{L^4}^4 \|a_{00}\|_{L^2}^2 \neq 2\pi \|a_{00}\|_{L^6}^6. \tag{11.24}
$$

This condition can be checked numerically or even analytically, e.g. for those solutions of NLS which are available in an explicit form. To illustrate usability, we present an example where we numerically verify (11.24) for the cnoidal wave solutions of NLS. Let us define

$$
\phi(x, k) = \sqrt{2}k \frac{2\,K(k)}{\pi} \mathrm{cn}\left(\frac{2\,K(k)}{\pi} x, k \right),
$$

where $\mathrm{cn}(\cdot, k)$ is the Jacobi elliptic function, $K(k)$ is the complete elliptic integral of the first kind, and $k \in (0, 1)$ is the elliptic modulus. We refer to [9] for precise definitions. Then, $\phi(\cdot, k)$ is a 2π-periodic solution of

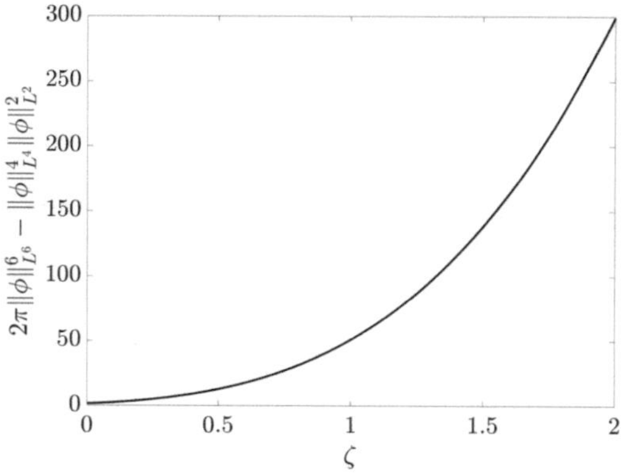

Fig. 11.4 Continuablity condition (11.24) for cnoidal waves as a function of $\zeta = \zeta(k)$

$$-a'' + \zeta(k)a - a^3 = 0, \quad \text{where} \quad \zeta(k) = \frac{4\,K^2(k)(2k^2-1)}{\pi^2}.$$

Numerical simulations in Fig. 11.4 show that (11.24) is satisfied at least for all values of $\zeta \in [0, 2]$ and thus solutions a_0 of (11.22) close to the cnoidal waves can be continued to solutions of (11.21).

11.4 Conclusion

To summarize, we have studied dynamical and stationary aspects of the Lugiato-Lefever equation with two-photon absorption and free carriers. First, we obtained a rigorous analytical result for modulation instability of constant steady states which is illustrated by numerical simulation of the instability region. Second, we gave an existence result for non-constant solutions in the range of small free-carrier parameters σ and μ. The results were illustrated by numerical simulations and the continuability condition was verified in the small forcing, small damping, and small TPA limit. The following questions remain open:

(i) Does modulation instability lead to the existence of stationary solutions? This is particularly interesting in the case $\mu\sigma > 0$ for which numerical results are not (yet) available. Note that the sign of $\mu\sigma$ is not determined by Theorem 11.3.
(ii) What are the stability properties of the solutions obtained in Theorem 11.3?

Acknowledgements Funded by the Deutsche Forschungsgemeinschaft (DFG, German Research Foundation)—Project-ID 258734477—SFB 1173.

Competing Interests The authors have no conflicts of interest to declare that are relevant to the content of this chapter.

References

1. Chembo, Y.K., Menyuk, C.R.: Spatiotemporal Lugiato-Lefever formalism for Kerr-comb generation in whispering-gallery-mode resonators. Phys. Rev. A **87**, 053852 (2013). https://doi.org/10.1103/PhysRevA.87.053852. https://link.aps.org/doi/10.1103/PhysRevA.87.053852
2. Chembo, Y.K., Yu, N.: Modal expansion approach to optical-frequency-comb generation with monolithic whispering-gallery-mode resonators. Phys. Rev. A **82**, 033801 (2010). https://doi.org/10.1103/PhysRevA.82.033801. https://link.aps.org/doi/10.1103/PhysRevA.82.033801
3. Delcey, L., Haragus, M.: Periodic waves of the Lugiato-Lefever equation at the onset of Turing instability. Philos. Trans. of the Roy. Soc. A **376**(2117), 20170188 (2018). https://doi.org/10.1098/rsta.2017.0188. https://doi.org/10.1098/rsta.2017.0188
4. Diddams, S.A., Udem, T., Bergquist, J.C., Curtis, E.A., Drullinger, R.E., Hollberg, L., Itano, W.M., Lee, W.D., Oates, C.W., Vogel, K.R., Wineland, D.J.: An optical clock based on a single trapped 199Hg+ ion. Science **293**(5531), 825–828 (2001). https://doi.org/10.1126/science.1061171. https://www.science.org/doi/abs/10.1126/science.1061171
5. Gärtner, J., Mandel, R., Reichel, W.: The Lugiato-Lefever equation with nonlinear damping caused by two photon absorption. J. Dynam. Differential Equations **34**(3), 2201–2227 (2022). https://doi.org/10.1007/s10884-021-09943-x. https://doi.org/10.1007/s10884-021-09943-x
6. Gershgorin, S.A.: Über die Abgrenzung der Eigenwerte einer Matrix. Bulletin de l'Académie des Sciences de l'URSS. **6**, 749–754 (1931)
7. Godey, C.: A bifurcation analysis for the Lugiato-Lefever equation. The European Physical Journal D **71**(5), 131 (2017). https://doi.org/10.1140/epjd/e2017-80057-2. https://doi.org/10.1140/epjd/e2017-80057-2
8. Godey, C., Balakireva, I.V., Coillet, A., Chembo, Y.K.: Stability analysis of the spatiotemporal Lugiato-Lefever model for Kerr optical frequency combs in the anomalous and normal dispersion regimes. Phys. Rev. A **89**, 063814 (2014). https://doi.org/10.1103/PhysRevA.89.063814. http://link.aps.org/doi/10.1103/PhysRevA.89.063814
9. Gradshteyn, I.S., Ryzhik, I.M.: Table of integrals, series, and products, eighth edn. Elsevier/Academic Press, Amsterdam (2015). Translated from the Russian, Translation edited and with a preface by Daniel Zwillinger and Victor Moll
10. Haldar, R., Roy, A., Mondal, P., Mishra, V., Varshney, S.K.: Free-carrier-driven Kerr frequency comb in optical microcavities: Steady state, bistability, self-pulsation, and modulation instability. Phys. Rev. A **99**, 033848 (2019). https://doi.org/10.1103/PhysRevA.99.033848. https://link.aps.org/doi/10.1103/PhysRevA.99.033848
11. Hamerly, R., Gray, D., Rogers, C., Jamshidi, K.: Conditions for parametric and free-carrier oscillation in silicon ring cavities. J. Lightwave Technol. **36**(19), 4671–4677 (2018). https://opg.optica.org/jlt/abstract.cfm?URI=jlt-36-19-4671
12. Hansson, T., Modotto, D., Wabnitz, S.: Dynamics of thc modulational instability in microresonator frequency combs. Phys. Rev. A **88**, 023819 (2013). https://doi.org/10.1103/PhysRevA.88.023819. https://link.aps.org/doi/10.1103/PhysRevA.88.023819
13. Hansson, T., Wabnitz, S.: Dynamics of microresonator frequency comb generation: models and stability. Nanophotonics **5**(2), 231–243 (2016). https://doi.org/10.1515/nanoph-2016-0012. https://doi.org/10.1515/nanoph-2016-0012
14. Haragus, M., Johnson, M.A., Perkins, W.R., de Rijk, B.: Nonlinear modulational dynamics of spectrally stable Lugiato-Lefever periodic waves. Ann. Inst. H. Poincaré C Anal. Non Linéaire **40**(4), 769–802 (2023). https://doi.org/10.4171/aihpc/65. https://doi.org/10.4171/aihpc/65
15. Herr, T., Hartinger, K., Riemensberger, J., Wang, C.Y., Gavartin, E., Holzwarth, R., Gorodetsky, M.L., Kippenberg, T.J.: Universal formation dynamics and noise of Kerr-frequency combs in microresonators. Nature photonics **6**(7), 480–487 (2012)
16. Lau, R.K.W., Lamont, M.R.E., Okawachi, Y., Gaeta, A.L.: Effects of multiphoton absorption on parametric comb generation in silicon microresonators. Opt. Lett. **40**(12), 2778–2781 (2015). https://doi.org/10.1364/OL.40.002778. https://opg.optica.org/ol/abstract.cfm?URI=ol-40-12-2778

17. Mandel, R., Reichel, W.: A priori bounds and global bifurcation results for frequency combs modeled by the Lugiato-Lefever equation. SIAM J. Appl. Math. **77**(1), 315–345 (2017). https://doi.org/10.1137/16M1066221. https://doi.org/10.1137/16M1066221
18. Marin-Palomo, P., Kemal, J.N., Karpov, M., Kordts, A., Pfeifle, J., Pfeiffer, M.H., Trocha, P., Wolf, S., Brasch, V., Anderson, M.H., et al.: Microresonator-based solitons for massively parallel coherent optical communications. Nature **546**(7657), 274–279 (2017)
19. Miyaji, T., Ohnishi, I., Tsutsumi, Y.: Bifurcation analysis to the Lugiato-Lefever equation in one space dimension. Phys. D **239**(23–24), 2066–2083 (2010). https://doi.org/10.1016/j.physd.2010.07.014. http://dx.doi.org/10.1016/j.physd.2010.07.014
20. Parra-Rivas, P., Gomila, D., Gelens, L., Knobloch, E.: Bifurcation structure of localized states in the Lugiato-Lefever equation with anomalous dispersion. Phys. Rev. E **97**(4), 042204 (2018). https://doi.org/10.1103/PhysRevE.97.042204. https://journals.aps.org/pre/abstract/10.1103/PhysRevE.97.042204
21. Parra-Rivas, P., Gomila, D., Matías, M.A., Coen, S., Gelens, L.: Dynamics of localized and patterned structures in the Lugiato-Lefever equation determine the stability and shape of optical frequency combs. Phys. Rev. A **89**, 043813 (2014). https://doi.org/10.1103/PhysRevA.89.043813. https://link.aps.org/doi/10.1103/PhysRevA.89.043813
22. Parra-Rivas, P., Knobloch, E., Gomila, D., Gelens, L.: Dark solitons in the Lugiato-Lefever equation with normal dispersion. Phys. Rev. A **93**(6), 1–17 (2016). https://doi.org/10.1103/PhysRevA.93.063839. https://journals.aps.org/pra/abstract/10.1103/PhysRevA.93.063839
23. Sahoo, A., Roy, S.: Stability and variational analysis of cavity solitons under various perturbations. Phys. Rev. A **100**, 053814 (2019). https://doi.org/10.1103/PhysRevA.100.053814. https://link.aps.org/doi/10.1103/PhysRevA.100.053814
24. Stanislavova, M., Stefanov, A.G.: Asymptotic stability for spectrally stable Lugiato-Lefever solitons in periodic waveguides. J. Math. Phys. **59**(10), 101502, 12 (2018). https://doi.org/10.1063/1.5048017. https://doi.org/10.1063/1.5048017
25. Suh, M.G., Vahala, K.J.: Soliton microcomb range measurement. Science **359**(6378), 884–887 (2018). https://doi.org/10.1126/science.aao1968. https://www.science.org/doi/abs/10.1126/science.aao1968
26. Trocha, P., Gärtner, J., Marin-Palomo, P., Freude, W., Reichel, W., Koos, C.: Analysis of Kerr comb generation in silicon microresonators under the influence of two-photon absorption and fast free-carrier dynamics. Phys. Rev. A **103**, 063515 (2021). https://doi.org/10.1103/PhysRevA.103.063515. https://link.aps.org/doi/10.1103/PhysRevA.103.063515
27. Trocha, P., Karpov, M., Ganin, D., Pfeiffer, M.H.P., Kordts, A., Wolf, S., Krockenberger, J., Marin-Palomo, P., Weimann, C., Randel, S., Freude, W., Kippenberg, T.J., Koos, C.: Ultrafast optical ranging using microresonator soliton frequency combs. Science **359**(6378), 887–891 (2018). https://doi.org/10.1126/science.aao3924. https://www.science.org/doi/abs/10.1126/science.aao3924
28. Udem, T., Holzwarth, R., Hänsch, T.W.: Optical frequency metrology. Nature **416**(6877), 233–237 (2002)
29. Uecker, H., Wetzel, D., Rademacher, J.D.M.: pde2path–a Matlab package for continuation and bifurcation in 2D elliptic systems. Numer. Math. Theory Methods Appl. **7**(1), 58–106 (2014). https://doi.org/10.4208/nmtma.2014.1231nm. https://doi.org/10.4208/nmtma.2014.1231nm

Chapter 12
The Reversible Simulator a Data-Driven Approach for Solving Forward and Inverse Problems

Tristan van Leeuwen

Abstract Inverse problems are prevalent across various fields of science and engineering, with applications spanning medical imaging, materials science, non-destructive testing, astrophysics, climate science, and seismology. These problems share a common goal: estimating a quantity of interest from measurements obtained under specific experimental conditions. Traditionally, simulation and inference have been considered separate tasks. However, recent advancements in conditional variational inference offer a promising approach to integrate these tasks within a single framework. This paper reviews some of these advancements in the context of linear inverse problems, focusing on the use of straightforward generative models for inference and experimental design in computed tomography.

Keywords Inverse problems · Bayesian inference · Uncertainty quantification

12.1 Introduction

Inverse problems are ubiquitous in science and engineering, with applications including medical imaging, materials science, non-destructive testing, astrophysics, climate science, and seismology. What unifies these applications, is that the aim is to estimate a quantity of interest, u, from measurements, f^δ, which are obtained using certain experimental settings, s. We assume that the data-generating process can be simulated by some combination of deterministic (i.e., PDEs, integral operators, ...) and stochastic (i.e., Gaussian process) models. **Simulating the underlying process often entails solving large-scale PDEs, and is computationally expensive**.

A popular framework for modeling, analyzing and solving inverse problems is the Bayesian one [11, 12]. Here, both the underlying data-generating process and prior information are described by probability distributions. The likelihood, describing the data-generating process is denoted by $\pi_{\text{data}}(f^\delta|s, u)$, while prior information on u is encoded by $\pi_{\text{prior}}(u)$. Accordingly, the posterior is denoted by

T. van Leeuwen (✉)
Centrum Wiskunde & Informatica, Amsterdam, Netherlands
e-mail: t.van.leeuwen@cwi.nl

A. Ott et al. (eds.), *Applications of Mathematics in Sciences, Engineering, and Economics*, Springer Proceedings in Mathematics & Statistics 515,
https://doi.org/10.1007/978-3-032-01279-1_12

$$\pi_{\text{post}}(u|s, f^\delta) \propto \pi_{\text{data}}(f^\delta|s, u)\pi_{\text{prior}}(u). \tag{12.1}$$

In a way, this posterior is the answer to the inverse problem; it captures all relevant prior assumptions and allows one to make probabilistic statements about the corresponding quantity of interest.

In practical applications, one often only considers the maximum a-posteriori probability (MAP) estimate

$$\widehat{u}_{\text{MAP}} = \underset{u}{\text{argmin}} - \log \pi_{\text{post}}(u|s, f^\delta). \tag{12.2}$$

The resulting optimization problem is typically solved iteratively, where each iteration requires at least one forward simulation of the system. **This makes MAP-estimation at least an order of magnitude more expensive than forward simulation.**

The related uncertainty is sometimes quantified by considering a Gaussian approximation of the posterior with mean $\widehat{u}_{\text{MAP}}$ and covariance

$$\widehat{\Sigma}_{\text{MAP}} = -\left[\nabla_u^2 \log \pi_{\text{post}}(\widehat{u}_{\text{MAP}}|s, f^\delta)\right]^{-1}. \tag{12.3}$$

To obtain more accurate and more detailed information on the uncertainty, one would need to generate samples from the posterior (12.1). Despite recent developments in Markov Chain Monte Carlo (MCMC) methods, this is not feasible for large-scale applications. An alternative approach is the so-called randomize-then-optimize approach, which solves the inference problem (12.2) for a perturbed data set. Each sample from the posterior in this approach requires computing a MAP estimate. **This currently makes uncertainty quantification (UQ) for large-scale problems at least an order of magnitude more expensive than MAP-estimation, and hence at least two orders of magnitude more expensive than forward simulation.**

This has given rise to recent research on *variational inference* using generative models, which can efficiently generate samples from complex probability distributions, which were fitted to training data. Early work on this topic includes [1], while more recent approaches are described in [4, 5, 7, 8, 10].

These developments allow us to define a *reversible simulator* that can do forward simulation and inference at the same computational cost. Figure 12.1 shows a schematic depiction of this concept. In principle, this lowers the computational cost of UQ by at least an order of magnitude, though this comes at the cost of an off-line training step. Still, for applications where data are collected routinely under similar circumstances this may prove feasible.

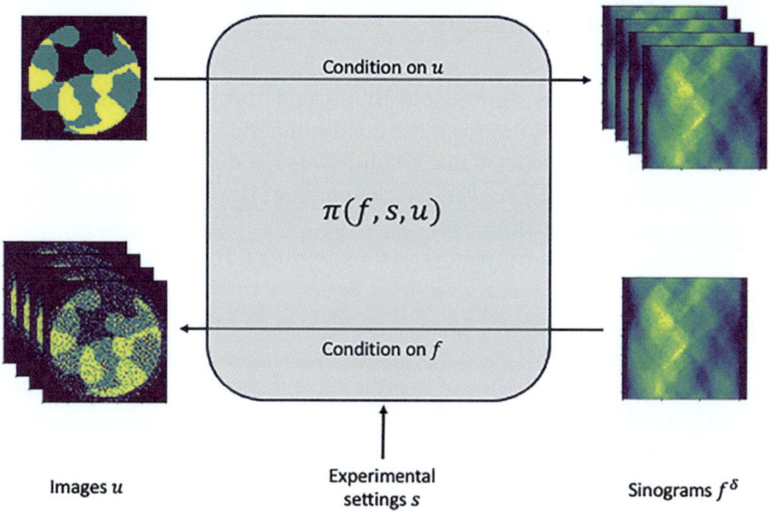

Fig. 12.1 In the forward mode, the Reversible Simulator produces samples from the likelihood; in this case simulating noisy sinograms for a given input image. In the reverse mode, it produces samples from the posterior from which we can compute, for example, the posterior mean and variance

12.1.1 Approach and Challenges

To be able to access the posterior and the likelihood, we need to learn a joint probability density function (pdf) $\pi(f, s, u)$ from examples $\{(f_i, s_i, u_i)\}_{i=1}^N$. This then allows us to simulate measurements by conditioning on (s, u) and to generate samples from the posterior by conditioning it on (f^δ, s).

The main challenges include *(i)* including the underlying physics to enable effectively learning the underlying distribution from a limited number of samples, *(ii)* parametrization of π to allow for conditional sampling, and *(iii)* computational efficiency, to allow for fast simulation and inference.

12.1.2 Contributions and Outline

In this paper, we adopt conditional normalizing flows for simulation, inference, and experimental design arising from certain linear inverse problems. In particular, we propose simple architectures for treating linear inverse problems with additive Gaussian noise.

The paper is organized is as follows. In Sect. 12.2, we outline the underlying methodology in more detail. Some illustrative examples are presented in Sect. 12.3. Finally, Sect. 12.4 discusses the results and concludes the paper.

12.2 A Reversible Simulator

The reversible simulator is represented by an underlying probability distribution $\pi(f, s, u)$. By drawing samples from the conditional distribution for a given (s, u), we can simulate measurements, while conditioning the distribution on a given (f, s) allows us to do inference. Moreover, we can perform Bayesian experimental design by maximizing a utility function over s.

A well-known approach for parametrizing a pdf is through a change of variables

$$\pi(x) = F^{\sharp}\pi_0(x) \equiv \pi_0\left(F(x)\right)|\nabla F(x)|, \tag{12.4}$$

where $F^{\sharp}\pi_0$ denotes the *pull-back* of the reference density π_0, and F is an invertible transport map with inverse $G = F^{-1}$.

Samples from π can be generated as $x = G(y)$, with y drawn from π_0. Here, we denote by x the collection of variables (f, s, u) while the corresponding latent variables are denoted by $y = (g, t, v)$. Likewise, we split the transform and its inverse in 3 blocks corresponding to the 3 variables, $F = (F_g, F_t, F_v)$, $G = (G_f, G_s, G_u)$. Throughout we will assume that the reference density can be written as the product of its marginals, which we will also denote by π_0 in a slight abuse of notation; $\pi_0(y) \equiv \pi_0(g)\pi_0(t)\pi_0(v)$.

Typically, one chooses π_0 to be a standard Normal distribution. In this case, we learn the transport map by minimizing the Kullback–Leibler (KL) divergence between π and $F^{\sharp}\pi_0$:

$$\min_F \mathbb{E}_{x\sim\pi}\ \tfrac{1}{2}\|F(x)\|_2^2 - \log|\nabla F(x)|. \tag{12.5}$$

Having parametrized the mapping in a suitable manner we pose (12.5) as a non-linear optimization problem which we can solve with stochastic gradient descent (SGD). Certain triangular architectures will allow for efficient computation of the required Jacobian, while the invertibility of the transform can be exploited to efficiently compute the gradient of the objective w.r.t. its parameters. See [6] for a recent overview.

The resulting mapping $\widehat{F}$ now represents our reversible simulator. For simulation, we need to sample from the conditional distribution

$$\widehat{\pi}_{\text{like}}(f|s, u) = \frac{\pi_0(\widehat{F}(f, s, u))|\nabla\widehat{F}(f, s, u)|}{\int \pi_0(\widehat{F}(f, s, u))|\nabla\widehat{F}(f, s, u)|\mathrm{d}f}.$$

The procedure for doing this is sketched in Algorithm 1. Note that sampling from this distribution does not require us to compute the normalizing constant. Likewise, samples from the posterior

$$\widehat{\pi}_{\text{post}}(u|s, f) = \frac{\pi_0(\widehat{F}(f, s, u))|\nabla\widehat{F}(f, s, u)|}{\iint \pi_0(\widehat{F}(f, s, u))|\nabla\widehat{F}(f, s, u)|\mathrm{d}s\mathrm{d}u}$$

can be generated according to Algorithm 2.

In both algorithms, the non-linear system of equations can be solved by applying a Newton-like method to $H_{\text{simulation}}(f, t, v) = F(f, s, u) - (g, t, v)$ and $H_{\text{inference}}(g, t, u) = F(f, s, u) - (g, t, v)$ respectively. Alternatively, (accelerated) fixed point iterations can be used if we have access to the inverse map G [3].

Algorithm 1 Simulation

Require: (s, u), transport map F
Ensure: f drawn from the conditional distribution $\widehat{\pi}(f|s, u)$
  Draw g from π_0
  Solve for (f, t, v) from $(g, t, v) = F(f, s, u)$
  return f, (t, v)

Algorithm 2 Inference

Require: (f, s), transport map F
Ensure: u drawn from the posterior distribution $\widehat{\pi}(u|s, f)$
  Draw v from π_0
  Solve for (g, t, u) from $(g, t, v) = F(f, s, u)$
  return u, (t, g)

Based on these capabilities we can perform Bayesian experimental design (BED) by optimizing some utility function. A well-known example is *D-optimal* design, which aims to maximize the expected information gain (EIG)

$$\max_{s} \mathbb{E}_{(f,u)\sim\widehat{\pi}_{\text{like}}(f|s,u)\pi_{\text{prior}}(u)} \left[\log \widehat{\pi}_{\text{post}}(u|s, f) - \log \pi_{\text{prior}}(u)\right]. \tag{12.6}$$

where π_{prior} denotes the prior distribution over which we want to optimize the design. An alternative is *A-optimal* design, which minimizes the expected risk:

$$\min_{s} \mathbb{E}_{\widehat{u}\sim\pi_{\text{post}}(u|s,f), f\sim\pi_{\text{like}}(f|s,u), u\sim\pi_{\text{prior}}} \|\widehat{u} - u\|_2^2. \tag{12.7}$$

In practice, we would estimate the required quantities using a Monte Carlo approximation [9].

12.2.1 Practical Implementation with Triangular Maps

The Algorithms can be greatly simplified if we construct triangular maps (so-called Knothe–Rosenblatt arrangements) [2]. To this end, we introduce an upper and a lower triangular transport map

$$\overrightarrow{F}(f, s, u) = \left(\overrightarrow{F}_g(f, s, u), \overrightarrow{F}_t(s, u), \overrightarrow{F}_v(u)\right),$$

$$\overleftarrow{F}(f,s,u) = \left(\overleftarrow{F}_g(f), \overleftarrow{F}_t(f,s), \overleftarrow{F}_v(f,s,u)\right).$$

A nice feature is that the variational problem (12.5) now separates over the individual components of the triangular maps. Since we are only interested in $\overrightarrow{F}_g$ (for simulation), $\overleftarrow{F}_v$ (for inference), and $\overrightarrow{F}_v$ (for prior sampling), we only need to learn these components by solving

$$\min_{\overrightarrow{F}_g} \mathbb{E}_{(f,s,u)\sim\pi} \tfrac{1}{2}\|\overrightarrow{F}_g(f,s,u)\|_2^2 - \log|\nabla_f \overrightarrow{F}_g(f,s,u)|,$$

$$\min_{\overleftarrow{F}_v} \mathbb{E}_{(f,s,u)\sim\pi} \tfrac{1}{2}\|\overleftarrow{F}_v(f,s,u)\|_2^2 - \log|\nabla_u \overleftarrow{F}_v(f,s,u)|.$$

$$\min_{\overrightarrow{F}_v} \mathbb{E}_{u\sim\pi} \tfrac{1}{2}\|\overrightarrow{F}_v(u)\|_2^2 - \log|\nabla_u \overleftarrow{F}_v(u)|.$$

The estimated conditional distributions needed for simulation and inference are now given by

$$\widehat{\pi}_{\text{like}}(f|s,u) = \pi_0(\overrightarrow{F}_g(f,s,u))|\nabla_f \overrightarrow{F}_g(f,s,u)|,$$

$$\widehat{\pi}_{\text{post}}(u|s,f) = \pi_0(\overleftarrow{F}_v(f,s,u))|\nabla_u \overleftarrow{F}_v(f,s,u)|.$$

Likewise, the prior is given by

$$\widehat{\pi}_{\text{prior}}(u) = \pi_0(\overrightarrow{F}_v(u))|\nabla_u \overrightarrow{F}_v(u)|.$$

The upper triangular map can be used for simulation and reduces Algorithm 1 to solving f from

$$\overrightarrow{F}_g(f,s,u) = g, \quad \text{with } g \text{ drawn from } \pi_0,$$

since this arrangement ensures that the map $f \mapsto \overrightarrow{F}_g(f,s,u)$ is invertible. Likewise, the lower triangular map can be used for inference, reducing Algorithm 2 to solving u from

$$\overleftarrow{F}_v(f,s,u) = v, \quad \text{with } v \text{ drawn from } \pi_0,$$

which is facilitated by the fact that the map $u \mapsto \overleftarrow{F}_v(f,s,u)$ is guaranteed to be invertible.

12.2.2 An Architecture for Linear Inverse Problems

A template for linear inverse problems is the additive Gaussian noise model with a Gaussian prior:

$$f \sim \mathcal{N}(Ku, \Sigma_{\text{like}}), \quad u \sim \mathcal{N}(\mu_{\text{prior}}, \Sigma_{\text{prior}}),$$

with $K \in \mathbb{R}^{m\times n}$ denoting the forward operator. The resulting posterior is a normal distribution with mean and covariance

$$\mu_{\text{post}} = \mu_{\text{prior}} + \Sigma_{\text{post}} K^T \Sigma_{\text{like}}^{-1} \left(f - K\mu_{\text{prior}}\right), \quad \Sigma_{\text{post}} = \left(K^T \Sigma_{\text{like}}^{-1} K + \Sigma_{\text{prior}}^{-1}\right)^{-1}.$$

The parameter s appears implicitly in Σ_{like} and/or K. The *true* transport maps would in this case be given by

$$\overrightarrow{F}_g(f, s, u) = \Sigma_{\text{like}}^{-1/2} \left(f - Ku\right), \tag{12.8}$$

$$\overleftarrow{F}_v(f, s, u) = \Sigma_{\text{post}}^{-1/2} \left(u - \mu_{\text{prior}}\right) + \Sigma_{\text{post}}^{1/2} K^T \Sigma_{\text{like}}^{-1}(f - K\mu_{\text{prior}}), \tag{12.9}$$

$$\overrightarrow{F}_v(u) = \Sigma_{\text{prior}}^{-1/2} \left(u - \mu_{\text{prior}}\right). \tag{12.10}$$

We propose the following architecture for the transport maps, which preserve these linear relationships.

$$\overrightarrow{F}_g(f, s, u) = \overrightarrow{A}_{gf}(s) f + \overrightarrow{A}_{gu}(s) u + \overrightarrow{b}_g(s), \tag{12.11}$$

$$\overleftarrow{F}_v(f, s, u) = \overleftarrow{A}_{vf}(s) f + \overleftarrow{A}_{vu}(s) u + \overleftarrow{b}_v(s), \tag{12.12}$$

$$\overrightarrow{F}_v(u) = \overrightarrow{A}_{vu} u + \overrightarrow{b}_v. \tag{12.13}$$

Here, $\overrightarrow{A}_{vu} \in \mathbb{R}^{n\times n}$, $\overrightarrow{A}_{gf}(s) \in \mathbb{R}^{m\times m}$ are invertible upper triangular matrices, $\overleftarrow{A}_{vu}(s) \in \mathbb{R}^{n\times n}$ is an invertible lower triangular matrix. The dependency of the weights on s is adapted to the application at hand, a few examples of which are shown in a later section.

12.3 Numerical Examples

The following examples illustrate the workings of the reversible simulator.

12.3.1 A Gaussian Example

For the first example, we consider the following data generating process

$$f^\delta \sim \mathcal{N}(Ku, s^2 I),\ u \sim \mathcal{N}(\mu_{\text{prior}}, \Sigma_{\text{prior}}), \tag{12.14}$$

with

$$K = \begin{pmatrix} 1 & 0.5 & 0 & 0 \\ 0.5 & 1 & 0.5 & 0 \\ 0 & 0.5 & 1 & 0.5 \\ 0 & 0 & 0.5 & 1 \end{pmatrix}, \quad \mu_{\text{prior}} = \mathbf{1}, \quad \Sigma_{\text{prior}} = 0.1I.$$

12.3.1.1 Parametrizing the Transport Maps

We define the transport maps as

$$\begin{aligned} \overrightarrow{F}_g(f, s, u) &= s^{-1}(\overrightarrow{A}_{gf} f + \overrightarrow{A}_{gu} u + \overrightarrow{b}_g), \\ \overleftarrow{F}_v(f, s, u) &= s^{-1}(\overleftarrow{A}_{vf} f + \overleftarrow{A}_{vu} u + \overleftarrow{b}_v), \\ \overrightarrow{F}_v(u) &= \overrightarrow{A}_{vu} u + \overrightarrow{b}_v, \end{aligned}$$

which matches the expected behavior as $s \downarrow 0$ (cf. (12.11)-(12.12)).

12.3.1.2 Training

We train the transport maps on training data ($N = 10^6$) generated for a fixed value of $s = 0.03$, allowing us to compute the corresponding transport maps directly from the mean and co-variance of the training data. To stabilize the process, we use a shrinkage estimator of the covariance with $\alpha = 10^{-6}$ (see the Appendix for more details).

12.3.2 Validation

To validate the transport maps, we apply it to test data ($N = 10^6$) generated for $s \in [0.01, 0.05]$. The KL-divergence between the estimated and true likelihood, posterior and prior are shown in Fig. 12.2. The errors behave as expected; the prior and likelihood fit with the chosen parametrization of the mappings and give a small error while the error in the posterior is generally larger, with a minimum for $s = 0.03$.

A different view on the error in the likelihood and posterior is shown in Fig. 12.3. Here, we show the mean and standard deviation of samples of the likelihood and

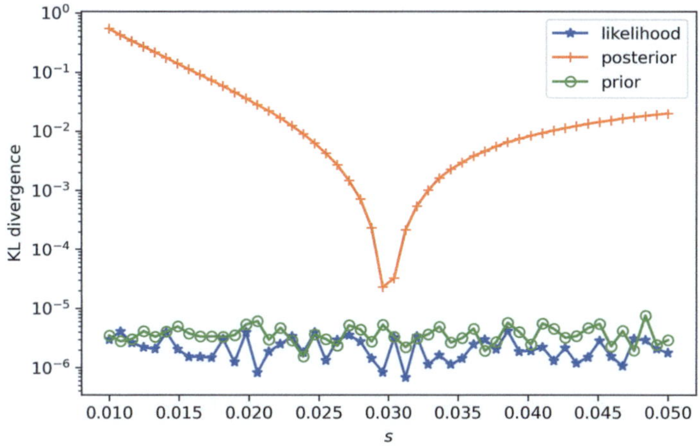

Fig. 12.2 KL-divergence between the estimated and true densities

posterior corresponding to $u = \mu_{\text{prior}}$ and $f = K\mu_{\text{prior}}$ respectively. We note that the main source of error is the standard deviation of the posterior samples, stemming from the chosen parametrization of the mappings.

12.3.3 Computed Tomography I

For a more realistic example, we consider computed tomography. Here, the measurements are given by

$$f^\delta = -\log\left(s^2 I^\delta\right), \; I^\delta \sim \mathcal{P}(s^{-2}\exp(-Ku)), \tag{12.15}$$

where $u \in \mathbb{R}^{n^2}$ are randomly generated $n \times n$ binary images, $K \in \mathbb{R}^{n^2 \times n^2}$ is a discretization of the Radon transform for a fixed acquisition geometry with n detector pixels and n equispaced angles, and $\mathcal{P}$ denotes the Poisson distribution. The parameter s in this case controls the intensity of the X-ray beam: $I_0 = s^{-2}$. An example of the corresponding data is shown in Fig. 12.4.

In the high-dose limit, we can approximate the corresponding likelihood with a Normal distribution with mean Ku and covariance $\text{diag}(s^2\exp(Ku))$. We define the transport map accordingly in the same way as in the previous example.

12.3.3.1 Training

We train the transport maps on data ($n = 32$, $N = 10^4$) for $s = 0.055$ with $\alpha = 10^{-6}$ using the procedure described in the Appendix.

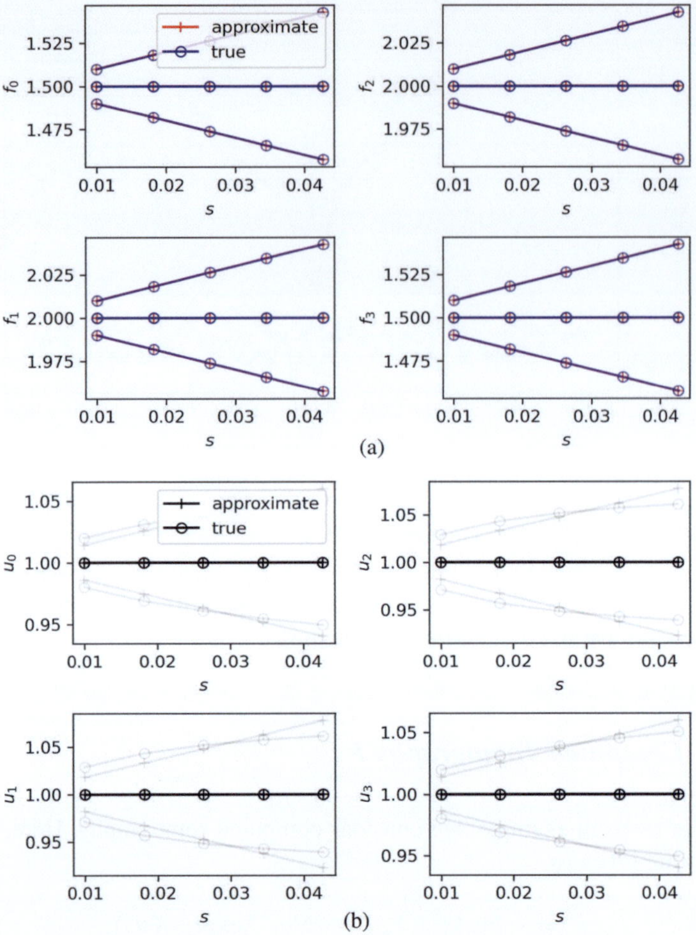

Fig. 12.3 Mean and standard deviation of samples generated from the likelihood (**a**) and posterior (**b**)

12.3.4 Validation

We apply the trained mapping to validation data generated for $s \in [0.01, 0.1]$. The KL divergence is shown in Fig. 12.5. We note that the error in the prior is smallest, and that the error in the likelihood gets smaller for larger s.

To get more insight in the learned map, we compare samples from the learned likelihood and prior to samples of the true likelihood and prior. Figure 12.6 shows examples samples drawn from the estimated prior. Figure 12.7 shows the mean and variance of samples from the estimated and true likelihood for a given phantom. We see a good correspondence of the mean, while the intensity-dependent variance of the true likelihood is not well-captured as expected. Figure 12.8 shows a summary

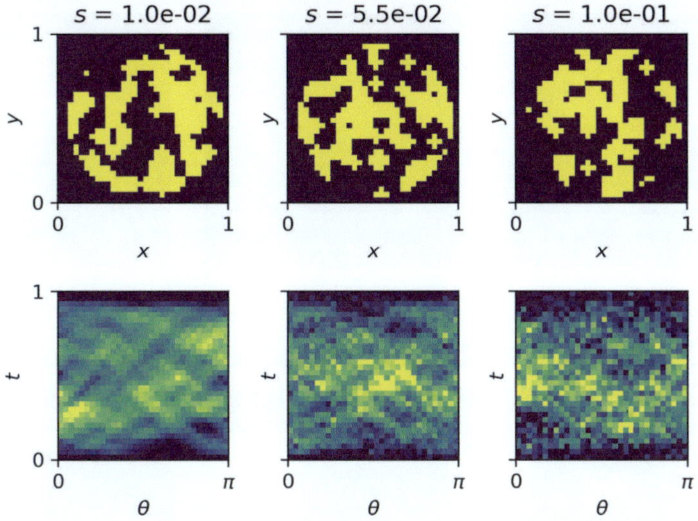

Fig. 12.4 The data consists of images images (top) and their corresponding sinograms (bottom) for $s \in [0.01, 0.1]$

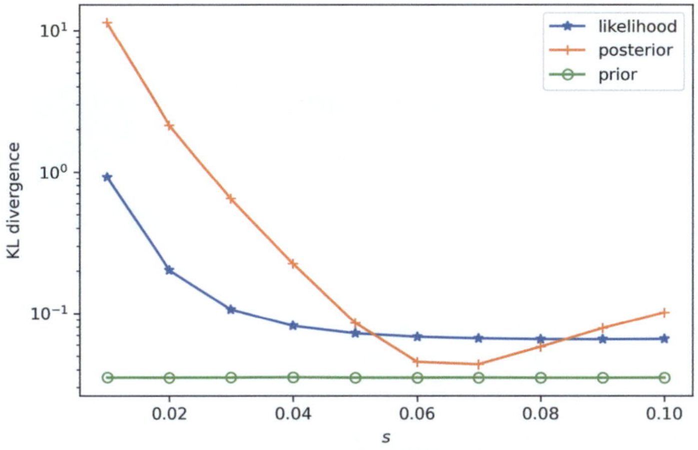

Fig. 12.5 KL divergence between the estimated and true densities

statistic of the samples drawn from the estimated and true likelihood for varying values of $s \in [0.01, 0.1]$, showing a good correspondence with increasing error for larger s.

In Fig. 12.9 we show an example of the posterior mean and variance for a given sinogram. Compared to the true phantom the standard deviation provides a useful indication of the error as shown in Fig. 12.10.

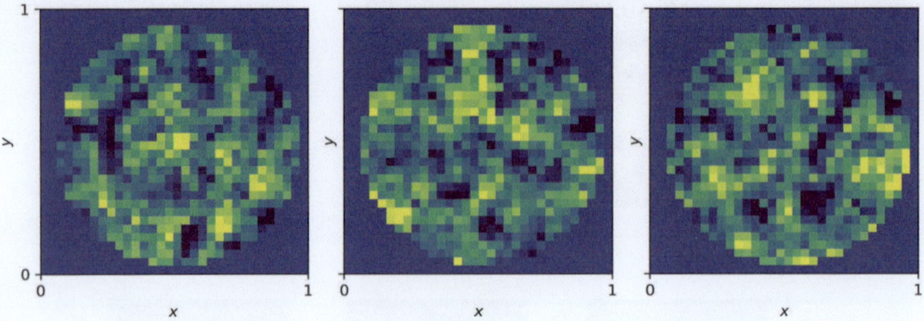

Fig. 12.6 A few generated prior samples

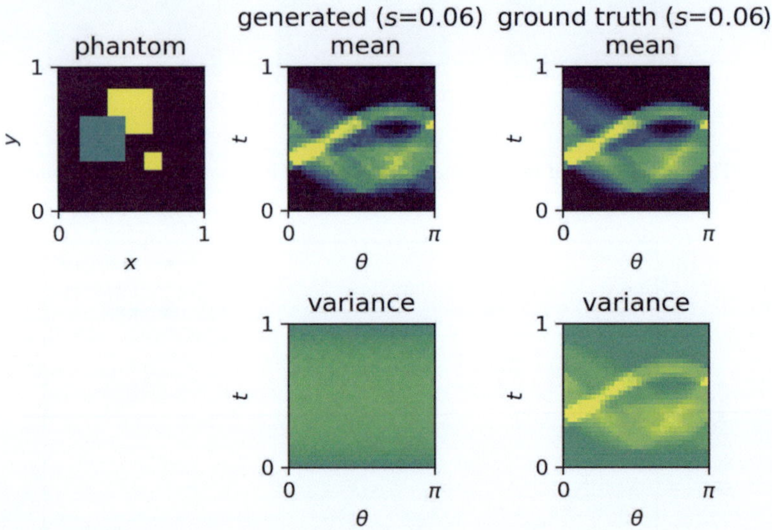

Fig. 12.7 Comparison in terms of mean and variance of the simulated sinograms to samples of the true likelihood for a phantom consisting of 3 squares

12.3.5 Computed Tomography II

We follow a similar setup as the previous example, except here we fix the noise level and let $s \in \mathbb{R}^m$ denote the $m \leq n$ angles at which the data are acquired:

$$f^\delta = -\log\left(I^\delta/I_0\right), \; I^\delta \sim \mathcal{P}(I_0 \exp(-K(s)u)). \tag{12.16}$$

For moderate noise levels, the likelihood again can be well-approximated by a normal distribution. To account for the non-trivial dependence on s we propose the following parametrization of the transport maps:

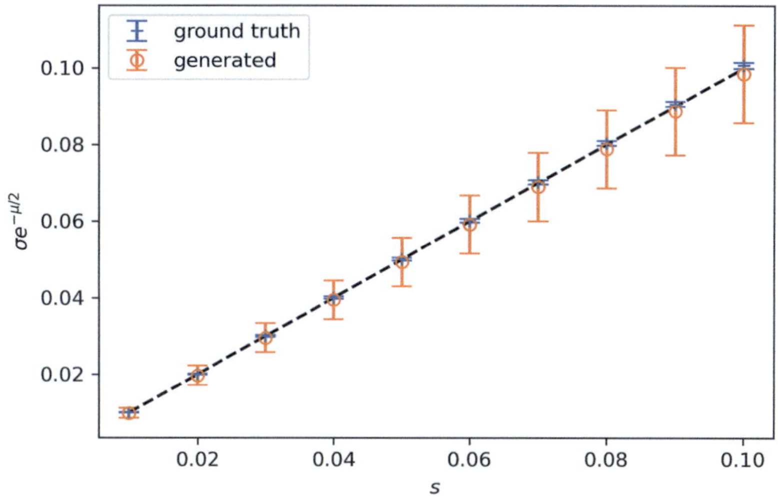

Fig. 12.8 Here we plot $\sigma e^{-\mu/2}$, with μ, σ denoting the mean and variance of the likelihood samples as a function of s. For the used likelihood, we expect this summary statistic to follow a straight line

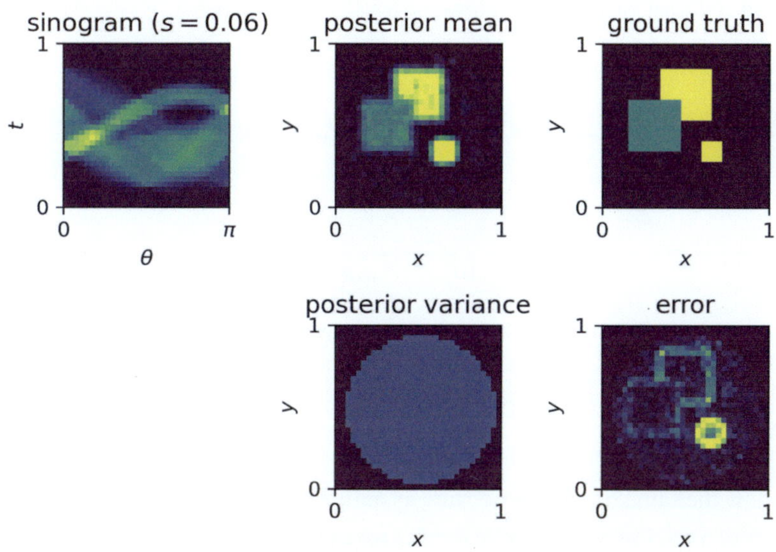

Fig. 12.9 Example of the posterior mean and variance for a given sinogram, as compared to the underlying true phantom and the corresponding reconstruction error

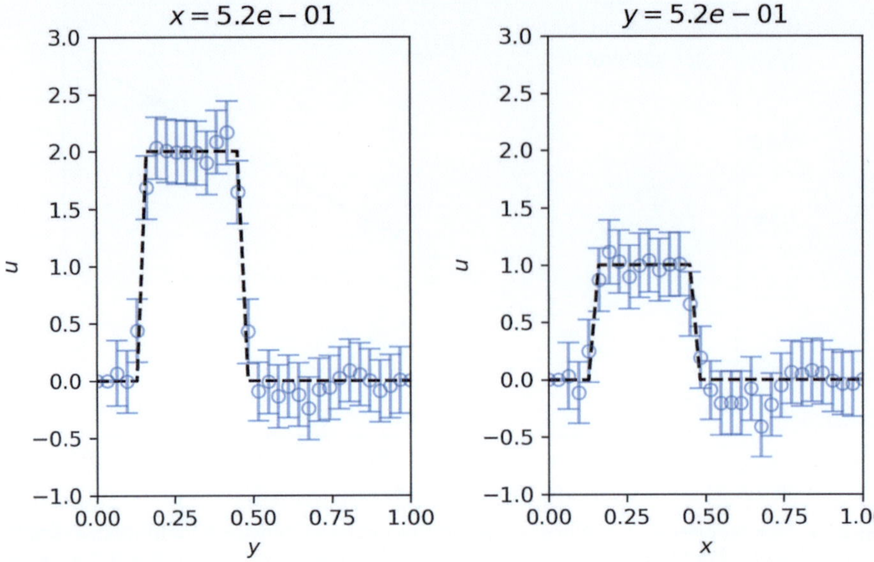

Fig. 12.10 Detailed view of the posterior mean and standard deviation as compared to the true phantom

$$\overrightarrow{F}_g(f, s, u) = \overrightarrow{A}_{gf} f + \overrightarrow{A}_{gu} K(s)u + \overrightarrow{b}_g,$$

$$\overleftarrow{F}_v(f, s, u) = \overleftarrow{A}_{vf} K(s)^T f + \overleftarrow{A}_{vf} u + \overleftarrow{b}_v.$$

$$\overrightarrow{F}_v(u) = \overrightarrow{A}_{vu} u + \overrightarrow{b}_v.$$

In effect, in known physics is included explicitly through K and K^T [5].

12.3.5.1 Training

We follow the same training procedure as outline above, except for $\overrightarrow{F}_g$ we use pairs $(f_i, K(s)u_i)$ (cf. Fig. 12.11) while for $\overleftarrow{F}_v$ we use pairs $(K(s)^T f_i, u_i)$ (cf. Fig. 12.12). For the training we fix s to represent n regularly spaced angles in $[0, \pi]$.

12.3.5.2 Validation

To highlight the ability to perform experimental design with $\overleftarrow{F}_v$, we use it to compute the reconstruction error between a ground-truth image (cf. Fig 12.9) and the posterior mean, as a function of s. This corresponds to A-optimal experimental design (cf. Eq. 12.7). The chosen architecture allows to evaluate the posterior

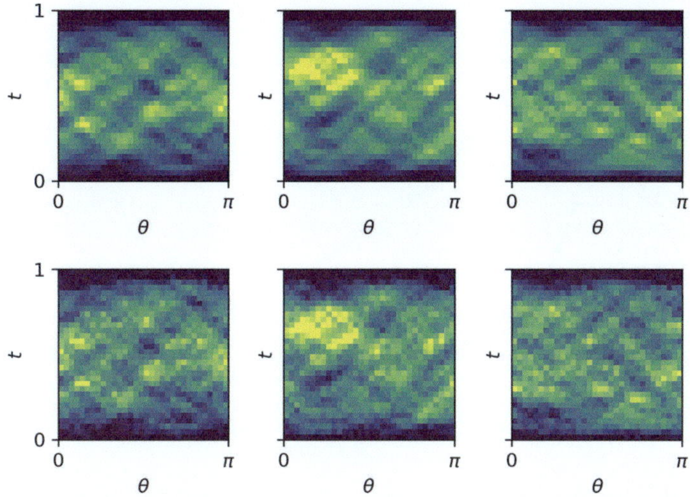

Fig. 12.11 The data consists of forward projected images (top) and their corresponding noisy sinograms (bottom)

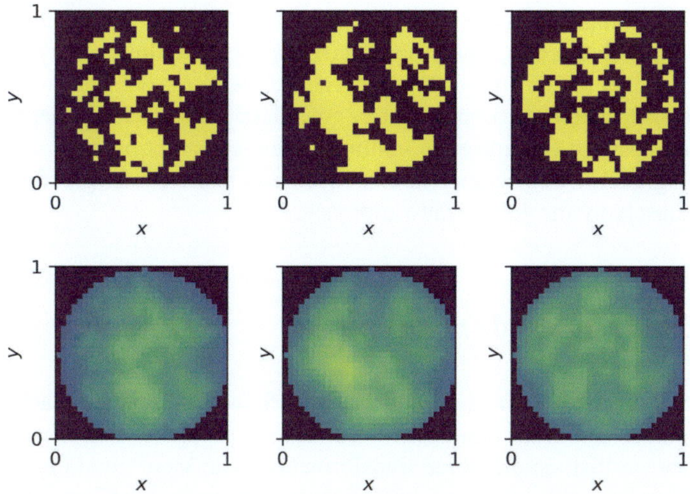

Fig. 12.12 The data consists of images (top) and their corresponding backprojected sinograms (bottom)

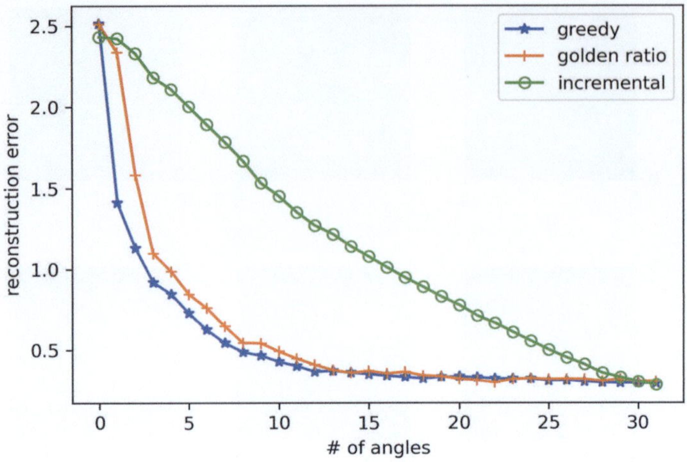

Fig. 12.13 Reconstruction error as a function of the number of angles for three angle-selection procedures

mean for any number of angles, so we can plot the error as we increase the number of angles. In particular, we consider three scenarios: adding angles sequentially (i.e., $s = (0)$, $s = (0, \Delta s)$, $s = (0, \Delta s, 2\Delta s)$, etc.) with $\Delta s = \pi/n$ (incremental) or $\Delta s = 0.618\pi$ (golden ratio), and a greedy procedure where each next angle is chosen amongst all possible candidates to minimize the reconstruction error. The result is shown in Fig. 12.13. We note that the error decreases fasted for the greedy procedure, followed closely by the golden ratio approach.

12.4 Conclusion and Discussion

We reviewed the use of (triangular) normalizing flows for simulation, inference, and experimental design in the context of linear inverse problems. In a few case studies we have shown that simple affine transformations can yield good approximations for linear inverse problems, even when we depart from additive Gaussian noise. The proposed architectures involve affine transformations and are amenable to application on large-scale inverse problems. In particular, tools from numerical linear algebra can be brought to bear on this problem and yield efficient ways to approximate the affine transformations. We expect that further refinement of the results can be achieved by composing the affine transformations with (point-wise) non-linear operations.

Acknowledgements I want to express my gratitude to Gabrio Rizutti, Felix Herrmann and Rafael Orozco with whom I have had numerous fruitful discussions on this topic in the past years.

Appendix

Training and Validation

The triangular maps arise by fitting a Gaussian to the data $\{(f_i, u_i)\}_{i=1}^N$, and computing a Cholesky decomposition of the Schur complements of the resulting covariance matrix. The core of this involves estimating the mean and covariance of a set of samples, and validating the fit.

Training

Given a set of $n-$dimensional training data $\{x_i\}_{i=1}^N$ drawn from π, we aim to fit a Normal distribution to it. The resulting mean and (shrinkage) covariance estimates are

$$\widehat{\mu} = N^{-1} \sum_{i=1}^{N} x_i,$$

$$\widehat{\Sigma}_\alpha = (1-\alpha)\widehat{\Sigma} + \alpha n^{-1} \text{trace}(\widehat{\Sigma}) I,$$

with

$$\widehat{\Sigma} = N^{-1} \sum_{i=1}^{N} (x_i - \widehat{\mu})(x_i - \widehat{\mu})^T.$$

The corresponding mapping is then defined as

$$F(x) = \widehat{\Sigma}_\alpha^{-1/2}(x - \widehat{\mu}),$$

where the matrix square root is to be interpreted in terms of a Cholesky decomposition, leading an upper or lower triangular matrix.

Validation

To validate the trained mapping we would ideally like to compute the difference between the true distribution π and its estimated counter part.

If we have access to the underlying true pdf π we could compute the KL-divergence

$$\text{KL}(\pi, F^\sharp \pi_0) \approx N^{-1} \sum_{i}^{N} \log \pi(x_i) - \log(\pi_0(F(x_i))|\nabla F(x_i)|),$$

with $x_i \sim \pi$ representing the validation data. We can alternatively compute this as

$$\mathrm{KL}(\pi, F^{\sharp}\pi_0) = \mathrm{KL}(F_{\sharp}\pi, \pi_0) \approx N^{-1} \sum_i^N \log \pi(G(y_i)|\nabla G(y_i)|) - \log \pi_0(y_i),$$

with $y_i = F(x_i)$. When the distributions involved are Gaussian (i.e., $\pi = \mathcal{N}(\mu, \Sigma)$, $\pi_0 = \mathcal{N}(0, I)$), we can compute this as

$$\mathrm{KL}(\pi, F^{\sharp}\pi_0) \approx \tfrac{1}{2} \left(\|\widehat{\mu}_0\|_2^2 + \mathrm{trace}\left(\widehat{\Sigma}_0\right) - n - \log|\widehat{\Sigma}_0| \right), \tag{12.17}$$

where $\widehat{\mu}_0$, $\widehat{\Sigma}_0$ are the sample mean and covariance of the normalized samples $\{y_i\}_{i=1}^N$. This effectively tests how far (in distribution) the normalized samples y_i are to π_0.

Even when π is not a normal distribution, this is a useful way of measuring how well the normalized samples fit π_0 as in that case we expect $\widehat{\mu}_0 \approx \mathbf{0}$, $\widehat{\Sigma}_0 \approx I$.

References

1. Ardizzone, L., Kruse, J., Rother, C., Köthe, U.: Analyzing inverse problems with invertible neural networks. In: International Conference on Learning Representations (2018)
2. Marzouk, Y., Moselhy, T., Parno, M., Spantini, A.: An Introduction to Sampling via Measure Transport, Feb. 2016
3. Moens, V., Sootla, A, Ammar, H.B., Wang, J.: Viscos Flows: Variational Schur Conditional Sampling With Normalizing Flows, pp. 1–18 (2021)
4. Orozco, R., Louboutin, M., Siahkoohi, A., Rizzuti, G., van Leeuwen, T., Herrmann, F.: Amortized Normalizing Flows for Transcranial Ultrasound with Uncertainty Quantification, vol. 1–18, 2023 (2020)
5. Orozco, R., Siahkoohi, A., Rizzuti, G., van Leeuwen, T., Herrmann, F.: Adjoint Operators Enable Fast and Amortized Machine Learning Based Bayesian Uncertainty Quantification, pp. 56 (2023)
6. Papamakarios, G., Nalisnick, E., Rezende, D.J., Mohamed, S., Lakshminarayanan, B.: Normalizing flows for probabilistic modeling and inference. J. Mach. Learn. Res. **22**(57), 1–64 (2021)
7. Radev, S.T., Schmitt, M., Pratz, V., Picchini, U., Köthe, U., Bürkner, P.C.: Jointly amortized neural approximation of complex bayesian models JANA. In: Proceedings of Machine Learning Research, vol. 216, pp. 1695–1706 (2023)
8. Radev, S.T., Voss, A., Mertens, U.K., Ardizzone, L.: Learning complex stochastic models with invertible neural networks and Ullrich Köthe BayesFlow (2020)
9. Rainforth, T., Foster, A., Ivanova, D.R., Smith, F.B.: Mod. Bayesian Exp. Des. (1) (2023)
10. Strutz, D., Curtis, A.: Variational Bayesian Experimental Design for Geophysical Applications, July 2023
11. Stuart, A.M.: Inverse problems: a Bayesian perspective. Acta Numerica **19**, 451–559 (2010)
12. Tarantola, A.: Inverse problem theory and methods for model parameter estimation. Soc. Ind. Math. (2005)

Chapter 13
The Effect of Protest Management Strategies on Protesting Activity and Social Tension: A Mathematical Perspective

Christina Wuyan Wang, Timothy Wessler, P. Jeffrey Brantingham, and Nancy Rodriguez

Abstract Protest activity, a constitutionally protected right in the United States under the First Amendment, serves as a key tool for individuals with limited individual influence to unite collectively and amplify their impact. Despite its legal recognition, the historical interaction between the government and protesters has been complex. In this study, we extend a model proposed in [2] to incorporate the influence of protest management strategies. Our research establishes the existence and stability of traveling wave solutions, supported by both theoretical analyses and numerical simulations. We delve into the impact of two distinct protest management approaches on the qualitative and semi-quantitative characteristics of these traveling waves. These metrics can aid in categorizing different types of protests.

Keywords Social science modeling · Reaction-diffusion systems · Traveling waves

C. W. Wang (✉) · N. Rodriguez
University of Colorado, Boulder, USA
e-mail: wuyan.wang@colorado.edu

N. Rodriguez
e-mail: rodrign@colorado.edu

T. Wessler
University of North Carolina, Chapel Hill, USA
e-mail: wessler@live.unc.edu

P. J. Brantingham
University of California, Los Angeles, USA
e-mail: branting@ucla.edu

A. Ott et al. (eds.), *Applications of Mathematics in Sciences, Engineering, and Economics*, Springer Proceedings in Mathematics & Statistics 515,
https://doi.org/10.1007/978-3-032-01279-1_13

13.1 Introduction

A protest is a form of collective behavior used by groups to affect political change [16, 20]. In many cases, protests are spontaneous or loosely organized [21] and exhibit interesting spatial and temporal patterns [13]. A self-reinforcing reaction-diffusion model to describe the interplay between the intensity of activity and the level of social tension was introduced in [2] by Berestycki and collaborators. This model exhibited intricate behavior and coherent structures, such as traveling wave solutions. We extend this model to include the global effects of policing strategies. We are motivated by the fact that many societies promote protest activity as a right but have fraught relationships with protest events and protesters when it comes to practical matters of policing [8, 15]. We therefore aim to understand the effect of different protest management strategies on the dynamics of protesting activity and social tension. The model we will analyze is the system for the interplay of (u, v, P) functions of physical space, x, and time t:

$$\begin{cases} u_t = d_1 \Delta u + f(u, v, P), & x \in \mathbb{R}^2,\ t > 0, \\ v_t = d_2 \Delta v + g(u, v, P), & x \in \mathbb{R}^2,\ t > 0, \\ P_t = s(u, v, P), & x \in \mathbb{R},\ t > 0, \\ (u(x,0), v(x,0), p(x,0)) = (u_0(x), v_0(x), p(x)). \end{cases} \tag{13.1}$$

In (13.3), the scalar fields u, v, and P measure the levels of social activity, social tension, and police density, respectively, with prescribed initial conditions $u_0(x)$, $v_0(x)$, and $p(x)$. The reaction functions, which model the kinetics between these three unknowns, will be detailed in the modeling section (see Sect. 13.1.1).

13.1.1 Model Motivation and Background

While the factors that lead individuals to make any given decision widely vary, in certain situations many individuals converge in space and time to generate collective behavior. The macroscopic patterns produced by the collective exhibit regularities that can be studied using mathematical models. A notable example is that of protests. Over the years, social scientists have explored the factors contributing to the convergence of people with a shared objective in both time and space. This research has revealed a crucial element for this collective action to occur: heightened social tension within a group, which often arises from a prolonged situation that deviates significantly from the group's expectations [18]. Additionally, a triggering event typically initiates the collective action [10, 21].

Spatial-temporal patterns in protest activities manifest in numerous instances. For instance, data analysis from the 2005 French riots revealed the emergence of traveling wave-like solutions originating from the point of the triggering event [5]. Similarly, during the Revolution of Dignity in Ukraine from late 2013 to 2014, a temporal pattern of self-excitation followed by relaxation was observed, with diffusion influenced by the political implications for Oblasts [1].

An essential aspect of the evolution of protest activities is the use of police protest management strategies. A key question arises regarding how these varied management strategies impact the overall dynamics [8, 14, 15]. Research indicates that, in certain situations, participating in a protest can alleviate stress for participants [6]. Conversely, in other scenarios, engagement in protesting may result in mental and physical trauma [11, 12, 17]. Mathematical models offer a means to investigate the influence of different protest management strategies, aiming to identify approaches that lead to "productive action"—where protest participants experience reduced stress levels and diminished social tension within the group.

13.1.1.1 The Model Without Protest Management

In [2], Berestycki and collaborators introduced a model for the interplay between the level of social activity and social tension

$$\begin{cases} u_t &= d_1 \Delta u + r(v)G(u) - \omega u, \quad x \in \mathbb{R}^n,\ t > 0, \\ v_t &= d_2 \Delta v + 1 - h(u)v, \quad x \in \mathbb{R}^n,\ t > 0, \\ u(x,0) &= u_0(x) \quad \text{and} \quad v(x,0) = v_0(x), \end{cases} \tag{13.2}$$

where u denotes the *level of activity* and v the *social tension*. The functions

$$r(v) = \frac{1}{1 + e^{-\beta(v-1)}}, \quad G(u) = \Gamma u(1-u) \quad \text{and} \quad h(u) = \theta(1+u)^{-k_0}$$

are based on various sociological theories and interviews with protest participants. Please refer to Table 13.3 for the physical interpretation of these parameters. The model is rooted in socio-psychological theories and insights gathered from interviews with participants in the 2005 French riots (Fig. 13.1). It is widely accepted that social tension, coupled with a triggering event, can give rise to collective action [10, 21]. Social tension is characterized as a negative emotional state within a group of individuals due to a prolonged, insoluble situation involving a misalignment between social expectations, interests, and the overall needs of the population [18, 19]. If social tension reaches a critical level, a triggering event—a swift catalyst capable of reshaping the political landscape and attracting or intensifying participation—can spark a protest. A useful analogy is that social tension creates combustible material, and a triggering event acts as the match that ignites the fire.

Data, sociological theories, and interviews with protest participants indicate that protesting activity can be self-exciting. In other words, action begets more action, up to a certain carrying capacity. This phenomenon is captured by the function G, which resembles a Fisher-KPP-like term. However, this self-excitation effect only occurs when there is sufficiently high tension in the system; the function r

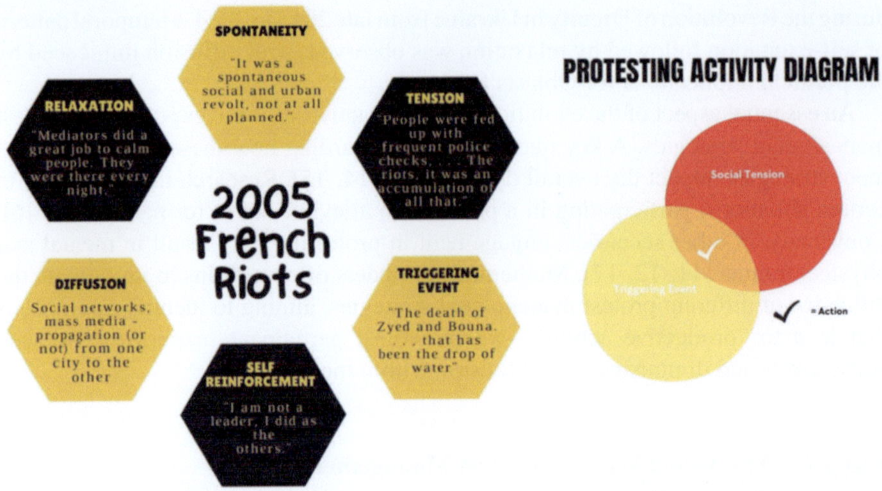

(a) Interviews (b) Protesting Activity

Fig. 13.1 Diagrams illustrating the factors that influence the rise of protesting activity. **a** Insights from interviews with participants of the 2005 French Riots. **b** Key elements necessary for a social outburst of protest

represents a switching mechanism that modulates the self-excitation effect depending on the social tension. The influence of the level of action on social tension introduces intriguing variations in the dynamics of the model. The modulation of v by u is characterized by the function h. The parameter k_0 determines whether the relationship between the two variables is enhancing or inhibiting. When $k_0 > 0$, action leads to a slowdown in tension; conversely, when $k_0 < 0$, action decreases the tension. As mentioned earlier, both scenarios have been observed in real life. On the one hand, many protests induce mental and physical trauma [11, 12, 17]; on the other, some individuals have reported a release of stress after participating in peaceful protests. A situation with the presence of counter-protesters or when a peaceful protest turns violent may lead to $k_0 > 0$.

Prior studies have developed mathematical approaches to studying the k_0 positive and k_0 negative cases. In [24, 25], Yang and Rodríguez have studied the traveling wave solutions both numerically and analytically for both cases. Such solutions in the tension-enhancing case have also been explored in [3] in heterogeneous environments where the restriction of information varies by region. The inhibitive case has been observed in [5] by fitting the 2005 French riots dataset with a Susceptible-Infected-Recovered (SIR)-type model. Moreover, a model similar to the one we are considering with a time-periodic source term has been studied in [4] where convergence to periodic solutions are proved for both the tension-enhancing and tension-inhibiting case.

13.1.1.2 Incorporating Protest Management Strategies

While the original system described in Eq. (13.2) exhibits intriguing behavior, it lacks integration of the significant impact that protest management has on the overall activity. In the United States, the government frequently employs law enforcement to handle protest activities, employing various protest management strategies. Chapter 2 of [8] introduces five crucial characteristics of protest policing practices, outlined in Table 13.1. To summarize, these characteristics encompass the degree of police concern for the First Amendment rights of protesters and their obligations to uphold and safeguard those rights, the level of police tolerance for community disruption, the dynamics of communication between police and demonstrators, the extent and method of arrests as a means of managing demonstrators, and the extent and method of using force, either independently or in conjunction with arrests, to control demonstrators.

In the 1960s, the prevailing management model in the United States was an **escalated force model**, characterized by a robust and forceful approach based on deterrence theory [7, 8, 14]. Deterrence theory posits that individuals seek to minimize pain and maximize pleasure, leading to the belief that the use of force and arrests would discourage action. The second column of Table 13.1 presents the different characteristics outlined earlier in the context of the escalated force model. By contrast, in the 1980s and 1990s, the **negotiated management model** emerged as the dominant U.S. approach [8]. In stark opposition to the escalated force model, the primary goals of the negotiated management model were to safeguard protesters' First Amendment rights and preserve lives. The third column of Table 13.1 illustrates the various characteristics within the framework of the Negotiated Management model. While there are other models, including several variations of the Escalated Force model, a more detailed description of the various protest management strategies employed in the United States can be found in [14]. We will concentrate on the two models highlighted above and presented in Table 13.1.

As is often the case, intriguing behaviors frequently stem from conflicting influences. While police presence may dissuade certain individuals from participating [22], an inappropriate protest management strategy could result in heightened tensions or increased participation [9]. Our objective is to refine model (13.2) to encompass these opposing effects. Initially, we examine the direct impact of police presence

Table 13.1 The main characteristics of the escalated force and the negotiated management models

Characteristic	Escalated force	Negotiated management
Respect for 1st Amend. Rights	Low	High
Tolerance for disruption	Low	High
Communication	Low	High
Arrest	Quick	Slow
Use of force	Quick	Slow

on u. We posit that police presence can act as a deterrent or, conversely, motivate people to participate. This effect is represented by modifying the first evolution equation as follows: $u_t = d_1\Delta u + H(P)r(v)G(u) - \omega u$, where $H(P) = (1+P)^{k_1}$. Note that in this case when $k_1 > 0$ the presence of police enhances the level of activity when $k_1 < 0$ it inhibits the action, and when $k_1 = 0$ the equation reduces to the original evolution equation.

The presence of police and their management style can also indirectly influence social tension, impacting the dynamics of v. This is accomplished by adjusting the second equation as follows:

$$v_t = d_2\Delta v + 1 - h(u)T(P)v,$$

where $T(P) = (1+P)^{-k_2}$. It is important to note that when $k_2 > 0$, police presence enhances social tension, whereas when $k_2 < 0$, it inhibits social tension. As mentioned earlier, when $k_2 = 0$, our equation reverts to the original one. Throughout these modifications, we adhere to the convention that a positive k_i $(i = 0, 1, 2)$ signifies enhancement, while a negative parameter implies inhibition. See Table 13.2 for a summary of the interpretation of the sign of these parameters.

Finally, we assume that the police are deployed at a proportional rate to the level of activity u up to some carrying capacity. This leads to the following equation:

$$P_t = u(1-P),$$

where the level of activity serves as the growth rate of the police presence, and the carrying capacity of policing is assumed to be one. Note that we assume no police diffusion.

Putting everything together, the system that we are interested in is

$$\begin{cases} u_t = d_1 u_{xx} + f(u, v, P), \\ v_t = d_2 v_{xx} + g(u, v, P), \\ P_t = s(u, v, P), \end{cases} \tag{13.3}$$

where the source functions are:

Table 13.2 Summary of the physical interpretation of k_i $(i = 0, 1, 2)$

Parameter	Negative value	Positive value
k_0	Protesting action is tension-inhibiting	Protesting action is tension-enhancing
k_1	Police presence has a deterrence effect on Protesting	Police presence has an incendiary effect on Protesting
k_2	Escalating force	Negotiated management model

Table 13.3 The description of parameters in system (13.3)

Parameter	Description
d_1	Diffusivity of protest activity
d_2	Diffusivity of social tension
α	Critical level of social tension to transition from without to with self-reinforcement
ω	Rate of decay of protest activity
θ	Scaling factor for rioting level that scales social tension
Γ	Rate of rioting self-reinforcement
β	Rate of transition of without to with self-reinforcement

$$\begin{cases} f(u, v, P) = (1 + P)^{k_1} \dfrac{1}{1 + e^{-\beta(v-\alpha)}} \Gamma u\,(1 - u) - \omega u, \\ g(u, v, P) = 1 - (1 + P)^{-k_2}\, \theta\, (1 + u)^{-k_0}\, v, \\ s(u, v, P) = u(1 - P). \end{cases} \tag{13.4}$$

The descriptions of model parameters are given in Table 13.3. For the remainder of the paper, we assume $\theta = 1$. For physical reasons, we limit the domains for the parameters to be $\alpha, \beta, \Gamma, \omega > 0$.

13.1.1.3 Main Results

In this work, we investigate the impact of various police management strategies on traveling wave solutions' quantitative and qualitative features. These solutions depict a mobile front where a high level of protest activity continuously spreads into regions with no activity. The traveling wave solutions to system (13.2) consistently leave behind a significantly elevated level of activity and/or social tension. Our focus involves proving the existence and stability of traveling wave solutions for system (13.3). The examination of wave existence is addressed in Sect. 13.3.1. We handle cases of monostable and bistable sources separately. We establish the existence of a monotone traveling wave satisfying system (13.3) with a unique wave speed in bistable regimes. For monostable regimes, we demonstrate the existence of monotone traveling waves as long as the wave speed surpasses a specified threshold. The stability analysis is provided in Sect. 13.3.2, where we also present asymptotic approach rates to the steady-state solutions. Our findings indicate that traveling waves with a bistable source are asymptotically stable with a shift according to the sup-norm. In contrast, waves with a monostable source are asymptotically stable according to a weighted norm, as detailed in the upcoming theorem.

We complement the theoretical results with numerical experiments across various parameter combinations and analyze the outcomes. We present key metrics to assess how parameter alterations influence the system's dynamics quantitatively.

These metrics encompass the maximum and cumulative levels of activity and tension experienced and the level of protest activity at the interface infiltrating the region not yet exposed to the protest traveling wave. Additionally, we partition the parameter space into distinct regions based on specific simulation outcomes. We identify regions in the parameter space where, after the traveling wave has long passed, there are either (1) persistent nonzero levels of protesting with the maximum density of police present, or (2) protest levels have diminished to near zero, and police densities are below maximum levels. Furthermore, we pinpoint regions where protesting and policing strategies have elevated tension following the dynamics of the traveling wave and regions where these strategies have successfully alleviated tensions—whether the tension relief is immediate following the protesting or occurs after a brief rise but ultimately results in a decline in tension.

13.2 Background Matters

In this section, we discuss the spatially homogeneous system, its equilibrium solutions, and the linearized stability of the system about these equilibrium solutions. Moreover, we introduce some relevant definitions and notations for our analysis of traveling wave solutions.

13.2.1 The Spatially Homogeneous System

First, we find and analyze the stationary points of the spatially homogeneous system

$$\begin{cases} u_t = H(P)r(v)G(u) - \omega u, \\ v_t = 1 - h(u)T(P)v, \\ P_t = u(1-P). \end{cases} \tag{13.5}$$

The nullclines for the P-equation are $u = 0$ or $P = 1$. If $u = 0$, then we have a hyperplane of fixed points $(0, (1+P)^{k_2}, P)$ for $0 < P < 1$, which correspond to the trivial steady states. If $P = 1$, then we have fixed points given by the intersection of the three nullclines

$$\begin{cases} 2^{k_1} r(v)G(u) - \omega u = 0, \\ 2^{k_2}(1+u)^{k_0} = v, \\ P = 1, \end{cases} \tag{13.6}$$

which correspond to the non-trivial steady states. From these equations of nullclines, we can see that the u value at non-trivial steady states should always be strictly between zero and one.

The Jacobian of $F = [f, g, s]^T$ is given by

$$DF(\boldsymbol{u}) = \begin{bmatrix} H(P)r(v)G'(u) - \omega & H(P)r'(v)G(u) & H'(P)r(v)G(u) \\ -h'(u)T(P)v & -h(u)T(P) & -h(u)T'(P)v \\ 1-P & 0 & -u \end{bmatrix}. \tag{13.7}$$

Then, the characteristic polynomial of the Jacobian is

$$\begin{aligned} &-\lambda^3 + (f_u + g_v + s_P)\lambda^2 + (-f_u(g_v + s_P) - g_v s_P + f_v g_u + f_P s_u)\lambda \\ &+ f_u g_v s_P - f_v g_u s_P + f_v g_P s_u - f_P s_u g_v = 0. \end{aligned} \tag{13.8}$$

We find the eigenvalues and evaluate them at different steady states.

At the trivial steady state, $f_v, f_P, s_P = 0$, so the characteristic polynomial (13.8) reduces to $-\lambda^3 + (f_u + g_v)\lambda^2 - f_u g_v \lambda = 0$, yielding the eigenvalues $0, f_u, g_v$. Evaluated at the trivial steady state, the three eigenvalues become

$$\begin{cases} \lambda_1 = 0, \\ \lambda_2 = (1+P)^{k_1} \dfrac{\Gamma}{1 + e^{-\beta(v^* - \alpha)}} - \omega, \\ \lambda_3 = -(1+P)^{-k_2}, \end{cases} \tag{13.9}$$

where $v^* = (1+P)^{k_2}$. Notice that $\lambda_3 < 0$ always, and the sign of λ_2 depends on the parameters.

At the nontrivial steady states, the eigenvalues are

$$\begin{cases} \lambda_1 = \dfrac{f_u + g_v + \sqrt{(f_u + g_v)^2 + 4 f_v g_u - 4 f_u g_v}}{2}, \\ \lambda_2 = \dfrac{f_u + g_v - \sqrt{(f_u + g_v)^2 + 4 f_v g_u - 4 f_u g_v}}{2}, \\ \lambda_3 = -u^*, \end{cases} \tag{13.10}$$

where the partial derivatives are evaluated at the nontrivial steady states $(u^*, v^*, 1)$ which can be obtained by solving (13.6). Note that λ_3 evaluated at the nontrivial steady states is always real and negative. For the monotone system, it turns out that all eigenvalues are real. For λ_1, λ_2 to be negative, we need $f_v g_u - f_u g_v < 0$ and $f_u + g_v < 0$, the latter of which can be shown using some algebra, the former of which simplifies to

$$\omega \beta k_0 e^{-\beta(v^* - \alpha)} \frac{v^*}{1 + u^*} < 2^{k_1} \Gamma. \tag{13.11}$$

Condition (13.11) will be checked numerically for the proofs later in the paper.

For the theoretical results, we focus on analyzing the case when system (13.3) is monotone. However, we will explore more general parameter sets numerically. By definition, our parabolic system (13.3) is monotone if for any $\boldsymbol{u} \in \mathbb{R}^3$,

$$\frac{\partial F_i}{\partial u_j} \geq 0, \quad i, j = 1, 2, 3, \quad i \neq j,$$

where $F_i = f, g, s$ for $i = 1, 2, 3$, respectively. One can verify that system (13.3) is monotone when $k_0, k_1, k_2, \beta, \Gamma \geq 0, 0 \leq u, P \leq 1$.

We are interested in studying the system with different k_0, k_1, k_2 values. Given that changing the rest of the parameters has essentially the same effect on the number of steady states as changing k_0, k_1, k_2, we will fix $\alpha = 1, \beta = 1, \Gamma = 1, \omega = 0.5$ from now on unless specified otherwise. For example, varying α changes the number of steady states the same way as varying k_1 does.

Now, we divide the $k_1 - k_2$ space into different regions depending on how many steady states there are. We consider the tension-enhancing ($k_0 \geq 0$) and inhibiting ($k_0 < 0$) cases separately.

13.2.1.1 Tension-Enhancing Case: $k_0 > 0$

When $k_0 \geq 0$, the v-nullcline is monotone increasing in the positive $u - v$ region. Since we assume $\omega, \Gamma, \beta > 0$, the u-nullcline is always monotone increasing. In this case, there are either zero, one, or two non-trivial steady states. Figure 13.2 shows the number of steady states in the $k_1 - k_2$ regime. Given any constant initial condition for P, P evolves to reach a specific constant value between zero and one at the trivial steady state, so the red regions indeed represent the regimes with one steady state instead of a hyperplane of them.

13.2.1.2 Tension-Inhibiting Case: $k_0 < 0$

When $k_0 < 0$, the v-nullcline is always monotone decreasing in the positive $u - v$ region. Again, the u-nullcline is always monotone increasing. In this case, there is either zero or one non-trivial steady state. Figure 13.3 shows a plot separating the

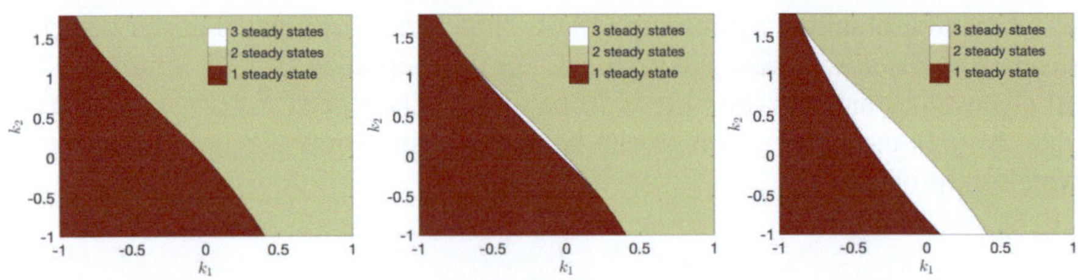

(a) $k_0 = 1$ (b) $k_0 = 2.5$ (c) $k_0 = 4.5$

Fig. 13.2 The $k_1 - k_2$ space separated into regimes with different numbers of steady states with positive k_0, where $\alpha = 1, \beta = 1, \Gamma = 1, \omega = 0.5$

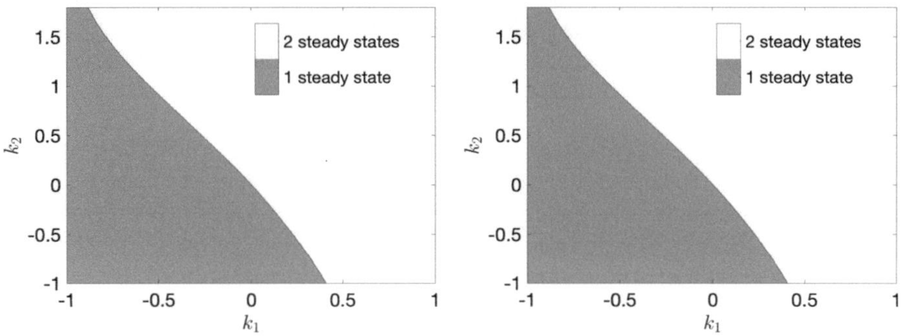

(a) $k_0 = -0.5$ (b) $k_0 = -4.5$

Fig. 13.3 The $k_1 - k_2$ space separated into regimes with different numbers of steady states with negative k_0, where $\alpha = 1, \beta = 1, \Gamma = 1, \omega = 0.5$

regions with different numbers of steady states in the $k_1 - k_2$ regime. Notice that the magnitude of k_0 does not change the diagram as long as $k_0 < 0$.

13.2.2 Existence and Stability of Traveling Waves

We are concerned with the existence and stability of traveling wave solutions to our system. Traveling wave solutions to our reaction-diffusion system (13.3) have the vector form $\boldsymbol{u}(x, t) = \hat{\boldsymbol{u}}(z)$ where $z = x - ct$ for some wave speed $c \in \mathbb{R}$ and satisfy

$$\begin{cases} A\hat{\boldsymbol{u}}'' + C\hat{\boldsymbol{u}}' + F(\hat{\boldsymbol{u}}) = 0, \\ \lim_{z\to\pm\infty} \hat{\boldsymbol{u}}(z) = \hat{\boldsymbol{u}}_{\pm}, \end{cases} \tag{13.12}$$

where $F = \begin{bmatrix} f \\ g \\ s \end{bmatrix}$ with f, g, s defined in (13.3), $A = \begin{bmatrix} 1 & 0 & 0 \\ 0 & 1 & 0 \\ 0 & 0 & 0 \end{bmatrix}$ is the diffusion matrix, and $C = cI$. Depending on the stability of the stationary points $\hat{\boldsymbol{u}}_{\pm}$, three types of sources are possible. The source is bistable if both the vectors $\hat{\boldsymbol{u}}_{+}$ and $\hat{\boldsymbol{u}}_{-}$ are stable. The source is monostable if one of $\hat{\boldsymbol{u}}_{+}$ and $\hat{\boldsymbol{u}}_{-}$ is stable, and the other is unstable. The source is unstable if both $\hat{\boldsymbol{u}}_{+}$ and $\hat{\boldsymbol{u}}_{-}$ are unstable. It can be seen in Sect. 13.3.1 that the existence and uniqueness of traveling wave solutions depend on the source type. Note that after the change of variable, $z = x - ct$, system (13.3) is a special case of system (13.12).

We are interested in monotone waves, as non-monotone waves for monotone systems are unstable [23]. For definiteness, we assume that the waves are monotonically decreasing. Consequently, we have $\hat{\boldsymbol{u}}_{+} < \hat{\boldsymbol{u}}_{-}$ (this inequality should be interpreted component wise).

Let E be a Banach space to study the stability of waves, and we consider small perturbations from a Banach space H, lying in E. We denote the norm in H by $\|\cdot\|_H$. These spaces will be made precise in each theorem.

Definition 13.1 ([23]) A wave $\hat{\boldsymbol{u}}(x)$ is *asymptotically stable with shift* according to the norm $\|\cdot\|_H$, if there exists a positive number ϵ such that for an arbitrary real vector-valued function $\boldsymbol{u}_0 \in E$ with $\boldsymbol{u}_0 - \hat{\boldsymbol{u}} \in H$ and $\|\boldsymbol{u}_0 - \hat{\boldsymbol{u}}\|_H < \epsilon$, the solution $\boldsymbol{u}(x,t)$ of system (13.3) with initial condition $\boldsymbol{u}(x,0) = \boldsymbol{u}_0(x)$ exists for all $t > 0$, is unique, $\boldsymbol{u}(x,t) - \hat{\boldsymbol{u}}(x) \in H$, and satisfies the estimate

$$\|\boldsymbol{u}(x,t) - \hat{\boldsymbol{u}}(x+h)\|_H \leq Me^{-bt}, \tag{13.13}$$

where h is a number depending on $\boldsymbol{u}_0(x)$; $M > 0$ and $b > 0$ are independent of t, h and $\boldsymbol{u}_0(x)$.

Definition 13.2 ([23]) A wave $\hat{\boldsymbol{u}}(x)$ is *asymptotically stable* according to the norm $\|\cdot\|_H$, if there exists a positive number ϵ such that for an arbitrary real vector-valued function $\boldsymbol{u}_0 \in E$ with $\boldsymbol{u}_0 - \hat{\boldsymbol{u}} \in H$ and $\|\boldsymbol{u}_0 - \hat{\boldsymbol{u}}\|_H < \epsilon$, the solution $\boldsymbol{u}(x,t)$ of system (13.3) with initial condition $\boldsymbol{u}(x,0) = \boldsymbol{u}_0(x)$ exists for all $t > 0$, is unique, $\boldsymbol{u}(x,t) - \hat{\boldsymbol{u}}(x) \in H$, and satisfies the estimate

$$\|\boldsymbol{u}(x,t) - \hat{\boldsymbol{u}}(x)\|_H \leq Me^{-bt}, \tag{13.14}$$

where $M > 0$ and $b > 0$ are independent of t and $\boldsymbol{u}_0(x)$.

13.3 Traveling Waves in the Monotone System: $k_0, k_1, k_2 \geq 0, 0 \leq u \leq 1$

13.3.1 Existence of Traveling Wave Solutions

To prove the existence of traveling wave solutions to (13.3), we need to determine the signs of the eigenvalues of Jacobian given in (13.7). Let us first look at the eigenvalues at the trivial steady states. Recall that the eigenvalues are given by (13.9), where $\lambda_1 = 0$ and $\lambda_3 < 0$. It is left to determine the sign of λ_2. Let us denote by $k_2(k_1, P)$ the boundary separating $\lambda_2 > 0$ from $\lambda_2 < 0$. By plotting in the $k_1 - k_2$ plane, we find that the curve $k_2(k_1, 1)$, represented by the blue dotted line in Fig. 13.4, coincides with the curve separating the region with two steady states from other numbers of steady states, which we denote by $k_2(k_1)$. More precisely, we see in Fig. 13.4 that, for $P = 1$, $\lambda_2 > 0$ when there are two steady states, and $\lambda_2 < 0$ when there are one or three steady states. For convenience, we denote the boundary separating the region with one steady state from other numbers of steady states by $k_2(k_0, k_1)$. Note that $k_2(k_1)$ coincides with $k_2(k_0, k_1)$ when k_0 is small (see Figs. 13.4a and 13.5a).

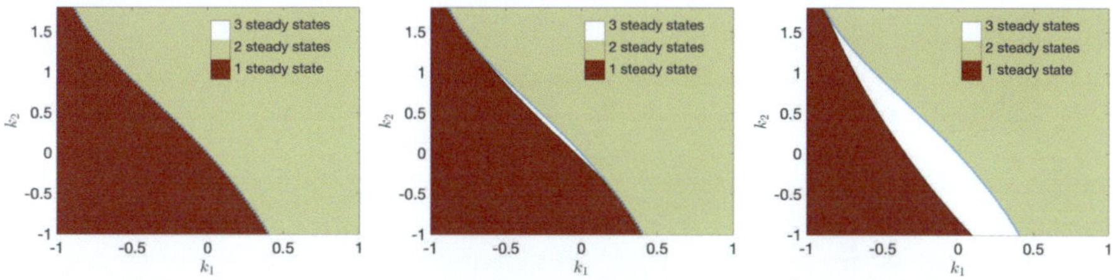

(a) $k_0 = 1, P = 1$ (b) $k_0 = 2.5, P = 1$ (c) $k_0 = 4.5, P = 1$

Fig. 13.4 The numbers of steady states depending on k_1 and k_2. The blue dotted line represents when $\lambda_2 = 0$, denoted by $k_2(k_1, 1)$, which coincides with $k_2(k_1)$. When $k_2 > k_2(k_1, 1)$, $\lambda_2 > 0$, and vice versa. For plot **a–c**, $\omega = 0.5, \alpha = 1, \Gamma = 1, \beta = 1$

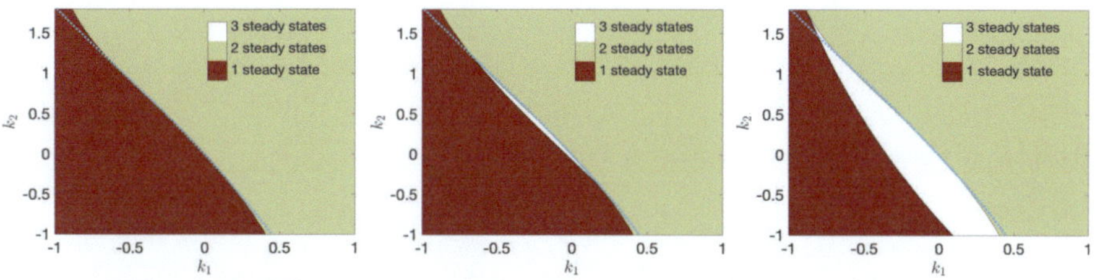

(a) $k_0 = 1, P = 0.5$ (b) $k_0 = 2.5, P = 0.5$ (c) $k_0 = 4.5, P = 0.5$

Fig. 13.5 The numbers of steady states depending on k_1 and k_2. The blue dotted line represents when $\lambda_2 = 0$, denoted by $k_2(k_1, 0.5)$. When $k_2 > k_2(k_1, 0.5)$, $\lambda_2 > 0$, and vice versa. For plot **a–c**, $\omega = 0.5, \alpha = 1, \Gamma = 1, \beta = 1$

When $P \neq 1$, the curve $k_2(k_1, P)$ does not necessarily coincide with $k_2(k_1)$ (see Fig. 13.5). In this more general case, we have up to six possible regions to analyze (see Fig. 13.6).

Depending on the parameter set, regions five and six may be empty (see Fig. 13.7).

The signs of eigenvalues at non-trivial steady states need to be checked separately for different values of k_1, k_2.

We now present the results for the existence of the traveling wave solutions.

Theorem 13.1 (Monostable) *Let $k_0, k_1, k_2 \geq 0$. For $k_2 > k_2(k_0, k_1)$ and $k_2 > k_2(k_1, P)$, there exists a positive constant c_* such that for all $c \geq c_*$, there exist monotone traveling wave solutions to* (13.12). *When $c < c_*$, such waves do not exist.*

Proof In the region where $k_0, k_1, k_2 \geq 0$, $k_2 > k_2(k_0, k_1)$ and $k_2 > k_2(k_1, P)$, there are either two or three steady states, corresponding to region ③ or ⑥, respectively. We denote by $\hat{\boldsymbol{u}}_+, \hat{\boldsymbol{u}}^1, \hat{\boldsymbol{u}}_-$ the trivial, middle non-trivial, and largest non-trivial steady state, respectively, so that $\hat{\boldsymbol{u}}_+ < \hat{\boldsymbol{u}}^1 < \hat{\boldsymbol{u}}_-$. Since $k_2 > k_2(k_1, P)$, the second eigenvalue of $DF(\hat{\boldsymbol{u}}_+)$ given in (13.9) lies in the right half-plane in this parameter regime, as shown in Figs. 13.5 or 13.4. For the nontrivial steady states, one can check numerically that all three eigenvalues of $DF(\hat{\boldsymbol{u}}_-)$ are real and negative, and, if we are in region ⑥, that at least one eigenvalue of $DF(\hat{\boldsymbol{u}}^1)$ lies in the right half plane. Finally, by Theorem 2.2 in Introduction of [23], we obtain the existence of waves. □

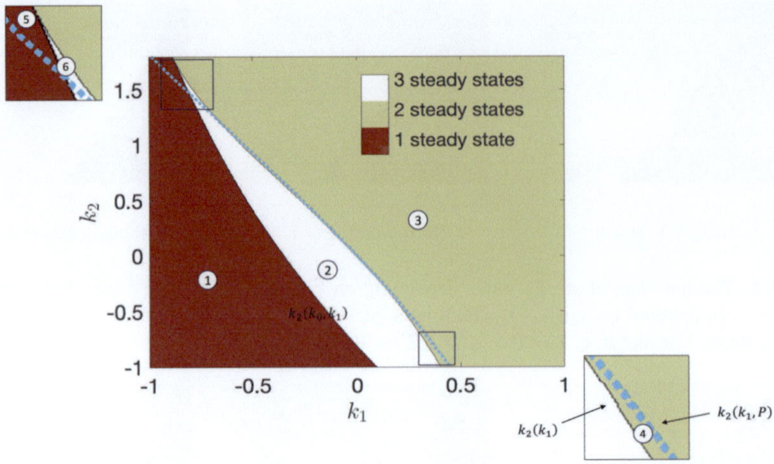

Fig. 13.6 Six possible regions in the $k_1 - k_2$ regime for $k_0 \geq 0$

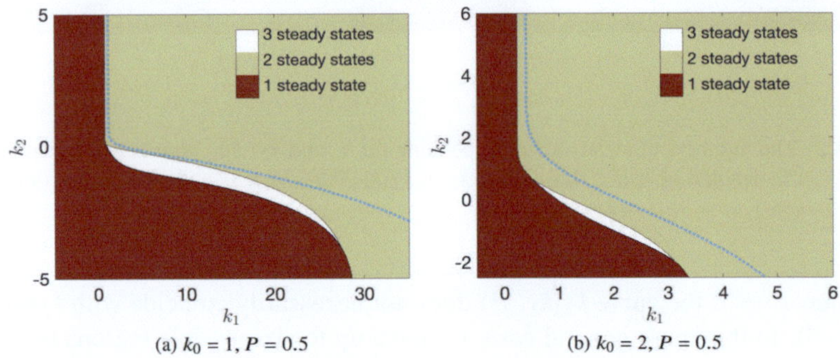

Fig. 13.7 The numbers of steady states depending on k_1 and k_2 focusing on the situations where regions five and six do not exist. For plot **a**, $\alpha = 1, \beta = 20, \Gamma = 160, \omega = 250$. For plot **b**, $\alpha = 1, \beta = 2.5, \Gamma = 210, \omega = 245$

Theorem 13.2 (Bistable) *Let $k_0, k_1, k_2 \geq 0$. Additionally, let $k_0 > k_0^*$ for some $k_0^* \geq 0$ such that $k_2(k_1)$ and $k_2(k_0, k_1)$ do not coincide, guaranteeing a region with three steady states. For $k_2(k_0, k_1) < k_2 < k_2(k_1, P)$, there exists a monotonically decreasing traveling wave solution to* (13.12) *with a unique wave speed c if either there are two steady states, or there are three steady states and one of the two following conditions is satisfied:*

1. $k_2 \neq 0, u_1 \neq 1,$
2. $k_2 = 0, k_1 \neq 0, u_1 \neq 1,$

where u_1 is the u-value at the middle stationary point.

Proof In the region where $k_0, k_1, k_2 \geq 0$ and $k_2(k_0, k_1) < k_2 < k_2(k_1, P)$, there are either two or three steady states, corresponding to region ④ or ②, respectively. Same as in the proof above, we denote the three steady states by $\hat{\boldsymbol{u}}_+, \hat{\boldsymbol{u}}^1, \hat{\boldsymbol{u}}_-$. We want to prove that all the eigenvalues of the matrices $DF(\hat{\boldsymbol{u}}_+)$ and $DF(\hat{\boldsymbol{u}}_-)$ lie in the left half-plane. For the trivial steady state, recall that the first and third eigenvalue of $DF(\hat{\boldsymbol{u}}_+)$ given in (13.9) are both negative. Since $k_2 < k_2(k_1, P)$, the second eigenvalue of $DF(\hat{\boldsymbol{u}}_+)$ lies in the left half-plane, as shown in Figs. 13.5 or 13.4. For the nontrivial steady state $\hat{\boldsymbol{u}}_-$, one can check numerically that all three eigenvalues of $DF(\hat{\boldsymbol{u}}_-)$ are real and negative. If we are in region ④, we are done. If we are in region ②, then it remains to show that there exists a non-negative vector $\boldsymbol{q}_1 = \begin{bmatrix} q_1 & q_2 & q_3 \end{bmatrix}$ such that $\boldsymbol{q}_1 DF(\hat{\boldsymbol{u}}^1) > 0$. If $\hat{\boldsymbol{u}}^1 = \begin{bmatrix} u_1 & v_1 & P_1 \end{bmatrix}$, then the inequality can be written as

$$\begin{cases} -q_1|f_u| + q_2|g_u| > 0, \\ q_1|f_v| - q_2|g_v| > 0, \\ q_1|f_P| + q_2|g_P| - q_3|s_P| > 0, \end{cases} \tag{13.15}$$

where we have used the fact that $s_u(\hat{\boldsymbol{u}}^1) = 0$ and $f_u(\hat{\boldsymbol{u}}^1), g_v(\hat{\boldsymbol{u}}^1), s_P(\hat{\boldsymbol{u}}^1) < 0$. Note that if $k_0 = 0$ then there are only two stationary points, so we only need to consider the case where $k_0 \neq 0$, which guarantees $g_u(\hat{\boldsymbol{u}}^1) \neq 0$. Since $k_2 = 0$ implies and $g_P(\hat{\boldsymbol{u}}^1) = 0$, we want to discuss by cases depending on whether k_2 is zero or not. Note that $g_v(\hat{\boldsymbol{u}}^1) \neq 0$ for any parameter values.

Case 1: if $k_2 > 0$, then inequality (13.15) can be further written as

$$\begin{cases} q_2 > \dfrac{q_1|f_u|}{|g_u|}, \\ q_2 < \dfrac{q_1|f_v|}{|g_v|}, \\ q_2 > -\dfrac{q_1|f_P| - q_3|s_P|}{|g_P|}. \end{cases}$$

Such $\boldsymbol{q}_1$ exists if there are q_1, q_3 such that

$$\begin{cases} \dfrac{q_1|f_v|}{|g_v|} > \dfrac{q_1|f_u|}{|g_u|}, \\ \dfrac{q_1|f_v|}{|g_v|} > -\dfrac{q_1|f_P| - q_3|s_P|}{|g_P|}, \end{cases}$$

which leads to

$$\begin{cases} q_1\left(\dfrac{|f_v|}{|g_v|} - \dfrac{|f_u|}{|g_u|}\right) > 0, \\ q_3 < q_1\dfrac{|g_P|}{|s_P|}\left(\dfrac{|f_v|}{|g_v|} + \dfrac{|f_P|}{|g_P|}\right). \end{cases}$$

Table 13.4 Six regions and their source types

Label	Region	Number of steady states	Source type
①	$k_2 < k_2(k_0, k_1), k_2 < k_2(k_1, P)$	1	Near-pulse
②	$k_2(k_0, k_1) < k_2 < k_2(k_1), k_2 < k_2(k_1, P)$	3	Bistable
③	$k_2 > k_2(k_1), k_2 > k_2(k_1, P)$	2	Monostable
④	$k_2(k_1) < k_2 < k_2(k_1, P)$	2	Bistable
⑤	$k_2(k_1, P) < k_2 < k_2(k_0, k_1)$	1	Near-pulse
⑥	$k_2(k_0, k_1) < k_2 < k_2(k_1), k_2 > k_2(k_1, P)$	3	Monostable

Note that such q_1, q_3 exist only if

$$\begin{cases} \dfrac{|f_v|}{|g_v|} - \dfrac{|f_u|}{|g_u|} > 0, \\ \dfrac{|g_P|}{|s_P|}\left(\dfrac{|f_v|}{|g_v|} + \dfrac{|f_P|}{|g_P|}\right) > 0, \end{cases}$$

yielding the conditions

$$\begin{cases} f_v g_u - f_u g_v > 0, \\ u_1 \neq 1. \end{cases}$$

Case 2: if $k_2 = 0$, then a similar derivation as above yields the conditions

$$\begin{cases} f_v g_u - f_u g_v > 0, \\ k_1 \neq 0, u_1 \neq 1. \end{cases}$$

One can check that the above strict inequality for both cases is satisfied at $\hat{\boldsymbol{u}} = \hat{\boldsymbol{u}}^1$.
Therefore, we obtain the existence and uniqueness of monotone traveling waves by Theorem 3.2 in Chap. 3 of [23]. □

In Table 13.4 we present a summary of different regions in the $k_1 - k_2$ regime with the according numbers of steady states and source types. The regions correspond to the ones labeled in Fig. 13.6.

Figure 13.8 illustrates traveling wave profiles in the regimes discussed in Theorems 13.1 and 13.2.

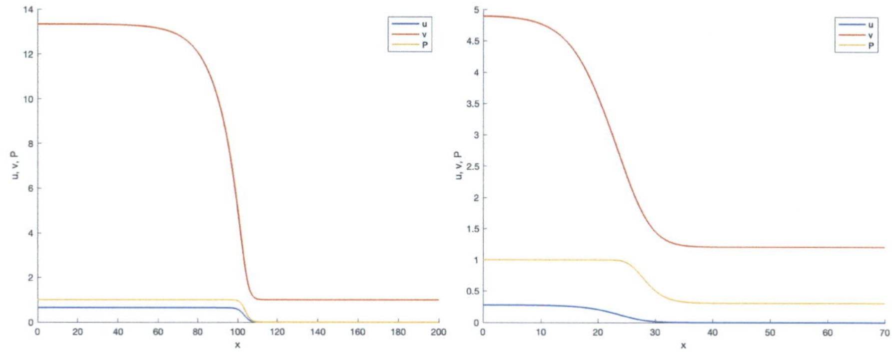

(a) The monostable region with $k_2 = 0.5$ and $\Gamma = 1$.

(b) The bistable region with $k_2 = 0.7$ and $\Gamma = 0.5$.

Fig. 13.8 Examples of traveling wave profiles with $k_0 = 4.5$, $k_1 = 0.5$, $\alpha = 1$, $\beta = 1$, and $\omega = 0.5$

13.3.2 Stability and Asymptotic Decay Rates of Traveling Wave Solutions

In this section, we explore the stability of the traveling wave solutions whose existence is proved in the previous section. We want to determine the types of perturbations allowed to obtain stable waves in different parameter regimes.

13.3.2.1 Asymptotic Decay Rates

We first present some preliminary results on the asymptotic decay rates of the waves, which will assist us in proving the stability of the waves in the monostable region. These rates show that the traveling waves and their derivatives respectively approach the steady states and zero exponentially fast as $|z| \to \infty$. Exact proofs for the following propositions can be found in Appendix 13.6.

Proposition 13.1 (Asymptotic approach rates for $k_0 \geq 0$) *Let* $\hat{\boldsymbol{u}}(z) = \begin{bmatrix} \hat{u}(z) \\ \hat{v}(z) \\ \hat{P}(z) \end{bmatrix}$ *be the traveling wave solution to* (13.3) *that satisfies*

$$\begin{cases} \lim_{z\to+\infty} \hat{\boldsymbol{u}}(z) = \hat{\boldsymbol{u}}_+, \quad \lim_{z\to-\infty} \hat{\boldsymbol{u}}(z) = \hat{\boldsymbol{u}}_-, \\ \hat{u}(z), \hat{v}(z) \geq 0, \quad 0 \leq \hat{P}(z) \leq 1, \\ \hat{u}'(z), \hat{v}'(z) < 0, \end{cases}$$

where $\hat{\boldsymbol{u}}_+ = \begin{bmatrix} 0 \\ (1+P)^{-k_2} \\ P \end{bmatrix}$, $\hat{\boldsymbol{u}}_- = \begin{bmatrix} u^* \\ v^* \\ 1 \end{bmatrix}$. *Define*

$$\mu_{1,2} = -(f_u(\hat{\boldsymbol{u}}_-) + g_v(\hat{\boldsymbol{u}}_-)) \\ \pm \sqrt{(f_u(\hat{\boldsymbol{u}}_-) + g_v(\hat{\boldsymbol{u}}_-))^2 + 4 f_v(\hat{\boldsymbol{u}}_-) g_u(\hat{\boldsymbol{u}}_-) - 4 f_u(\hat{\boldsymbol{u}}_-) g_v(\hat{\boldsymbol{u}}_-)},$$

where μ_1, μ_2 take the plus, minus sign, respectively.

(a) There are nonnegative vectors $\boldsymbol{k}_c, \boldsymbol{K}_c, \boldsymbol{k}_\epsilon, \boldsymbol{K}_\epsilon, \boldsymbol{m}_c, \boldsymbol{M}_c, \boldsymbol{m}_\epsilon, \boldsymbol{M}_\epsilon$, negative numbers $\lambda_5, \lambda_3, \lambda_2$, and positive numbers $\epsilon, \tilde{\lambda}_1, \tilde{\lambda}_2$ such that the nonzero elements of $\boldsymbol{K}_\epsilon, \boldsymbol{M}_\epsilon$ go to ∞ as $\epsilon \to 0$, and

$$\begin{cases} \boldsymbol{k}_c e^{\lambda_5 z} \le \hat{\boldsymbol{u}}(z) - \hat{\boldsymbol{u}}_+ \le \boldsymbol{K}_c e^{\lambda_5 z}, & (z \ge 0),\ f_u(\hat{\boldsymbol{u}}_+) < g_v(\hat{\boldsymbol{u}}_+), \\ \boldsymbol{k}_\epsilon e^{(\lambda_5 - \epsilon) z} \le \hat{\boldsymbol{u}}(z) - \hat{\boldsymbol{u}}_+ \le \boldsymbol{K}_\epsilon e^{(\lambda_5 + \epsilon) z}, & (z \ge 0),\ f_u(\hat{\boldsymbol{u}}_+) = g_v(\hat{\boldsymbol{u}}_+), \\ \boldsymbol{k}_c e^{\lambda_3 z} \le \hat{\boldsymbol{u}}(z) - \hat{\boldsymbol{u}}_+ \le \boldsymbol{K}_c e^{\lambda_3 z}, & (z \ge 0),\ g_v(\hat{\boldsymbol{u}}_+) < f_u(\hat{\boldsymbol{u}}_+) < 0, \\ \boldsymbol{k}_c e^{\lambda_2 z} \le \hat{\boldsymbol{u}}(z) - \hat{\boldsymbol{u}}_+ \le \boldsymbol{K}_c e^{\lambda_2 z}, & (z \ge 0),\ f_u(\hat{\boldsymbol{u}}_+) > 0,\ c > 2\sqrt{f_u(\hat{\boldsymbol{u}}_+)}, \\ \boldsymbol{k}_\epsilon e^{(\lambda_2 - \epsilon) z} \le \hat{\boldsymbol{u}}(z) - \hat{\boldsymbol{u}}_+ \le \boldsymbol{K}_\epsilon e^{(\lambda_2 + \epsilon) z}, & (z \ge 0),\ f_u(\hat{\boldsymbol{u}}_+) > 0,\ c = 2\sqrt{f_u(\hat{\boldsymbol{u}}_+)}, \\ \boldsymbol{m}_c e^{\tilde{\lambda}_1 z} \le \hat{\boldsymbol{u}}_- - \hat{\boldsymbol{u}}(z) \le \boldsymbol{M}_c e^{\tilde{\lambda}_1 z}, & (z \le 0),\ \frac{u^*}{c} < \frac{-c+\sqrt{c^2+2\mu_2}}{2}, \\ \boldsymbol{m}_\epsilon e^{(\tilde{\lambda}_1 + \epsilon) z} \le \hat{\boldsymbol{u}}_- - \hat{\boldsymbol{u}}(z) \le \boldsymbol{M}_\epsilon e^{(\tilde{\lambda}_1 - \epsilon) z} & (z \le 0),\ \frac{u^*}{c} = \frac{-c+\sqrt{c^2+2\mu_2}}{2}, \\ \boldsymbol{m}_\epsilon e^{(\tilde{\lambda}_2 + \epsilon) z} \le \hat{\boldsymbol{u}}_- - \hat{\boldsymbol{u}}(z) \le \boldsymbol{M}_\epsilon e^{(\tilde{\lambda}_2 - \epsilon) z} & (z \le 0),\ \frac{-c+\sqrt{c^2+2\mu_2}}{2} < \frac{u^*}{c},\ \mu_1 = \mu_2, \\ \boldsymbol{m}_c e^{\tilde{\lambda}_2 z} \le \hat{\boldsymbol{u}}_- - \hat{\boldsymbol{u}}(z) \le \boldsymbol{M}_c e^{\tilde{\lambda}_2 z}, & (z \le 0),\ \frac{-c+\sqrt{c^2+2\mu_2}}{2} < \frac{u^*}{c},\ \mu_1 \ne \mu_2. \end{cases}$$

(b) There are nonpositive vectors $\boldsymbol{h}_c, \boldsymbol{H}_c, \boldsymbol{h}_\epsilon, \boldsymbol{H}_\epsilon$, nonnegative vectors $\boldsymbol{n}_c, \boldsymbol{N}_c, \boldsymbol{n}_\epsilon, \boldsymbol{N}_\epsilon$, negative numbers $\lambda_5, \lambda_3, \lambda_2$, and positive numbers $\epsilon, \tilde{\lambda}_1, \tilde{\lambda}_2$ such that the nonzero elements of $\boldsymbol{h}_\epsilon$ and $\boldsymbol{N}_\epsilon$ respectively go to $-\infty$ and ∞ as $\epsilon \to 0$, and

$$\begin{cases} \boldsymbol{h}_c e^{\lambda_5 z} \le \hat{\boldsymbol{u}}'(z) \le \boldsymbol{H}_c e^{\lambda_5 z}, & (z \ge 0),\ f_u(\hat{\boldsymbol{u}}_+) < g_v(\hat{\boldsymbol{u}}_+), \\ \boldsymbol{h}_\epsilon e^{(\lambda_5 + \epsilon) z} \le \hat{\boldsymbol{u}}'(z) \le \boldsymbol{H}_\epsilon e^{(\lambda_5 - \epsilon) z}, & (z \ge 0),\ f_u(\hat{\boldsymbol{u}}_+) = g_v(\hat{\boldsymbol{u}}_+), \\ \boldsymbol{h}_c e^{\lambda_3 z} \le \hat{\boldsymbol{u}}'(z) \le \boldsymbol{H}_c e^{\lambda_3 z}, & (z \ge 0),\ g_v(\hat{\boldsymbol{u}}_+) < f_u(\hat{\boldsymbol{u}}_+) < 0, \\ \boldsymbol{h}_c e^{\lambda_2 z} \le \hat{\boldsymbol{u}}'(z) \le \boldsymbol{H}_c e^{\lambda_2 z}, & (z \ge 0),\ f_u(\hat{\boldsymbol{u}}_+) > 0,\ c > 2\sqrt{f_u(\hat{\boldsymbol{u}}_+)}, \\ \boldsymbol{h}_\epsilon e^{(\lambda_2 + \epsilon) z} \le \hat{\boldsymbol{u}}'(z) \le \boldsymbol{H}_\epsilon e^{(\lambda_2 - \epsilon) z}, & (z \ge 0),\ f_u(\hat{\boldsymbol{u}}_+) > 0,\ c = 2\sqrt{f_u(\hat{\boldsymbol{u}}_+)}, \\ \boldsymbol{n}_c e^{\tilde{\lambda}_1 z} \le -\hat{\boldsymbol{u}}'(z) \le \boldsymbol{N}_c e^{\tilde{\lambda}_1 z}, & (z \le 0),\ \frac{u^*}{c} < \frac{-c+\sqrt{c^2+2\mu_2}}{2}, \\ \boldsymbol{n}_\epsilon e^{(\tilde{\lambda}_1 + \epsilon) z} \le -\hat{\boldsymbol{u}}'(z) \le \boldsymbol{N}_\epsilon e^{(\tilde{\lambda}_1 - \epsilon) z} & (z \le 0),\ \frac{u^*}{c} = \frac{-c+\sqrt{c^2+2\mu_2}}{2}, \\ \boldsymbol{n}_\epsilon e^{(\tilde{\lambda}_2 + \epsilon) z} \le -\hat{\boldsymbol{u}}'(z) \le \boldsymbol{N}_\epsilon e^{(\tilde{\lambda}_2 - \epsilon) z} & (z \le 0),\ \frac{-c+\sqrt{c^2+2\mu_2}}{2} < \frac{u^*}{c},\ \mu_1 = \mu_2, \\ \boldsymbol{n}_c e^{\tilde{\lambda}_2 z} \le -\hat{\boldsymbol{u}}'(z) \le \boldsymbol{N}_c e^{\tilde{\lambda}_2 z}, & (z \le 0),\ \frac{-c+\sqrt{c^2+2\mu_2}}{2} < \frac{u^*}{c},\ \mu_1 \ne \mu_2. \end{cases}$$

13.3.2.2 Stability

We now present the stability results for the bistable and monostable regions separately. We will see that the stability in the monostable regions allows for perturbations from a family of Banach spaces embedded with exponentially weighted norms, which requires a faster decay rate at infinity compared to the bistable case.

Theorem 13.3 (Bistable) *Let $k_0, k_1 \geq 0, k_2 > 0$, or $k_0, k_1 > 0, k_2 \geq 0$. For $k_2(k_0, k_1) < k_2 < k_2(k_1, P)$, the traveling waves whose existence is guaranteed by Theorem 13.2 are asymptotically stable with shift in the sup-norm $\|\cdot\|$ defined by $\|\boldsymbol{u}\| = \sup_{x\in\mathbb{R}} |\boldsymbol{u}(x)|$ for all vector-valued, continuous and bounded function $\boldsymbol{u}$ defined on $\mathbb{R}$.*

Proof Let us denote the monotone wave solution of system (13.3) by $\hat{\boldsymbol{u}}(z)$. Since the system is monotone for $k_0, k_1, k_2 \geq 0$, its Jacobian $DF(\hat{\boldsymbol{u}}(z))$ has nonnegative off-diagonal elements. One can also check that it is functionally irreducible so long as $\Gamma, \beta \neq 0$, and either $k_0, k_1 \geq 0, k_2 > 0$ or $k_0, k_1 > 0, k_2 \geq 0$. For the eigenvalue conditions, recall from the proof for Theorem 13.2 that matrices $DF(\hat{\boldsymbol{u}}_-)$ and $DF(\hat{\boldsymbol{u}}_+)$ have all their eigenvalues in the left half-plane, where, again, $\hat{\boldsymbol{u}}_-$ and $\hat{\boldsymbol{u}}_+$ are the two steady states that $\hat{\boldsymbol{u}}(z)$ approaches as $z \to -\infty$ and $z \to +\infty$, respectively. The last condition $\lim_{z\to\infty} \hat{\boldsymbol{u}}'(z) = 0$ results from the monotonicity assumption $\hat{\boldsymbol{u}}'(z) < 0$ and the definition $\lim_{z\to\infty} \hat{\boldsymbol{u}}(z) = \hat{\boldsymbol{u}}_+$. Therefore, the wave $\hat{\boldsymbol{u}}(z)$ is asymptotically stable with a shift in the sup-norm by Theorem 4.1 in Chap. 5 of [23]. □

Theorem 13.4 (Monostable) *Let $k_0, k_1 \geq 0, k_2 > 0$, or $k_0, k_1 > 0, k_2 \geq 0$. For $k_2 > k_2(k_0, k_1)$ and $k_2 > k_2(k_1, P)$, the traveling waves whose existence is guaranteed by Theorem 13.1 are asymptotically stable in the norm $\|\cdot\|_\sigma$ for some $\sigma \geq 0$. The norm $\|\cdot\|_\sigma$ is defined by $\|\boldsymbol{u}\|_\sigma = \sup_{x\in\mathbb{R}} |\boldsymbol{u}(x)(1 + e^{\sigma x})|$ for all vector-valued, continuous and bounded function $\boldsymbol{u}$ defined on $\mathbb{R}$ such that $\lim_{|x|\to\infty} \boldsymbol{u}(x)(1 + e^{\sigma x}) = 0$.*

Proof Again, the Jacobian $DF(\hat{\boldsymbol{u}}(z))$ is functionally irreducible in the given region where either $k_0, k_1 \geq 0, k_2 > 0$ or $k_0, k_1 > 0, k_2 \geq 0$. We have checked in the proof for 13.1 that $DF(\hat{\boldsymbol{u}}_-)$ has all its eigenvalues in the left half-plane. Now, we want to find a nonnegative σ that shifts all eigenvalues of the matrix $A\sigma^2 - C\sigma + DF(\hat{\boldsymbol{u}}_+)$ to the left half-plane. The eigenvalues of this matrix are

$$\lambda_1 = \sigma^2 - c\sigma + f_u(\hat{\boldsymbol{u}}_+), \quad \lambda_2 = \sigma^2 - c\sigma + g_v(\hat{\boldsymbol{u}}_+), \quad \lambda_3 = -c\sigma.$$

Since $g_v(\hat{\boldsymbol{u}}_+) < f_u(\hat{\boldsymbol{u}}_+)$ always holds in the monostable regions, we only need to require that $\lambda_1 < 0$. This is satisfied as long as

$$\frac{c - \sqrt{c^2 - 4f_u(\hat{\boldsymbol{u}}_+)}}{2} < \sigma < \frac{c + \sqrt{c^2 - 4f_u(\hat{\boldsymbol{u}}_+)}}{2},$$

with $c > 2\sqrt{f_u(\hat{\boldsymbol{u}}_+)}$.

The last part of the proof is to show that $\hat{\boldsymbol{u}}'(z)e^{\sigma z}$ diverges as $z \to \infty$. By part (b) of Proposition 13.1, we have

$$\boldsymbol{h}_c e^{(\lambda_2+\sigma)z} \leq \hat{\boldsymbol{u}}'(z)e^{\sigma z} \leq \boldsymbol{H}_c e^{(\lambda_2+\sigma)z}.$$

Since $\lambda_2 = \dfrac{-c + \sqrt{c^2 - 4f_u(\hat{\boldsymbol{u}}_+)}}{2}$, using the lower bound for σ found above, we see that $\lambda_2 + \sigma > 0$ and conclude divergence.

By Theorem 4.1 in Chap. 5 of [23], the waves are asymptotically stable in the norm $\|\cdot\|_\sigma$. □

Our stability results generalize the results in [25], but we note that the waves in the monostable region in the monotone system in [25] should be asymptotically stable without a shift due to the divergence condition.

13.4 Model Outcomes from Numerical Experiments

In this section, we explore parameter space to study important features that are difficult to analyze theoretically. Our emphasis is on studying the effect of k_0, k_1, and k_2 on these features; refer to Table 13.2 for a refresher on the physical interpretation of these parameters. The specific values for all parameters employed in these simulations are detailed in Table 13.6 within Appendix 13.7.

13.4.1 Quantitative Outcomes

We present various metrics to assess the outcomes of simulations conducted under different conditions. While many measurements can quantify the dynamics of u, v, and P in a system initiated with predominantly localized protests, a moderate level of social tension everywhere, and no police presence, we intended to provide a combination of local and global measurements. It is important to note that the speed of the traveling wave remains independent of the parameters. Additionally, by $t = 15$, we observed that the profiles for u with the parameters outlined in Table 13.6 had fully formed. The profile for v required more time to form, and by $t = 100$, most wave profiles had been reached.

As mentioned in the introduction, a wave of protests is initiated when significant social tension is coupled with a triggering event. Our numerical simulations assume an initial social tension of one uniformly across space. The triggering event is assumed to occur at the leftmost boundary of the domain, resulting in an initial condition of $u(x, 0) = e^{-5x}$. It is natural to conceptualize this event as a localized function with a decaying impact relative to the distance from the triggering event location. It is well-established that exponentially decaying initial data gives rise to traveling wave solutions. Furthermore, we assume an initial condition of zero throughout the domain for police presence.

13.4.1.1 Invading Protest Levels

A metric of interest for individuals is the degree of activity and social tension encountered as the protest wavefront passes through. Specifically, the invading protest wavefront serves as a distinct boundary separating no protest activity immediately to its right from a positive level of activity on it. Given the constant wave speed across all simulations, we observe a near-identical nature of this interface among simulations. To the left of this point, the behavior diverges from simulation to simulation for varying parameter values (refer to Fig. 13.9d–f). The x coordinate where the simulation-to-simulation behavior diverges is easily discernible in simulations with negative k_0 values, identified as the local maximum of the interface (e.g., in Fig. 13.9d, it occurs at approximately $x = 454$). We denote this x coordinate as the location of the invading interface because it marks the trailing edge of a sharp interface distinguishing the region unexposed to protesting activity from the exposed region. In Fig. 13.9a–c, we present the level of invading protest activity (i.e., the level at the invading interface) for various values of k_0, k_1, and k_2, demonstrating slight variations. This also underscores the point that the invading level does not necessarily correspond to the highest point of the traveling wave, as some profiles continue to grow beyond the invading interface (e.g., the profiles in Fig. 13.9d).

As depicted in Fig. 13.9a–c, an increase in any of the parameters k_0, k_1, or k_2 results in a higher level of u at the invading interface. In other words, the intensity of protesting activity strictly and monotonically rises as a function of any of the parameters k_i. While the final level of protesting activity after the front has passed cannot be predicted *a priori* solely based on the level of protest activity at the invading interface, it holds true that in regions unexposed to protesting, the highest levels of protesting activity at the invading interface occur when the values of k_i are at their maximum. An intriguing observation is that k_1 appears to exert a greater influence on the activity level than k_2. In Fig. 13.9a–c, we note that the bottom right quadrants, where $k_1 < 0$ and $k_2 > 0$, exhibit less activity compared to scenarios where $k_1 > 0$.

13.4.1.2 Residual Levels Left After the Front Interface Passes

Another noteworthy metric is the residual level of protest activity and social tension that persists after the interface traverses. Based on the prior sections' analysis, we know that the constant state at the fixed x coordinate must have either u approaching zero or P approaching one. In other words, the protest traveling wave can result in a sustained nonzero level of protest activity that no longer alters social tension levels, or it may exhibit a near-pulse profile of activity that essentially diminishes after passing.

We have measured the residual levels of protest activity, tension, and police presence across various values of k_0, k_1, and k_2, illustrating them in a heatmap presented in Fig. 13.10. It is crucial to highlight that in all experiments, when $k_2 < 0$, indicative of a negotiated management strategy, the residual values are either zero or close to zero, even when $k_1 > 0$. By examining these values, we gain insights into the

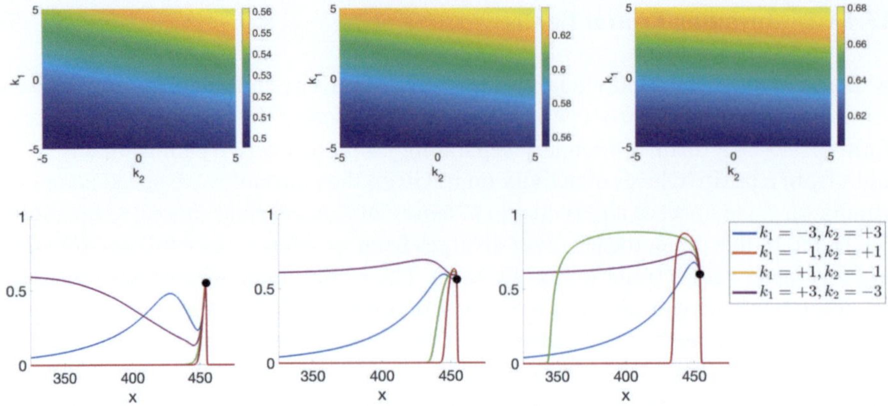

Fig. 13.9 **a–c** Heatmaps for the value of the invading protest levels for simulations run with various combinations of k_1 and k_2. **d–f** Sample profiles from different regions from the figures (**a**)–(**c**) with the legend providing the k_1 and k_2 coordinates for the respective profiles. The black dot signifies the location of the invading interface. Note: in (**d**) the green curve is plotted but is mostly covered by the red curve

contrasting roles that police presence can play in influencing protest activity. For instance, in scenarios where $k_2 > 0$, signifying an escalated force management strategy, we observe a positive level of activity remaining, even if police presence results in deterrence ($k_1 < 0$). Furthermore, there are specific regions where $k_1 < 0$ and $k_2 > 0$, showcasing a notable persistence of tension after the front interface has passed.

From these observations, it becomes apparent that a positive k_2 value outweighs an inhibitory k_1 value ($k_1 < 0$), suggesting that if police presence raises social tension, it will lead to more protest activity even if it discourages protest participation. Moreover, the parameter k_2 appears sensitive for small absolute values but less so for large magnitudes concerning the expected residual protest activity after the wave interface passes. This implies that the extent to which police presence enhances social tension may not matter as much as the fact that it does enhance it; a positive level of protest activity is expected to persist for a fixed k_1 value. A similar trend is observed for k_1, where a positive k_1 value leads to approximately the same residual activity regardless of the value, whereas, for negative values, there are more gradual changes in residual activity as k_1 decreases (for a fixed $k_2 > 0$).

In Fig. 13.11a, a zoomed-in view of Fig. 13.10d focusing on the third quadrant, reveals that residual values are not monotonic with respect to k_1. Within this range, for each fixed value of k_2, as k_1 increases from –4 to 1, the residual value of v ascends to a local maximum before decreasing. Figure 13.11b displays the profiles of v at $t = 95$ with k_2 fixed at five different values of k_1. A comparison of different $v(x, t; k_0, k_1, k_2)$ values shows that $v(0, 95; -1, -4, 5) < v(0, 95; -1, -3, 5) < v(0, 95; -1, -2, 5)$, while $v(0, 95; -1, -2, 5) > v(0, 95; -1, -1, 5) > v(0, 95; -1, 0, 5) > v(0, 95; -1, 1, 5)$. An interpretation of this behavior is that if a protest

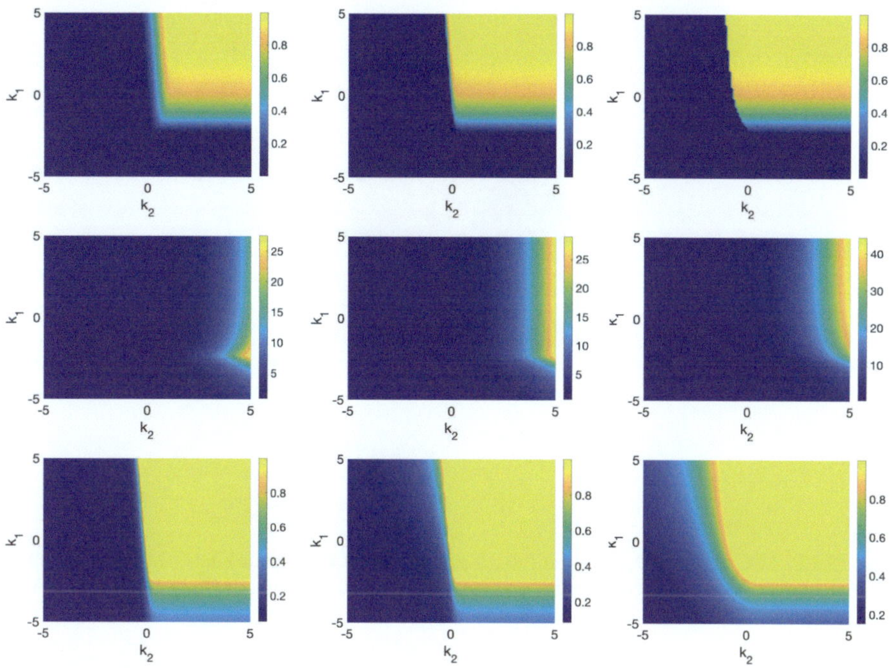

Fig. 13.10 Residual levels of protest activity **a–c**, social tension **d–f**, and police presence **g–i** for various k_0, k_1, and k_2 parameter combinations when $t = 100$

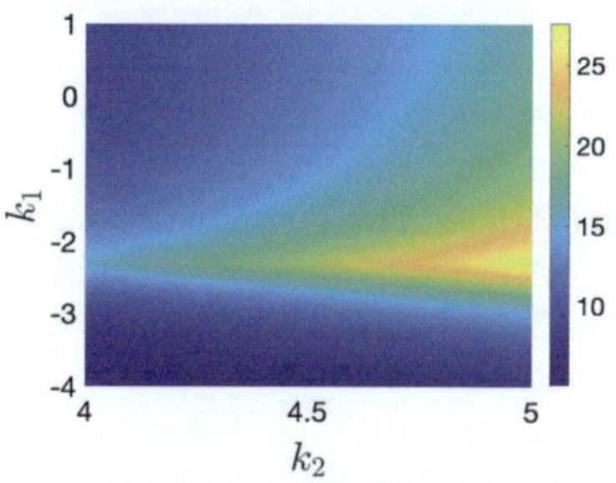

(a) Zoom in of Figure 13.10d

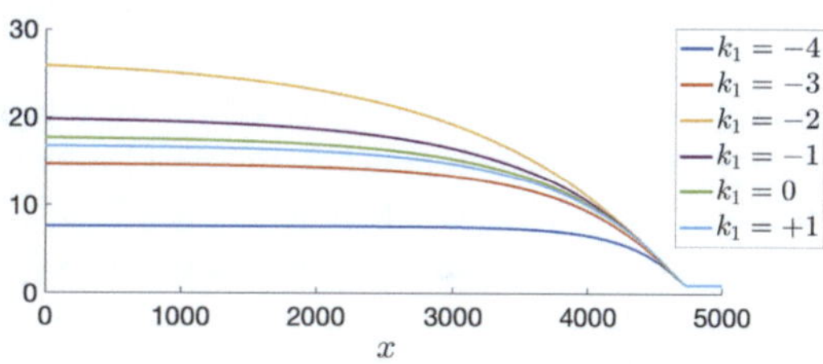

(b) v profiles with $k_0 = -1$ and $k_2 = +5$ (at $t = 95$)

Fig. 13.11 Zoom in on the residual levels of social tension and sample wave profiles

wave is traversing an environment in which protesting inhibits social tension ($k_0 < 0$) but police presence enhances social tension ($k_2 > 0$), then a policing policy aimed at discouraging protests ($k_1 < 0$) will result in stronger discouragement leading to higher social tension, up to a certain point when the discouragement becomes effective at reducing social tension.

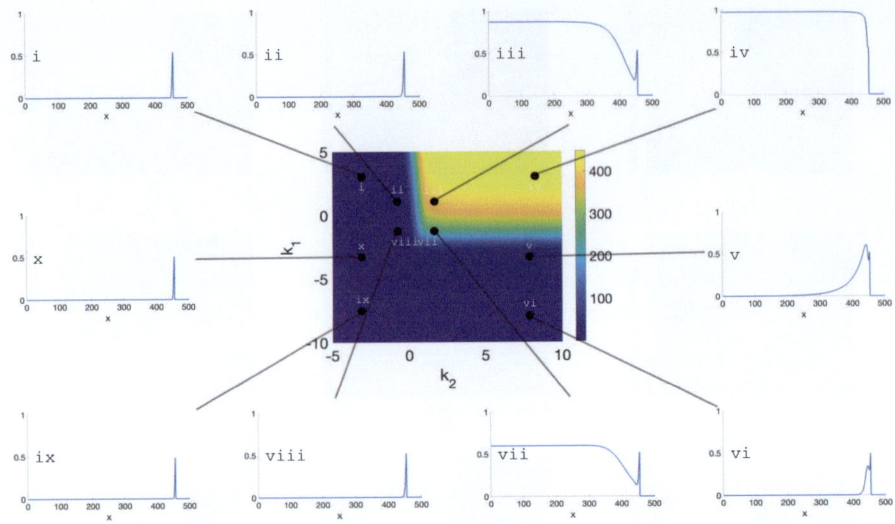

Fig. 13.12 AUC of u with $k_0 = -1$ at $t = 15$ and profiles at select coordinates. Moving clockwise from the top left corner the coordinates (k_1, k_2) are: i (3, –3), ii (1, –1), iii (1, 1), iv (3, 8), v (–3, 8), vi (–8, 8), vii (–1, 1), viii (–1, –1), ix (–8, –3), and x (–3, –3)

13.4.1.3 Cumulative Levels of Protest Activity and Social Tension

Perhaps one of the most important metrics is the total exposure of protesting activity and net social tension. To quantify this metric we look at Area Under the Curve (AUC). These were computed by integrating the area underneath the profiles across space, with the calculation of each profile taking place at the same fixed point in time when wave profiles that are formed for all relevant k parameter combinations. Due to the nature of the calculations, the AUC values are affected by maximum values as well residual values, but the plots in Figs. 13.12, 13.13, 13.14 and 13.15 more closely resemble their respective residual value plots than the maximum value plots.

In Fig. 13.12, we observe two predominant types of results: (1) a substantial AUC value due to u approaching a non-zero residual level, and (2) a more moderate AUC value as u approaches the $u = 0$ regime. For all $k_2 < 0$, the profiles for u demonstrate a narrow pulse-like behavior, indicating that protest activity is concentrated primarily at the invading interface. However, when $k_2 > 0$ and k_1 is sufficiently negative, protest activity extends beyond the invading interface, gradually diminishing to zero as the traveling wave progresses further away (see Fig. 13.12, profiles (v) and (vi)).

Figure 13.13 illustrates the social tension corresponding to the protest activity depicted in Fig. 13.12. When $k_0 = -1$ and $k_2 < 0$, which encompasses scenarios where the residual tension is less than one, the reductions in AUC are relatively gradual as both k_1 and k_2 decrease. In this parameter space, the profiles quickly approach the residual value, attributing the AUC decrease primarily to the diminish-

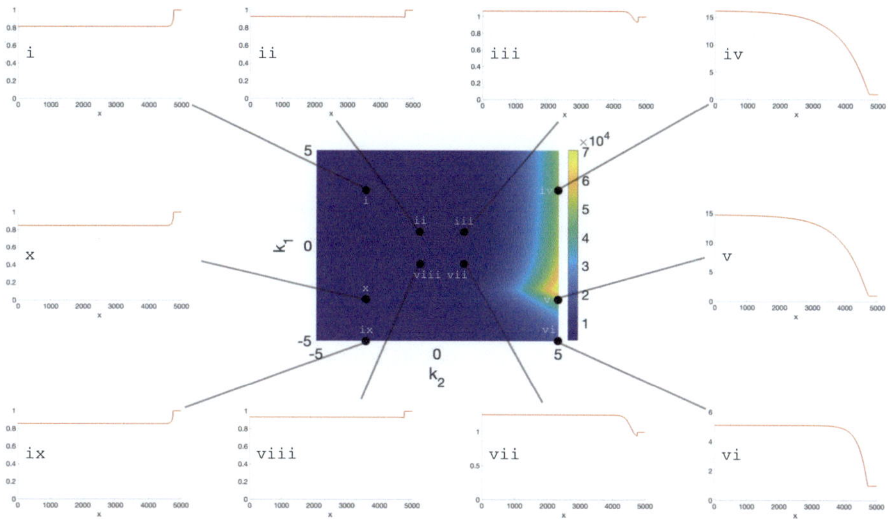

Fig. 13.13 AUC of v with $k_0 = -1$ at $t = 100$ and profiles at select coordinates (at $t = 95$). Moving clockwise from the top left corner the coordinates (k_1, k_2) are: i (3, –3), ii (1, –1), iii (1, 1), iv (3, 5), v (–3, 5), vi (–5, 5), vii (–1, 1), viii (–1, –1), ix (–5, –3), and x (–3, –3)

ing residual value as k_1 and k_2 decrease. Notably, the heatmap in Fig. 13.13 reveals a similar nonmonotonic trend in AUC as observed in the residual values in Fig. 13.10d.

The heatmap for $k_0 = 1$ in Fig. 13.14 reveals a more diverse range of behaviors than the predominantly observed two types (narrow pulse-like profiles of u and profiles trending towards non-zero steady states) in the $k_0 = -1$ heatmap depicted in Fig. 13.12. While profiles can still be categorized as near-pulse or non-pulse, the near-pulse waves, particularly when $k_1 > 0$, may exhibit plateaus near $u = 1$ for a significant portion of the x domain before dropping sharply towards zero. This plateauing behavior results in a relatively large AUC value, even though the residual value is roughly zero. Generally, in comparison to the $k_0 = -1$ scenario, protest activity is not confined solely to the invading interface when $k_0 = 1$, leading to a higher overall level of protest activity.

Figures 13.14 and 13.15 emphasize the significance of considering qualitative behaviors in conjunction with quantitative measurements. Solely focusing on maximum protest levels and global protest levels might overlook crucial insights into what is experienced as a protest wave approaches and passes a fixed point.

13.4.2 Qualitative Outcomes

In this section, we delve into the qualitative outcomes of simulations, considering the viewpoint of a resident fixed at a specific location. Taking the perspective of a town or

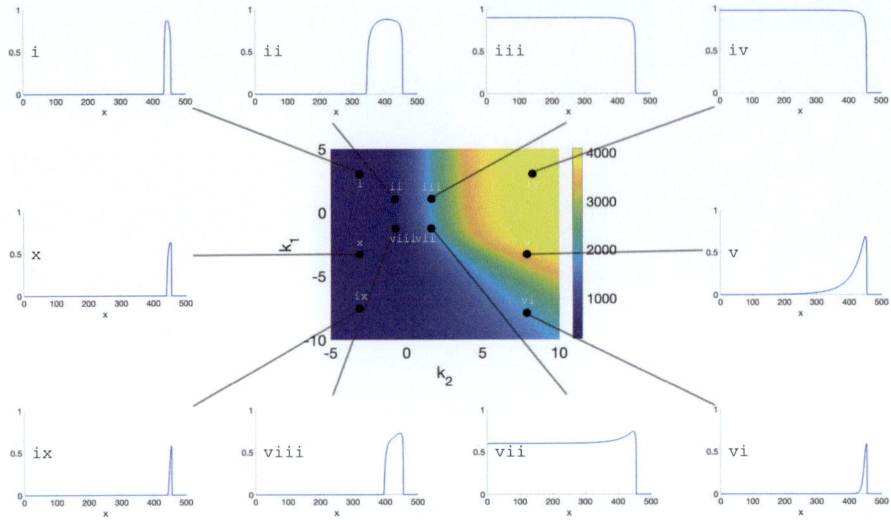

Fig. 13.14 AUC of u with $k_0 = 1$ at $t = 15$ and profiles at select coordinates. Moving clockwise from the top left corner the coordinates (k_1, k_2) are: i (3, –3), ii (1, –1), iii (1, 1), iv (3, 8), v (–3, 8), vi (–8, 8), vii (–1, 1), viii (–1, –1), ix (–8, –3), and x (–3, –3)

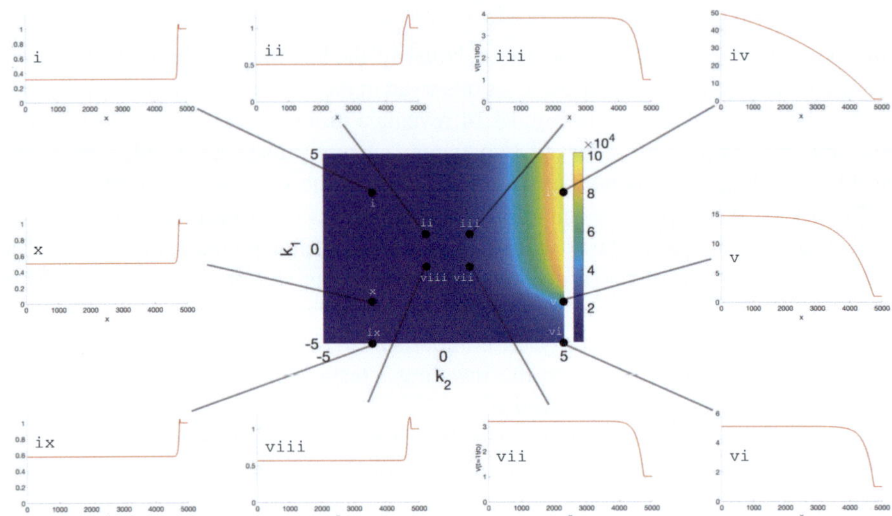

Fig. 13.15 AUC of v with $k_0 = 1$ at $t = 100$ and profiles at select coordinates (at $t = 95$). Moving clockwise from the top left corner the coordinates (k_1, k_2) are: i (3, –3), ii (1, –1), iii (1, 1), iv (3, 5), v (–3, 5), vi (–5, 5), vii (–1, 1), viii (–1, –1), ix (–5, –3), and x (–3, –3)

neighborhood, the encountered activity varies significantly based on the environment (parameters) through which the protesting wave passes. The fluctuations in protest activity, tension, and police presence play a crucial role in shaping the experience of residents. They may witness scenarios where there is no protest activity, followed by a sudden surge, only to witness a rapid and complete cessation of all activity. Alternatively, there may be instances of no protest activity, a sudden surge, followed by a swift decline before resurging and persisting at levels even higher than the initial invading levels. Although these experiences may initially appear similar, they can evolve into vastly different systems. Given the diverse outcomes, we have identified a few key features and the specific regions where these features manifest.

13.4.2.1 Productive and Unproductive Strategies for Relieving Tension

Arguably, the most crucial metric for assessing the effectiveness of protests and policing strategies involves comparing tension levels before and after the protests. In an environment characterized by a baseline tension level that experiences a wave of protest activity, evaluating the productivity of the protests and policing strategy in alleviating or exacerbating the underlying tension is vital. This assessment is made by comparing tension levels before and after the complete passage of the wave of protests through a region. We quantified residual tension and, in Figs. 13.16 and 13.17, juxtaposed these residual levels against the initial level ($v(x, 0) = 1$). Regions, where the wave of activity escalated social tension, are shaded purple, while those where tension decreased immediately and persistently are shaded green. If the tension increases at some point before ultimately being successfully reduced, the region is shaded light blue. We denote the quantity v_0 to be the initial tension and v_{ss} to be the difference between the residual tension and v_0.

In the scenario where $k_0 > 0$ (exemplified in Fig. 13.16), there is an immediate uptick in tension following the invading wave of protest activity. After this escalation, two outcomes emerge: (1) tension continues to rise monotonically until reaching a steady state with a higher tension in the system than the initial state (depicted in purple in Fig. 13.16a with example profiles in Fig. 13.16c and d), and (2) tension attains a local maximum, after which it decreases monotonically until reaching a steady state with lower tension than the initial state (illustrated in light blue in Fig. 13.16a with example profiles in Fig. 13.16b and e). If $k_2 > 0$, no value of k_1 can prevent a tension increase at the steady state. If k_2 is slightly negative, even with $k_1 > 0$, a residual tension increase occurs, and some $k_1 < 0$ values lead to residual tension increases as well. Figure 13.10f demonstrates that residual tension monotonically increases with both k_1 and k_2.

The system when $k_0 < 0$ (as seen in Fig. 13.17) appears to exhibit two possible outcomes: (1) immediate and ultimate relief of tension (indicated in green in Fig. 13.17a) or (2) initial abatement followed by an ultimate increase in tension compared to the initial levels (depicted in purple in Fig. 13.17a). When $k_2 < 0$, the value of k_1 becomes irrelevant; tension will invariably be relieved. Additionally, if $k_2 > 0$, tension will consistently increase unless the magnitude of k_2 is small and k_1 is posi-

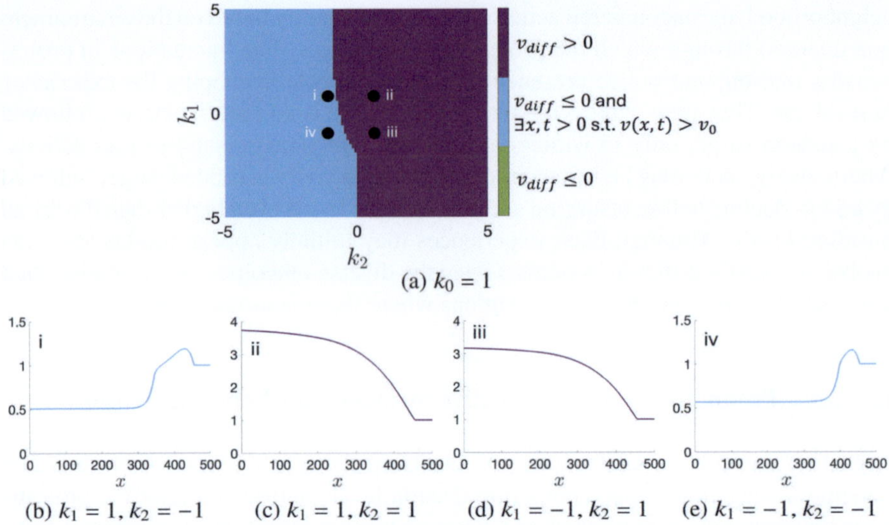

Fig. 13.16 **a** Categorization of regions in parameter space $(k_1, k_2) \in [-5, 5] \times [-5, 5]$ when $k = 1$ into two categories: (1) where tension is relieved ($v_{diff} \leq 0$–light blue) and (2) where it is exacerbated ($v_{diff} > 0$–purple). **b–e** illustrate sample tension profiles from (**a**) with the dots and labels indicating the (k_1, k_2) coordinates **b–e** are from. Note: there is no green region in this figure

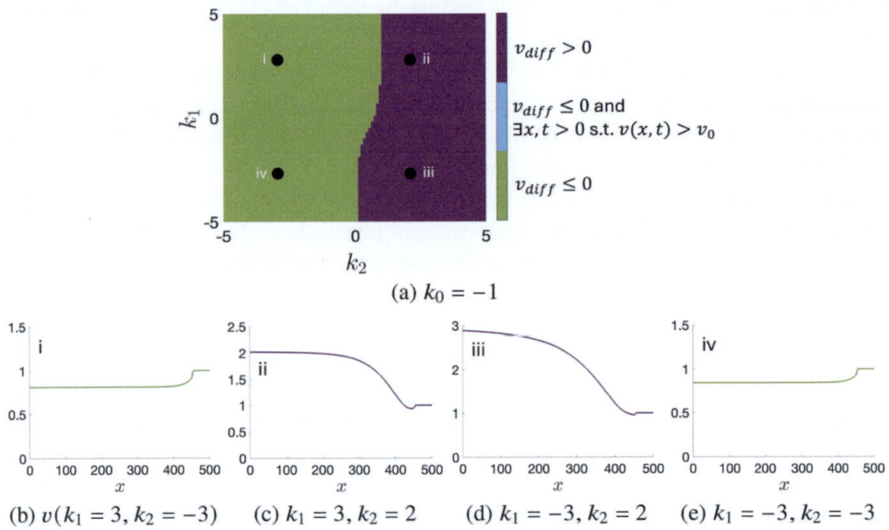

Fig. 13.17 **a** Categorization of parameter regions $(k_1, k_2) \in [-5, 5] \times [-5, 5]$ $k_0 = -1$ into two categories: (1) where tension is relieved ($v_{diff} \leq 0$–green) and where it is exacerbated ($v_{diff} > 0$–purple). **b–e** illustrate sample tension profiles from (**a**) with the dots and labels indicating the (k_1, k_2) coordinates (**b**)–(**e**) are from. Note: there is no light blue region in this figure

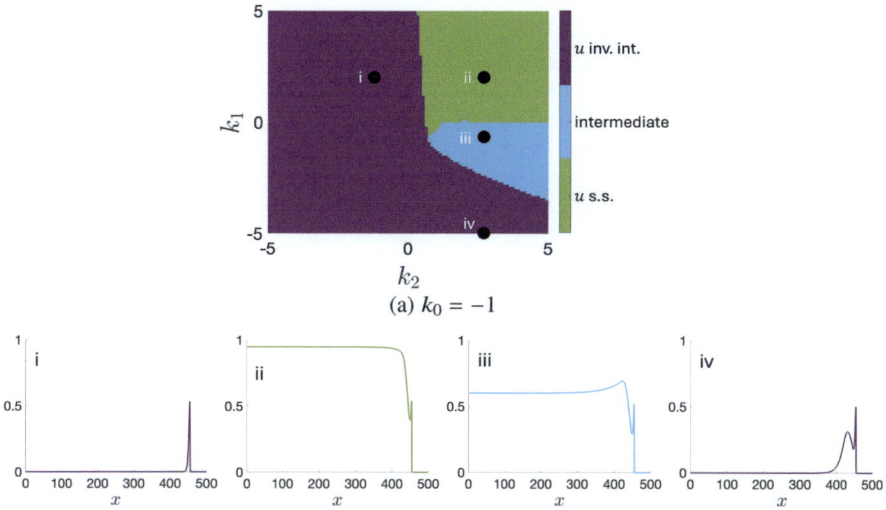

(a) $k_0 = -1$

(b) $k_1 = 2, k_2 = -1$ (c) $k_1 = 2, k_2 = 3$ (d) $k_1 = -1, k_2 = 3$ (e) $k_1 = -5, k_2 = 3$

Fig. 13.18 **a** The locations of global maximum levels of protesting for $k_0 = -1$ and $(k_1, k_2) \in [-5, 5] \times [-5, 5]$, while **b–e** are example protest activity profiles from (a) with the dots and labels indicating the (k_1, k_2) coordinates (**b**)–(**e**) are from. Purple indicates regions where the global maximum occurs at the invading interface; green indicates regions where the global maximum is the steady state (occurs far behind the invading interface); and light blue indicates regions where the global maximum occurs at some intermediate location

tive or nearly positive. Moreover, the profiles of v (shown in Fig. 13.17b–e) exhibit qualitative similarities in terms of their shapes, regardless of whether they belong to the green or purple region. The primary distinction lies in the residual values, as illustrated in Fig. 13.10d.

13.4.2.2 Location of Maximum Levels of Protest Activity

In Figs. 13.9 and 13.10, we are aware of the protest levels at the invading interface and the residual levels remaining. From a local standpoint, it becomes pertinent to discern whether the level observed at the invading interface will be the peak level experienced, with protesting reaching higher levels before diminishing, or if the protest activity will persistently escalate to a maximum as $t \to \infty$.

Based on the preceding discussion, it is established that all profiles in Fig. 13.18a fall within the receding regime. For very large coordinates of k_1 and/or k_2, the decrease in activity immediately behind the invading interface is subtle, while Fig. 13.18c and d exhibit a more visible decrease.

In Fig. 13.18a, when holding $k_1 > 0$ fixed and transitioning from some $k_2 < 0$ to some $k_2 > 0$, there is a shift from the global maximum being situated at the invading interface (depicted in purple, e.g., Fig. 13.18b) to being positioned far behind the

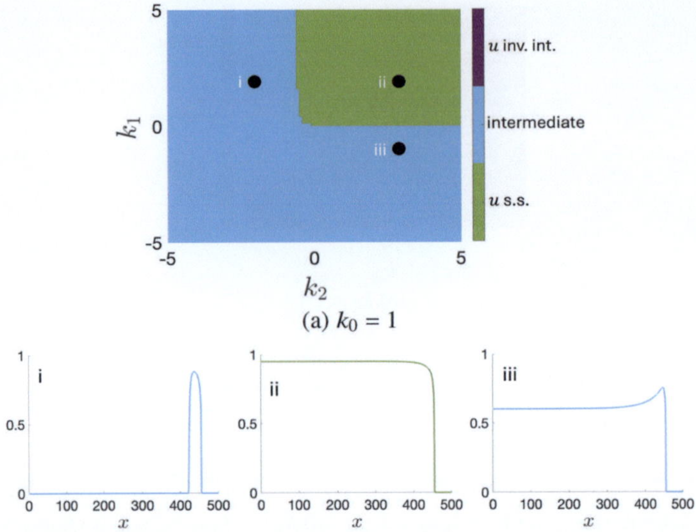

(a) $k_0 = 1$

(b) $k_1 = 2, k_2 = -2$ (c) $k_1 = 2, k_2 = 3$ (d) $k_1 = -1, k_2 = 3$

Fig. 13.19 **a** Shows locations of global maximum levels of protesting for $k_0 = -1$ and $(k_1, k_2) \in [-5, 5] \times [-5, 5]$, while **b–d** are example protest activity profiles from (**a**) with the dots and labels indicating the (k_1, k_2) coordinates (**b**)–(**d**) are from. Purple indicates regions where the global maximum occurs at the invading interface; green indicates regions where the global maximum is the steady state (occurs far behind the invading interface); and light blue indicates regions where the global maximum occurs at some intermediate location

wave (indicated in green, e.g., Fig. 13.18c). Examining profiles along this trajectory from $k_2 < 0$ to $k_2 > 0$, we observe pulse-like profiles that eventually transform into non-pulse profiles, yet with $u(x_1, 15) \approx u(x = 0, 15) \forall x_1 < 300$. As we progress to higher k_2 values, the residual level steadily increases until, at some point (dependent on the fixed $k_1 > 0$), the residual value surpasses the value at the invading interface. At this juncture, the system enters the green region.

Similarly, by fixing a $k_2 > 0$ in Fig. 13.18a and descending from some $k_1 > 0$ to some $k_1 < 0$, there is a transition from the global maximum being the residual value (indicated in green, e.g., Fig. 13.18c) to being positioned at some intermediary point between the invading interface and $u(x = 0)$ (depicted in light blue, e.g., Fig. 13.18d), and finally to the maximum being situated at the invading interface (depicted in purple, e.g., Fig. 13.18e). Examining profiles along this trajectory from $k_1 > 0$ to $k_1 \ll 0$, we observe the residual value decreasing progressively until another inflection point emerges, and a local maximum forms at some $x \in (400, 450)$. When this local maximum materializes, the system enters the light blue region, and this local maximum becomes the global maximum. With the continued reduction in the k_1 value, the residual value continues to decrease, gradually lowering the local maximum. Eventually, the local maximum falls below the value at the invading interface, marking the transition into the purple region.

In Fig. 13.19, for a fixed $k_1 > 0$ and moving from some $k_2 < 0$ to some $k_2 > 0$, we see a pulse-like profile (e.g., Fig. 13.19b), which as k_2 increases will get wider and wider (e.g., the green curve $k_1 = 1, k_2 = -1$ in Fig. 13.9f) until the plateau is as wide as the grid and hence does not drop to near-zero. At this point, the maximum level is the residual level (e.g., Fig. 13.19c) rather than occurring at some intermediate point. If starting at a fixed $k_2 > 0$ and moving from some $k_1 > 0$ down to some $k_1 < 0$, we observe a decrease in residual value, then eventually an emergence of an inflection point and formation of a local maximum at some $x \in (400, 450)$, at which point we have transitioned from the green region (e.g., Fig. 13.20c) to the light blue region (e.g., Fig. 13.19d). As we continue moving along the fixed $k_1 > 0$ to lower and lower k_2 values, the residual value continues decreasing until it is classified as a near-pulse profile. Eventually, the pulse-like profile will become narrow enough that the algorithm for detecting the location of the global maximum will determine whether it occurs at the invading interface.

It should be noted that for simulations $k_0 = 1$, $k_1 < -5$, and $k_2 < -5$, our algorithm found global maxima occurring at the invading interface in clear contradiction to our observation that when $k_0 > 0$ there is an increase in the level of protesting immediately behind the invading interface. We believe that this contradiction is purely due to limitations of numerical simulations: we believe that in these parameter regimes, there is a local maximum behind the invading interface, but that these maxima fall between points on our discretized grid so will not be found in our algorithm.

The heatmap Fig. 13.20a and the profiles behind it are qualitatively similar to the heatmap Fig. 13.19a and its profiles. One key difference is that pulse-like profiles for $k_1 < 0$ are much narrower in this $k_0 = 0$ case compared to the $k_0 = 1$ case above. Another key difference is that the observed expansion immediately behind the wavefront for $k_1 < 0$ and $k_2 > 0$ is also much narrower in the $k_0 = 0$ case compared to the $k_0 = 1$ case.

There is another obvious difference, which is the existence of a purple region. From earlier discussions, however, we know that for $k_0 > 0$, the protest activity is expanding immediately following the wavefront, so the global maximum cannot occur at the invading interface. When $k_0 = 0$, it is less clear whether there truly is a decrease immediately behind the invading interface or if there is an increase that is not observed in our numerical simulations. That is, it is unclear whether or not there should be a purple region, or if the purple region we see is entirely due to the limitations of numerical simulations. From visual inspection, Fig. 13.20e appears to show a decrease in u immediately behind the wavefront, but the actual global maximum value may occur between points on the discretized grid. Due to known limitations of the numerical methods, we cannot declare definitely that the purple regions have global maxima at their respective invading interfaces; rather, we can only state they are global maxima in our numerical simulations.

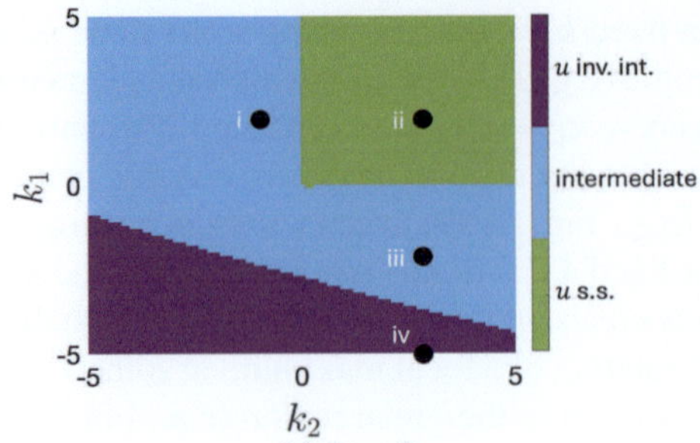

(a) $k_0 = 0$

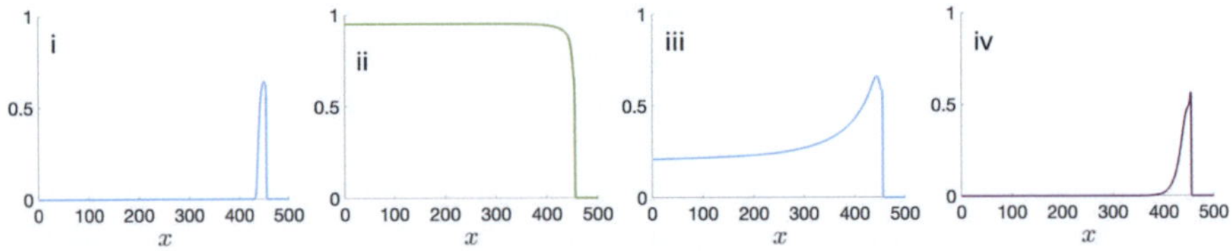

(b) $k_1 = 2, k_2 = -1$ (c) $k_1 = 2, k_2 = 3$ (d) $k_1 = -2, k_2 = 3$ (e) $k_1 = -5, k_2 = 3$

Fig. 13.20 **a** shows locations of global maximum levels of protesting for $k_0 = 0$ and $(k_1, k_2) \in [-5, 5] \times [-5, 5]$, while **b**–**e** are example protest activity profiles from (**a**) with the dots and labels indicating the (k_1, k_2) coordinates (**b**)–(**e**) are from. Purple indicates regions where the global maximum occurs at the invading interface; green indicates regions where the global maximum is the steady state (occurs far behind the invading interface); and light blue indicates regions where the global maximum occurs at some intermediate location

13.5 Discussion

We have presented a model to examine the dynamics of levels of social activity and social tension under the influence of policing. The parameters of the model were classified based on whether they led to a monotone or non-monotone system, and the stability of stationary solutions. In the monotone system, we established the existence of traveling waves, where the uniqueness of the wave speed depends on whether the source type is bistable or monostable. We have not established the existence or stability of traveling waves in the non-monotone system, a topic we aim to address in future work.

It is also intriguing and crucial to investigate solutions to system (13.3) in non-monotone regimes where at least one of the three k parameters has an inhibitory effect, as observed in real-world incident data (see [5]). Our numerical simulations confirm the existence of non-monotone wavefronts and near-pulse solutions within these regimes. Moreover, given that the classical theorems employed for asymptotic stability are inapplicable for non-monotone systems, exploring spectral stability for these regimes could be an initial step before delving into nonlinear stability. Additionally, it is natural to examine the solutions using bifurcation theory and identify and classify any potential bifurcations that may occur.

Another direction of interest is the study of system (13.3) as a control problem and optimize the level of social activity or tension, or a combination of both. This would be a more direct approach to obtaining optimal policing strategies for relieving tension or reducing activities. Moreover, in addition to the self-reinforcement accounted for in the u equation, exogenous factors can also be considered and incorporated into our model. For example, in the original model proposed in [2], continuing external events that happen at particular times and locations have been included in the source term for social tension.

A limitation of our study is that we cannot provide a comprehensive analytic derivation to classify the stability of the nontrivial stationary solutions due to the implicit expression for the nontrivial steady states. Also, our model has not been fitted to any real-world dataset. The application of our system to datasets containing measurements of the variables u, v, P is an intriguing possible course for future investigation.

From a numerical standpoint, we extensively investigate various parameter regimes and extract metrics with practical applications in mind. We envision that this research can contribute to categorizing different protesting activities as more data becomes available.

One crucial observation is that the effectiveness of a police management strategy can vary depending on the circumstances. For instance, employing an escalated force approach might initially reduce social tension, followed by an increase if $k_0 < 0$. Conversely, when $k_0 > 0$, tension and activity typically rise continuously. Regardless of the situation, the escalated force approach tends to amplify residual social tension, even if activity remains unchanged. However, there are instances where police presence is indispensable. For instance, in cases where $k_0 > 0$ and unrest escalates in the absence of police, adopting a negotiated management model can ameliorate protest outcomes. The condition $k_0 > 0$ arises when there are counterprotesters or agitators within a protest, possibly leading to violence regardless of police intervention. In such scenarios, we observe decreased social tension and activity, sometimes accompanied by traveling pulses.

The analysis suggests that the negotiated management model is generally preferable due to its ability to reduce social tension. However, it is important to note that the costs associated with action have not been considered. Accounting for these costs might warrant reducing action in certain situations, even if tension increases. However, relying on an escalated management model seldom yields positive outcomes, particularly since the system's state is often unknown. Developing methods to measure social tension and understand the demographics of potential participants can assist in selecting an appropriate strategy.

Acknowledgements Rodriguez partially funded by NSF-DMS-2042413 and AFOSR MURI FA9550-22-1-0380. Brantingham, Wang, Wessler were funded by AFOSR MURI FA9550-22-1-0380.

13.6 Appendix A. Asymptotic Decay Rates for $k_0 \geq 0$

Assume that the system has traveling wave solutions $\hat{\boldsymbol{u}}(z) = (\hat{u}(z), \hat{v}(z), \hat{P}(z))$, where we change the coordinate system to $z = x - ct$ and c is the wave speed. Then, system (13.3) becomes

$$\begin{cases} -cu_z = u_{zz} + f(u, v, P), \\ -cv_z = v_{zz} + g(u, v, P), \\ -cP_z = s(u, v, P), \end{cases}$$

which can be written as a five-dimensional, first-order system

$$\begin{cases} u' = w, \\ v' = y, \\ P' = -\frac{1}{c}s(u, v, P), \\ w' = -cw - f(u, v, P), \\ y' = -cy - g(u, v, P). \end{cases}$$

The linearization of this system near the fixed points is

$$U' = \begin{bmatrix} u \\ v \\ P \\ w \\ y \end{bmatrix}' = \begin{bmatrix} 0 & 0 & 0 & 1 & 0 \\ 0 & 0 & 0 & 0 & 1 \\ -\frac{1}{c}s_u(u, v, P) & 0 & -\frac{1}{c}s_P(u, v, P) & 0 & 0 \\ -f_u(u, v, P) & -f_v(u, v, P) & -f_P(u, v, P) & -c & 0 \\ -g_u(u, v, P) & -g_v(u, v, P) & -g_P(u, v, P) & 0 & -c \end{bmatrix} \begin{bmatrix} u \\ v \\ P \\ w \\ y \end{bmatrix} = AU. \tag{13.16}$$

Recall from the previous sections, that the system has the trivial steady states $(0, (1 + P)^{-k_2}, P, 0, 0)$ and, in certain parameter regimes, also the non-trivial steady states $(u^*, v^*, 1, 0, 0)$. The linearization matrix A at the trivial steady states is

$$A|_{(0,(1+P)^{-k_2},P,0,0)} = \begin{bmatrix} 0 & 0 & 0 & 1 & 0 \\ 0 & 0 & 0 & 0 & 1 \\ -\frac{1}{c}s_u(\hat{\boldsymbol{u}}_+) & 0 & 0 & 0 & 0 \\ -f_u(\hat{\boldsymbol{u}}_+) & 0 & 0 & -c & 0 \\ -g_u(\hat{\boldsymbol{u}}_+) & -g_v(\hat{\boldsymbol{u}}_+) & -g_P(\hat{\boldsymbol{u}}_+) & 0 & -c \end{bmatrix}$$

and has eigenvalues

$$\begin{cases} \lambda_1 = 0, \\ \lambda_{2,3} = \frac{-c \pm \sqrt{c^2 - 4f_u(\hat{\boldsymbol{u}}_+)}}{2}, \\ \lambda_{4,5} = \frac{-c \pm \sqrt{c^2 - 4g_v(\hat{\boldsymbol{u}}_+)}}{2}, \end{cases}$$

where the plus sign is for $\lambda_{2,4}$, and minus sign is for $\lambda_{3,5}$. This convention holds for the entire paper.

The matrix at the non-trivial steady states is

$$A|_{(u^*,v^*,1,0,0)} = \begin{bmatrix} 0 & 0 & 0 & 1 & 0 \\ 0 & 0 & 0 & 0 & 1 \\ 0 & 0 & -\frac{1}{c}s_P(\hat{\boldsymbol{u}}_-) & 0 & 0 \\ -f_u(\hat{\boldsymbol{u}}_-) & -f_v(\hat{\boldsymbol{u}}_-) & -f_P(\hat{\boldsymbol{u}}_-) & -c & 0 \\ -g_u(\hat{\boldsymbol{u}}_-) & -g_v(\hat{\boldsymbol{u}}_-) & -g_P(\hat{\boldsymbol{u}}_-) & 0 & -c \end{bmatrix}$$

with eigenvalues

$$\begin{cases} \lambda_1 & = -\frac{1}{c}s_P(\hat{\boldsymbol{u}}_-), \\ \lambda_{2,3} & = \frac{-c\pm\sqrt{c^2+2\mu_2}}{2}, \\ \lambda_{4,5} & = \frac{-c\pm\sqrt{c^2+2\mu_1}}{2}, \end{cases}$$

where

$$\mu_{1,2} = -(f_u(\hat{\boldsymbol{u}}_-) + g_v(\hat{\boldsymbol{u}}_-)) \pm \sqrt{(f_u(\hat{\boldsymbol{u}}_-) + g_v(\hat{\boldsymbol{u}}_-))^2 + 4f_v(\hat{\boldsymbol{u}}_-)g_u(\hat{\boldsymbol{u}}_-) - 4f_u(\hat{\boldsymbol{u}}_-)g_v(\hat{\boldsymbol{u}}_-)}.$$

13.6.1 As $z \to +\infty$

As $z \to +\infty$, $\hat{\boldsymbol{u}} \to \hat{\boldsymbol{u}}_+ = (0, (1+P)^{-k_2}, P)$. To avoid having complex eigenvalues, we restrict that $c^2 - 4f_u(\hat{\boldsymbol{u}}_+) \geq 0$ and $c^2 - 4g_v(\hat{\boldsymbol{u}}_+) \geq 0$. The second inequality holds automatically since $g_v(\hat{\boldsymbol{u}}_+) < 0$ always. The first inequality leads to the condition

$$c \geq 2\sqrt{l},$$

where we define $l := \max(f_u(\hat{\boldsymbol{u}}_+), 0)$. To find the principal eigenvalue, which is the negative eigenvalue with the smallest magnitude, we discuss by cases depending on the value of $f_u(\hat{\boldsymbol{u}}_+)$.

Case 1 If $f_u(\hat{\boldsymbol{u}}_+) < g_v(\hat{\boldsymbol{u}}_+)$, then $\lambda_3 < \lambda_5 < 0 < \lambda_4 < \lambda_2$, and λ_5 is the principal eigenvalue. The corresponding eigenvector is

$$\mathbf{y}_5 = \begin{bmatrix} 0 \\ -\dfrac{\lambda_5 + c}{g_v(\hat{\mathbf{u}}_+)} \\ 0 \\ 0 \\ 1 \end{bmatrix},$$

and the linear approximation about the steady state $\hat{\boldsymbol{u}}_+$ is

$$\begin{bmatrix} u(z) \\ v(z) - (1+P)^{-k_2} \\ P(z) - P \\ w(z) \\ y(z) \end{bmatrix} \approx A_1 \boldsymbol{y}_5 e^{\lambda_5 z},$$

as $z \to +\infty$, for some constant A_1. Since the second entry in $\boldsymbol{y}_5$ is negative, we know that $A_1 < 0$ so that the wave is decaying.

Case 2 If $f_u(\hat{\boldsymbol{u}}_+) = g_v(\hat{\boldsymbol{u}}_+)$, then $\lambda_3 = \lambda_5 < 0 < \lambda_2 = \lambda_4$, and $\lambda_3 = \lambda_5$ is the principal eigenvalue. The linear approximation about $\hat{\boldsymbol{u}}_+$ is

$$\begin{bmatrix} u(z) \\ v(z) - (1+P)^{-k_2} \\ P(z) - P \\ w(z) \\ y(z) \end{bmatrix} \approx A_2 z \boldsymbol{y}_5 e^{\lambda_5 z} + O(e^{\lambda_5 z}),$$

as $z \to +\infty$, for some constant A_2. Again, $A_2 < 0$ since we need the wave to be decaying.

Case 3 If $g_v(\hat{\boldsymbol{u}}_+) < f_u(\hat{\boldsymbol{u}}_+) < 0$, then $\lambda_5 < \lambda_3 < 0 < \lambda_2 < \lambda_4$, and λ_3 is the principal eigenvalue. The corresponding eigenvector is

$$\mathbf{y}_3 = \left.\begin{bmatrix} \dfrac{g_v - f_u}{-\lambda_3 g_u + \frac{1}{c} g_P s_u} \\ -\dfrac{\lambda_3 + c}{f_u} \\ \dfrac{s_u (g_v - f_u)}{\lambda_3 c(\lambda_3 g_u - \frac{1}{c} g_P s_u)} \\ \dfrac{\lambda_3 (g_v - f_u)}{-\lambda_3 g_u + \frac{1}{c} g_P s_u} \\ 1 \end{bmatrix}\right|_{\hat{\mathbf{u}}_+},$$

and the linear approximation about $\hat{\mathbf{u}}_+$ is

$$\begin{bmatrix} u(z) \\ v(z) - (1+P)^{-k_2} \\ P(z) - P \\ w(z) \\ y(z) \end{bmatrix} \approx A_3 \mathbf{y}_3 e^{\lambda_3 z},$$

as $z \to +\infty$, for some constant A_3. One can check that the first, second, and third entries in $\mathbf{y}_3$ are negative, and the fourth and fifth entries are positive. This implies that $A_3 < 0$ to guarantee the decay of the wave.

Case 4 If $f_u(\hat{\mathbf{u}}_+) = 0$, then $\lambda_5 < \lambda_3 < 0 = \lambda_2 < \lambda_4$. The linear approximation about $\hat{\mathbf{u}}_+$ is

$$\begin{bmatrix} u(z) \\ v(z) - (1+P)^{-k_2} \\ P(z) - P \\ w(z) \\ y(z) \end{bmatrix} \approx A_4 \mathbf{y}_2 + O(e^{\lambda_3 z}),$$

as $z \to +\infty$, for some constant A_4. Therefore, we cannot talk about asymptotic decay rates in this case.

Case 5 If $f_u(\hat{\mathbf{u}}_+) > 0$, then there are two subcases:

Case 5.1 If $c > 2\sqrt{f_u(\hat{\mathbf{u}}_+)}$, then $\lambda_5 < \lambda_3 < \lambda_2 < 0 < \lambda_4$, and λ_2 is the principal eigenvalue. The corresponding eigenvector is

$$\mathbf{y}_2 = \left. \begin{bmatrix} \dfrac{g_v - f_u}{-\lambda_2 g_u + \frac{1}{c} g_P s_u} \\ -\dfrac{\lambda_2 + c}{f_u} \\ \dfrac{s_u(g_v - f_u)}{\lambda_2 c(\lambda_2 g_u - \frac{1}{c} g_P s_u)} \\ \dfrac{\lambda_2 (g_v - f_u)}{-\lambda_2 g_u + \frac{1}{c} g_P s_u} \\ 1 \end{bmatrix} \right|_{\hat{\mathbf{u}}_+},$$

and the linear approximation about $\hat{\mathbf{u}}_+$ is

$$\begin{bmatrix} u(z) \\ v(z) - (1+P)^{-k_2} \\ P(z) - P \\ w(z) \\ y(z) \end{bmatrix} \approx A_5 \mathbf{y}_2 e^{\lambda_2 z},$$

as $z \to +\infty$, for some constant A_5. Similarly to the previous cases, the first three entries in $\mathbf{y}_2$ are negative and the last two are positive, implying that $A_5 < 0$.

Case 5.2 If $c = 2\sqrt{f_u(\hat{\mathbf{u}}_+)}$, then $\lambda_5 < \lambda_3 = \lambda_2 < 0 < \lambda_4$, and $\lambda_2 = \lambda_3$ is the principal eigenvalue. The linear approximation about $\hat{\mathbf{u}}_+$ is

$$\begin{bmatrix} u(z) \\ v(z) - (1+P)^{-k_2} \\ P(z) - P \\ w(z) \\ y(z) \end{bmatrix} \approx A_6 z \mathbf{y}_2 e^{\lambda_2 z} + O(e^{\lambda_2 z}),$$

as $z \to +\infty$, for some negative constant A_6.

13.6.2 As $z \to -\infty$

As $z \to -\infty$, $\hat{\boldsymbol{u}} \to \hat{\boldsymbol{u}}_- = (u*, v*, 1, 0, 0)$. The principal eigenvalue is the positive eigenvalue with the smallest magnitude. Note that λ_3 and λ_5 are always negative, and that λ_2 and λ_4 are nonnegative if and only if μ_2 and μ_1 are nonnegative, respectively. Therefore, we want to discuss by cases depending on the signs of μ_1, μ_2. We find numerically that μ_1, μ_2 are always real for $k_0 \geq 0$, and are not always real for $k_0 < 0$. As for their signs, recall from Sect. 13.2.1 that $f_u + g_v < 0$, implying that $\mu_1 > 0$ always holds, and that $\mu_2 > 0$ if and only if $f_v(\hat{\boldsymbol{u}}_-)g_u(\hat{\boldsymbol{u}}_-) - f_u(\hat{\boldsymbol{u}}_-)g_v(\hat{\boldsymbol{u}}_-) < 0$. This inequality condition can be numerically shown true for $k_0 \geq 0$ and as a result, $\mu_2 > 0$ always holds.

Since now we know that $\lambda_{1,2,4}$ are always positive and that $\lambda_2 \leq \lambda_4$, let us discuss by a few cases depending on what the principle eigenvalue is as $z \to -\infty$.

Case 1 If $\lambda_1 < \lambda_2 \leq \lambda_4$, then λ_1 is the principle eigenvalue. The corresponding eigenvector is

$$\mathbf{l}_1 = \left.\begin{bmatrix} \dfrac{c^3(f_v g_P + f_P(s_P - g_v)) - c f_P s_P^2}{\eta s_P} \\ \dfrac{1}{\lambda_1} \\ \dfrac{c^4(-f_v g_u + (f_u - s_P)(g_v - s_P)) + c^2(f_u + g_v - 2s_P)s_P^2 + s_P^4}{c\eta s_P} \\ \dfrac{-c^2(f_v g_P + f_P(s_P - g_v)) + f_P s_P^2}{\eta} \\ 1 \end{bmatrix}\right|_{\hat{\mathbf{u}}_-},$$

where $\eta = g_P s_P^2 - c^2(-f_u g_P + f_P g_u + g_P s_P)$. Therefore, the linear approximation about $\hat{\mathbf{u}}_-$ is

$$\begin{bmatrix} u^* - u(z) \\ v^* - v(z) \\ 1 - P(z) \\ -w(z) \\ -y(z) \end{bmatrix} \approx B_1 \mathbf{l}_1 e^{\lambda_1 z},$$

as $z \to -\infty$, for some positive constant B_1.

Case 2 If $\lambda_1 = \lambda_2 < \lambda_4$, then the linear approximation about $\hat{\boldsymbol{u}}_-$ is

$$\begin{bmatrix} u^* - u(z) \\ v^* - v(z) \\ 1 - P(z) \\ -w(z) \\ -y(z) \end{bmatrix} \approx B_2 z \boldsymbol{l}_1 e^{\lambda_1 z} + O(e^{\lambda_1 z}),$$

as $z \to -\infty$, for some negative constant B_2.

Case 3 If $\lambda_1 = \lambda_2 = \lambda_4$, then we have three repeated eigenvectors, and the linear approximation about $\hat{\boldsymbol{u}}_-$ is

$$\begin{bmatrix} u^* - u(z) \\ v^* - v(z) \\ 1 - P(z) \\ -w(z) \\ -y(z) \end{bmatrix} \approx B_3 z^2 \boldsymbol{l}_1 e^{\lambda_1 z} + O(z e^{\lambda_1 z}),$$

as $z \to -\infty$, for some positive constant B_3.

Case 4 If $\lambda_2 = \lambda_4 < \lambda_1$, then the corresponding eigenvector is

$$\boldsymbol{l}_2 = \begin{bmatrix} \dfrac{\left(c + \sqrt{c^2 + 2\mu_2}\right)(\mu_1 + 2 f_u)}{2 g_u \mu_2} \\ -\dfrac{2}{c - \sqrt{c^2 + 2\mu_2}} \\ 0 \\ \dfrac{\mu_1 + 2 f_u}{2 g_u} \\ 1 \end{bmatrix}\Bigg|_{\hat{\boldsymbol{u}}_-},$$

and the linear approximation about $\hat{\boldsymbol{u}}_-$ is

Table 13.5 Numerical parameter values that were used in simulations

Term	Value (short runs, long runs)	Description
$tEnd$	20,100	t at end of simulation
dt	0.0001	Temporal step size
$xMin$	0	x value at left of grid
$xMax$	500,5000	x value at right of grid
dx	0.1,1	Spatial step size
$u(x, 0)$	e^{-5x}	Level of u at start of simulation
$v(x, 0)$	1	Level of v at start of simulation
$P(x, 0)$	0	Level of P at start of simulation

$$\begin{bmatrix} u^* - u(z) \\ v^* - v(z) \\ 1 - P(z) \\ -w(z) \\ -y(z) \end{bmatrix} \approx B_4 z \boldsymbol{l}_2 e^{\lambda_2 z} + O(e^{\lambda_2 z}),$$

as $z \to -\infty$, for some negative constant B_4.

Case 5 If $\lambda_2 < \min(\lambda_1, \lambda_4)$, then λ_2 is the principle eigenvalue. In this case, we have

$$\begin{bmatrix} u^* - u(z) \\ v^* - v(z) \\ 1 - P(z) \\ -w(z) \\ -y(z) \end{bmatrix} \approx B_5 \boldsymbol{l}_2 e^{\lambda_2 z},$$

as $z \to -\infty$, for some positive constant B_5.

13.7 Appendix B. Values Used in Numerical Simulations

See Tables 13.5 and 13.6.

Table 13.6 Parameter values from system in system 13.3 that were used in simulations

Parameter	Value(s) in simulations
d_1	1
d_2	1
α	1
ω	100
θ	1
Γ	500
β	20
k_0	[–5, 5]
k_1	[–10, 10]
k_2	[–10, 10]

References

1. Bahid, Y., Kutsenko, O., Rodríguez, N., White, D.: The statistical and dynamic modeling of protests in Ukraine: the revolution of dignity and preceding times. Submitted to Plos One (2023)
2. Berestycki, H., Nadal, J.-P., Rodríguez, N.: A model of riot dynamics: shocks, diffusion, and thresholds. Netw. Heterog. Media **10**(3), 1–34 (2015)
3. Berestycki, H., Rodríguez, N.: Analysis of a heterogeneous model for riot dynamics: the effect of censorship of information. Eur. J. Appl. Math. **27**(3), 554–582 (2016)
4. Berestycki, H., Rossi, L., Rodríguez, N.: Periodic cycles of social outbursts of activity. J. Differ. Equ. **264**(1), 163–196 (2018)
5. Bonnasse-Gahot, L., Berestycki, H., Depuiset, M.-A., Gordon, M.B., Roché, S., Rodríguez, N., Nadal, J.-P.: Epidemiological modelling of the 2005 French riots: a spreading wave and the role of contagion. Sci. Rep. **8**(1), 107 (2018). Jan
6. Cosner, L.: The Function of Social Conflict. The Free Press, New York (1956)
7. D'arcy, J., Herath, T.: A review and analysis of deterrence theory in the is security literature: making sense of the disparate findings. Eur. J. Inf. Syst. **20**(6), 643–658 (2011)
8. Della Porta, D., Reiter, H.: Policing Protest: The Control of Mass Demonstrations in Western Democracies, vol. 6. University of Minnesota Press, ned - new edition (1998)
9. I. for Intergovernmental Research. After-action assessment of the police response to the: demonstrations in Ferguson, Missouri, p. 2015. Office of Community Oriented Policing Services, Technical Report August, COPS Office Crital Response Initiative, Washington DC (2014)
10. Gurr, T.: Why Men Rebel. Princeton University Press (1951)
11. Haar, R., Iacopino, V., Ranadive, N., Dandu, M., Weiser, S.: Death, injury and disability from kinetic impact projectiles in crowd-control settings: a systematic review. BMJ Open **7**(12), 1 9 (2017)
12. Haar, R., Iacopino, V., Ranadive, N., Weiser, S.D., Dandu, M.: Health impacts of chemical irritants used for crowd control: A systematic review of the injuries and deaths caused by tear gas and pepper spray. BMC Public Health **17**(1), 2020–2022 (2017)
13. Koopmans, R.: Protest in time and space: the evolution of waves of contention. In: The Blackwell Companion to Social Movements, pp. 19–46 (2004)
14. Maguire, E.R.: New directions in protest policing. St. Louis Univ. Public Law Rev. **35**(1) (2015)
15. Metcalfe, C., Pickett, J.T.: Public fear of protesters and support for protest policing: an experimental test of two theoretical models. Criminology **60**(1), 60–89 (2022)
16. Meyer, D.S.: Protest and political opportunities. Ann. Rev. Sociol. **30**(1), 125–145 (2004)

17. Ni, M.Y., Kim, Y., McDowell, I., Wong, S., Qiu, H., Wong, I., Galea, S., Leung, G.: Mental health during and after protests, riots and revolutions: a systematic review. Aust. N. Z. J. Psychiatry **54**(3), 232–243 (2020)
18. Parson, T.: The Social System. Free Press, Glencoe Ill (1951)
19. Prokazina, N., Kaira, Y., Kolmogorov, E.: Sociological analysis of social tension in the region (based on materials from the Orel region). Bull. Tula State Univ. Hum. **1**, 37–50 (2018)
20. Ratliff, T.N., Hall, L.L.: Practicing the art of dissent: toward a typology of protest activity in the united states. Hum. Soc. **38**(3), 268–294 (2014)
21. Snow, D.A., Moss, D.M.: Protest on the fly: toward a theory of spontaneity in the dynamics of protest and social movements. Am. Sociol. Rev. **79**(6), 1122–1143 (2014)
22. Taylor, M., Lewis, P., Clifton, H.: Why the riots stopped: fear, rain and a moving call for peace. Guardian (2011)
23. Volpert, A.I., Volpert, V.A., Volpert, V.A.: Traveling Wave Solutions of Parabolic Systems, vol. 140. American Mathematical Society (1994)
24. Yang, C., Rodríguez, N.: A numerical perspective on traveling wave solutions in a system for rioting activity. Appl. Math. Comput. **364**, 124646 (2020)
25. Yang, C., Rodríguez, N.: Existence and stability traveling wave solutions for a system of social outbursts. J. Math. Anal. Appl. **494**(1), 124583 (2021)

Chapter 14
Inferring a Cell Structure on the Space of Cyclooctane Conformations

Ulrich Bauer and Fabian Lenzen

Abstract The conformation space of cyclooctane, a ringlike organic molecule comprising eight carbon atoms, is a two-dimensional algebraic variety, which has been studied extensively for more than 90 years. We propose a cell structure representing this space, which arises naturally by partitioning the space into subsets of conformations that admit particular symmetries. We do so both for the labeled conformation space, in which the carbon atoms are considered as distinct, and for the actual, unlabeled, conformation space. The proposed cell structure is obtained by identifying subspaces of conformations based on symmetry patterns and studying the geometry and topology of these subsets using methods from dimensionality reduction and topological data analysis. Our findings suggest that, in contrast to the labeled variant, the conformation space of cyclooctane is contractible.

Keywords Cyclooctane · Molecule conformation space · Persistent homology

Ulrich Bauer—Funded by the Deutsche Forschungsgemeinschaft (DFG, German Research Foundation) through the Collaborative Research Center SFB/TRR 109 (EXC-2046/1, project ID: 195170736).
Fabian Lenzen—Funded by the Deutsche Forschungsgemeinschaft (DFG, German Research Foundation) under Germany's Excellence Strategy—The Berlin Mathematics Research Center MATH+ (EXC-2046/1, project ID: 390685689).

U. Bauer (✉)
Technical University of Munich, TUM School of Computation, Information and Technology, Department of Mathematics, Munich Data Science Institute (MDSI), Munich Center for Machine Learning (MCML), Munich, Germany
e-mail: ulrich.bauer@tum.de

F. Lenzen
Technische Universität Berlin, Faculty II Mathematics and Natural Sciences, Institute of Mathematics, Chair of Discrete Mathematics/Geometry, Berlin, Germany
e-mail: lenzen@math.tu-berlin.de

A. Ott et al. (eds.), *Applications of Mathematics in Sciences, Engineering, and Economics*, Springer Proceedings in Mathematics & Statistics 515,
https://doi.org/10.1007/978-3-032-01279-1_14

14.1 Introduction

The fact that most alkanes are not rigid molecules but admit some flexibility has caused a certain interest in understanding the geometry of the space of all "shapes", called *conformations*, that a given alkane can attain. Cycloalkanes are molecules forming a closed loop of carbon atoms, which restricts their flexibility considerably and gives their conformation spaces a particularly rich structure. In particular, conformation spaces of small cycloalkanes (with five to around twelve carbon atoms) have been the subject of various experimental and theoretical studies. In the present work, we consider cyclooctane, an alkane consisting of a cycle of eight carbon atoms, each of which binds to two hydrogen atoms.

The conformations that can be attained by a flexible molecule are governed by the potential energy required to form them, which results mostly from deviations of the bond lengths and bond angles from their equilibria, as well as steric and electronic interactions between hydrogen atoms bound to different carbon atoms. The terms representing bond lengths and bond angles grow by several orders of magnitude faster than the others; therefore, it is customary to regard these quantities as fixed at the equilibria [20, 24].

Since in the case of cyclooctane the bond angles do not allow for the formation of a planar, regular octagon, any conformation necessarily has to "pucker" in three dimensions, meaning that the molecule has to fold and twist into a non-flat shape in order to accommodate the bond angles in three dimensional space; see e.g. Figs. 14.1 and 14.6 for a few typical conformations. Because of this, conformations of this molecule are not rigid but have a certain amount of mobility, which gives rise to a rich structure of the conformation space.

14.1.1 *Representing Molecules as Linkages*

As a simplified model, we regard cyclooctane as an octagon embedded into three-dimensional Euclidean space with all edges having the same fixed length and all pairs of adjacent edges enclosing the same fixed angle. Such an object is an instance of a *linkage*:

Definition An *abstract linkage* is a pair (K, ℓ) consisting of a graph K with vertices K_0 and edges $K_1 \subseteq K_0 \times K_0$, and a length function $\ell\colon K_1 \to (0, \infty)$, $(i, j) \mapsto \ell_{ij}$ such that $\ell_{ij} = \ell_{ji}$ for all i, j. A *realization* of (K, ℓ) in the Euclidean space $\mathbf{R}^n$ is a mapping $x\colon K_0 \to \mathbf{R}^n$, $i \mapsto x_i$ such that for all edges $(i, j) \in K_1$ the Euclidean distance $\|x_i - x_j\|$ satisfies $d(x_i, x_j) = \ell_{ij}$.

An *abstract linkage with fixed joints* (K, ℓ, ϕ) is an abstract linkage endowed with a map $\phi\colon \{(i, j, k) \mid (i, j), (j, k) \in K_1\} \to \boldsymbol{R}/2\pi\mathbf{Z}$, $(i, j, k) \mapsto \phi_{ijk}$, such that $\phi_{ijk} = \phi_{kji}$ for all i, j, k. The map ϕ can be thought of as assigning an angle to each pair of adjacent edges. A *realization* of (K, ℓ, ϕ) is a realization x of (K, ℓ) such that

$\angle(x_i, x_j, x_k) = \phi_{ijk}$ for all $(i, j), (j, k) \in K_1$. We write $R^n(K, \ell, \phi)$ or just $R^n(K)$ for the space of realizations in $\mathbf{R}^n$.

The study of linkages goes back at least to [27]; their connection to algebraic sets has been studied in [26]. Note that an abstract linkage with fixed joints can equivalently be described as an abstract linkage (without joints), by additionally prescribing the distances $d(x_i, x_k)$ for $(i, j), (j, k) \in K_1$. Specifically, by the cosine law, realizations of (K, ℓ, ϕ) satisfy the quadratic equations

$$\|x_i - x_j\|^2 = \ell_{ij}^2 \qquad \text{for all } (i, j) \in K_1,$$
$$\|x_i - x_k\|^2 = \ell_{ij}^2 + \ell_{jk}^2 - 2\ell_{ij}\ell_{jk}\cos\phi_{ijk} \qquad \text{for all } (i, j), (j, k) \in K_1.$$

Therefore, $R^n(K, \ell, \phi)$ is a real algebraic variety (not necessarily connected, see [32]). The special Euclidean group SE(n), generated by translations and rotations in $\mathbf{R}^n$, acts on realizations by composition; given a realization x of (K, ℓ, ϕ), any $T \in \mathrm{SE}(n)$ yields another realization $Tx := T \circ x\colon K_0 \to \mathbf{R}^n$ of (K, ℓ, ϕ).

Definition The quotient $Cfg^n(K, \ell, \phi) := R^n(K, \ell, \phi)/\mathrm{SE}(n)$ (or just $Cfg^n(K)$) is called the *configuration space* of K. Its elements are called *configurations*. We write $\bar{x}$ for the equivalence class in $Cfg^n(K)$ of a configuration $x \in R^n(K)$.

To see that $Cfg^n(K, \ell, \phi)$ is also a variety, choose three vertices $i, j, k \in K_0$ and fix three points $x_i, x_j, x_k \in \mathbf{R}^3$ that satisfy $\|x_i - x_j\| = \ell_{ij}$, $\|x_j - x_k\| = \ell_{jk}$ and $\angle(x_i, x_j, x_k) = \phi_{ijk}$. Then every equivalence class in $Cfg^n(K, \ell, \phi)$ has a unique representative x with the prescribed images x_i, x_j and x_k.

14.1.2 Cyclooctane

From now on, we consider the closed loop linkage (K, ϕ) with eight vertices $K_0 = \{v_0, \ldots, v_7\}$, edges $e_0 = (v_0, v_1), \ldots, e_6 = (v_6, v_7), = e_7(v_7, v_0)$ of equal length $\ell = \ell_{i,i+1}$ and fixed joint angles $\phi = \phi_{i-1,i,i+1}$ for all $i \in K_0$, where we consider all indices modulo 8.

We are only interested in configurations in three dimensions, and write $R(K)$ for $R^3(K)$ and $Cfg(K)$ for $Cfg^3(K)$. For $\phi > \frac{3\pi}{4}$, $Cfg(K)$ is empty, and for $\phi = \frac{3\pi}{4}$,

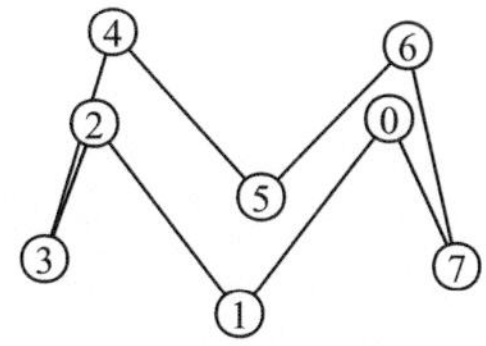

Fig. 14.1 A typical configuration of the cyclooctane linkage

$Cfg(K)$ has a single element, the embedding of a planar regular octagon into $\mathbf{R}^3$. For $\phi < \frac{3\pi}{4}$, a configuration necessarily folds and twists in all three spatial dimensions; see Fig. 14.1 and, for more examples, Fig. 14.6. An actual cyclooctane molecule has $\ell \approx 1.53$ A and $\phi \approx 115°$ [31, Table II][24, Table I]. This is the angle ϕ in the sample of $Cfg(K)$ used in our experiments [28].

The space $Cfg(K)$ has attracted a large attention in the literature because it is a two-dimensional variety. Hence, it is large enough to have interesting structure and topology, while still being manageably low-dimensional for manual scrutiny.

14.1.3 Labeled and Unlabeled Configurations

The map x keeps track of the individual images of the vertices in K; the space $Cfg(K)$ is in fact the space of all *labeled* configurations. However, as all carbon atoms in an actual molecule are indistinguishable, the labeling is only a technical artifact, and we are actually primarily interested in the topological structure of the quotient of $Cfg(K)$ by the symmetry group of the regular octagon (i.e., the dihedral group $D_8 = C_8 \rtimes C_2$, where C_8 acts by rotation and C_2 by reversal of the indexation), together with the quotient map $Cfg(K) \to Cfg(K)/D_8$. As an intermediate step, we also analyze the topology of the quotient of $Cfg(K)$ by the subgroup C_8 of D_8, together with the quotient maps $Cfg(K) \to Cfg(K)/C_8$ and $Cfg(K)/C_8 \to Cfg(K)/D_8$.

We note that, surprisingly, the unlabeled conformation space of cyclooctane seems not to have been discussed in the literature to the best of our knowledge, even though the study of equivalence classes of alkanes up to relabeling of carbon atoms is a classical topic dating back to Pólya [44].

14.1.4 Related Work

The conformation space of cyclooctane has been subject to a large number of studies. We discuss the literature most relevant to our work.

Conformational energies From the perspective of molecular chemistry, it is interesting to understand not only the space of realizable conformations, but also the function $E\colon Cfg \to [0, \infty)$ that assigns to each conformation its conformational energy.

In his influential work [24], Hendrickson discusses various terms contributing to the total conformational energy. With regard to their high stiffness, he assumes bond lengths to be fixed for all his calculations; see also [20]. In [25], the shapes of conformations of locally minimal conformational energy (there called *conformers*) are described in terms of their symmetries as well as their dihedral angles, and the names under which they are referred to. Hendrickson [25] also provides a nomenclature for certain symmetries in the conformations. He describes the symmetric, asymmetric and pseudo-rotational modes of interconversion between the conformations.

Since then, research has mostly focused on these conformers. Various different approaches for the description of conformers have been proposed, leading to the development of so-called *ring puckering* coordinates [12]; these describe a conformer by the out-of-plane displacement in terms of an amplitude and a phase coordinate w.r.t canonical conformations. Another approach [35] in terms of irreducible representations of dihedral groups (the symmetry groups of regular polygons), equivalent to ring puckering coordinates [6], have proven fruitful for conformational analysis [6, 16].

Sampling Sampling algebraic varieties such as $Cfg(K)$ is an involved problem in its own right. A feasible method should yield a uniform, sufficiently dense sample (i.e., solutions from the entire solution space are found) and allow for the estimation of the uncertainty of the solutions if necessary.

Cayley–Menger determinants A sufficient and necessary condition for the embedability of finitely many points into a Euclidean space with prescribed distances is given by the signs of the *Cayley–Menger-determinants* [5, Corollary 42.2]. A technique based on Cayley–Menger determinants is used in [36, 39] to develop a sampling technique that is used for sampling the configuration spaces describing conformations of cyclooctane and other short cyclic molecules.

Resultants Coutsias et al. develop a sampling algorithm using Sylvester and Dixon resultants [10]. These are successfully employed in [30] to compute a sample of the cyclooctane conformation space of approximately one million configurations. This sample, available online [28], also includes positions of the hydrogen atoms for the actual cyclooctane molecule; as there are two hydrogen atoms per carbon atom, the full conformations are thus given as a collection of points in $\mathbf{R}^{72}$ (eight carbon atoms sixteen hydrogen atoms with three spatial coordinates each). Note however that the positions of the hydrogen atoms are uniquely determined by those of the carbon atoms. All conformations in this sample have their center of mass at the origin and are aligned according to the Eckart condition (see Sect. 14.2.1). This is the sample employed in our analysis.

Homotopy continuation Recently, homotopy continuation has proven to be a worthwhile method to tackle computational algebro-geometrical problems [37]. Implementations for various environments are available; the Julia package by Breiding and Timme [7] provides code for uniformly sampling the cyclooctane conformation space.

Low-dimensional projections, surface reconstruction, topology Among the more recent work on cyclooctane, the investigations carried out in [9, 28–30] are of particular interest for us. Based on the incremental surface reconstruction algorithm in [18], which infers a triangulation of an orientable or non-orientable 2-manifold from a finite number of points, Martin and Watson presented an algorithm [29] that allows for the reconstruction of a 2-dimensional stratified space with singularities. The feasibility of their method is demonstrated both in terms of running times as well as quality of the computed triangulations, although neither [29] nor [18] provide complexity bounds or qualitative guarantees.

In [30], the method is applied to the cyclooctane sample from [28]. The result is compared to the image of the conformation space under a projection to three dimensions obtained using *Isomap*, a non-linear dimensionality reduction method (see Sect. 14.2.3 and [38]). In addition, the reconstruction of the actual surface enables the authors to identify the conformation space as a gluing of a sphere and a Klein bottle. As far as we know, [30] is the earliest work suggesting this description of the conformation space of cyclooctane. Their cyclooctane data set has also been studied using persistent homology in [42]. We extend the observations made there.

Persistent homology and Morse theory Persistent homology [13, 19, 40] is a tool from algebraic topology that describes topological features of a filtration. It is often applied to geometric filtrations such as the Vietoris–Rips filtration, which can be used to infer (under mild assumptions) the actual homology of a sampled topological space. Membrillo et al. undertake a two-step analysis of the conformation spaces of cyclooctane and other molecules, and the conformational energy distribution on these. In their analysis [32], which is also performed on the sample from [28], they use persistent homology to validate hypotheses about the homology of several conformation spaces. They further employ a numerical approximation of the Morse–Smale complex [14, 22, 23] of the conformational energy function based on a triangulation of the data set, applying topological simplification guided by persistent homology [4, 21] to extract the significant topological features of the energy function.

Symmetries An early study of conformation spaces of molecules in terms of actions of symmetry groups is carried out in [6, 35]. A thorough investigation of idealized configuration spaces of linkages (there called molecular graphs) and their quotients w.r.t. their symmetry groups is done in [32]. The authors observe that the symmetry group actions render the respective quotients of the configuration spaces into orbifolds. The work of [32] appears to be the first to introduce other metrics than the Euclidean metric on $Cfg(K)$, which are defined in terms of orbifold theory.

14.1.5 Our Contribution

We provide further insight into the structure of the conformation space of cyclooctane, supporting the description as the gluing of a sphere and a Klein bottle. Our findings are based on a partition of $Cfg(K)$ into certain subspaces that give rise to a cell structure on $Cfg(K)$. Each of these subspaces arises naturally as the set of all configurations in $Cfg(K)$ that exhibit a common *symmetry type*; see Sect. 14.3.1.

Symmetry types of cycloalkanes [24, 25] have already been seen to distinguish conformations from the sphere and the Klein bottle subspace of $Cfg(K)$ [30]. We extend this approach significantly, identifying more symmetry types and using their regimes for the segmentation of $Cfg(K)$ into a cell complex. As far as we know, no cell complex representing the conformation space of cyclooctane has been proposed so far.

Besides the labeled configuration space $Cfg(K)$, we also provide a description of the unlabeled configuration space $Cfg(K)/D_8$, which arises as a quotient of the labeled configuration space under the action of the symmetry group D_8. Interestingly, the latter space has a much simpler topological structure; namely, while $Cfg(K)$ is obtained as the gluing of a sphere and a Klein bottle, the space $Cfg(K)/D_8$ is contractible.

The experiments and the resulting plots have been made using a 6040-point subsample of the data set [28].[1] The subsample is provided as part of the data set, together with code to generate it, and subsampled to ensure a minimum distance between data points.

14.2 Methods

14.2.1 Metrics

We consider two metrics on $Cfg(K)$.

Angular metric Let $\bar{x} \in Cfg(K)$ be a configuration, represented by the realization $x \in R(K)$. We write e_i for the vector $e_i := x_{i+1} - x_i$. To x, we assign a sequence $\sigma(x) = (\sigma(x)_0, \dots, \sigma(x)_7) \subset (S^1)^8$ of oriented angles, called the *torsion angles* or *dihedral angles* of x, as follows.

Using $\|\cdot\|$ to denote the Euclidean norm and writing $n_i = \frac{e_i \times e_{i+1}}{\|e_i \times e_{i+1}\|}$ (with indices taken modulo 8), let $\sigma_i := \angle(n_{i-1}, n_i)$ be the oriented angle such that σ_i is positive if (n_{i-1}, n_i, e_i) is a right handed system; that is, if $\langle n_{i-1} \times n_i, e_i \rangle$ is positive. Therefore,

$$\sigma_i = \operatorname{atan2}\big(\langle n_{i-1} \times n_i, e_i \rangle, \langle n_{i-1}, n_i \rangle\big),$$

where atan2 is the continuous extension of $(a, b) \mapsto \arctan \frac{a}{b}$. Torsion angles are invariant under the action of SE(3), so the definition of $\sigma(x)$ passes down to a well-defined map $\sigma : Cfg(K) \to (S^1)^8, \bar{x} \mapsto \sigma(x)$. We call the sequence $\sigma(\bar{x})$ the *torsion* or *dihedral angles* of the configuration $\bar{x}$. Moreover, two configurations agree if and only if all their torsion angles do. Therefore, letting

$$d_\angle(\bar{x}, \bar{x}') = \sum_i \min\big(|\sigma(x)_i - \sigma(x')_i|, 2\pi - |\sigma(x)_i - \sigma(x')_i|\big)$$

defines a metric on $Cfg(K)$, called the *angular metric*.

---

[1] See the download "cyclo-octane example" in section "NMTRI: Non-Manifold Surface Reconstruction" on their website. For details, consider the file `readme.txt` in the download.

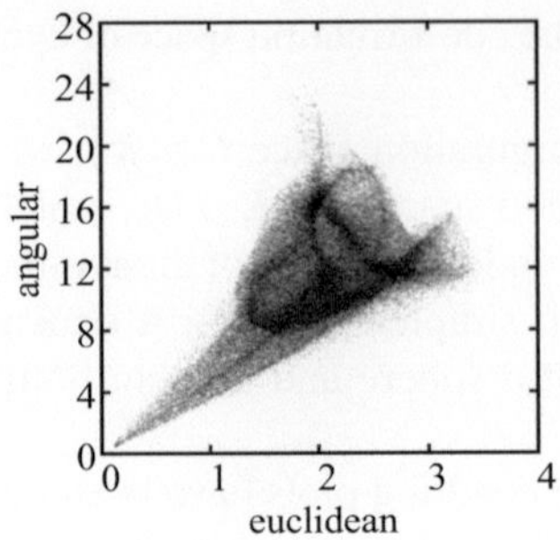

Fig. 14.2 The Euclidean distance $d_{\|}$ and the angular distance $d_{\angle}$ are equivalent. The picture has been obtained by plotting 100k randomly chosen pairwise distances (out of the $\approx 18M$ possible ones)

Euclidean metric In the sample [28] that we used in our analysis, each equivalence class in $Cfg(K)$ is represented by a *standard representative* that is aligned as follows. Let x_i^{pln} be the coordinates of the planar regular octagon in the xy-plane with circumradius 1, given by $x_i^{\mathrm{pln}} = (\cos(2i-1)\frac{\pi}{8}, \sin(2i-1)\frac{\pi}{8}, 0)^{\mathrm{T}}$. For the equivalence class $\bar{x}$ of a realization $x \in R(K)$ of K, Martin [28] chooses the representative x^{std} of $\bar{x}$ that satisfies the *Eckart condition*

$$\sum_i x_i^{\mathrm{std}} = 0, \qquad \sum_i x_i^{\mathrm{pln}} \times x_i^{\mathrm{std}} = 0 \tag{14.1}$$

w.r.t. x^{pln} [43]. The *Euclidean metric* is the metric $d_{\|}$ on $Cfg(K)$ defined as

$$d_{\|}(x, x')^2 = \sum_i \|x_i^{\mathrm{std}} - x_i'^{\mathrm{std}}\|^2$$

for two configurations $\bar{x}$ and $\bar{x}'$ in $Cfg(K)$, where $\|\cdot\|$ denotes the usual Euclidean metric on $\boldsymbol{R}^3$.

The angle sequence $(\sigma_0, \ldots, \sigma_7)$ uniquely determines a equivalence class in $Cfg(K)$. A method for sampling $Cfg(K)$ in terms of squared distances $\|x_{i-1} - x_{i+2}\|^2$, and thus in terms of dihedral angles, is described in [36].

Since $d_{\angle}$ is invariant under the action of SE(3) and involves no alignment to a reference orientation, one could consider it a more "natural" metric than $d_{\|}$. However, since previous investigations have considered $d_{\|}$ as the only choice of a metric on $Cfg(K)$, we include it in our investigations nevertheless. As it turns out, both metrics are equivalent with low metric distortion, as can be seen from Fig. 14.2.

14.2.2 Persistent Homology

We use *persistent homology* [13] as a central tool in our analysis. In the following, we give a quick account of persistent homology; for details, see [13, 34]. Given a finite metric space $(X = \{x^1, \ldots, x^k\}, d)$, the associated *Vietoris–Rips complex* $VR_*(X)$ assigns to each $r \in [0, \infty)$ the simplicial complex $VR_r(X)$ with $VR_r(X) = \{\sigma \subseteq X \mid \emptyset \neq \sigma,\ \forall x, x' \in \sigma : d(x, x') \leq r\}$, and to every pair $r \leq s$ in $[0, \infty)$ the inclusion map $\rho_{rs} : VR_r(X) \hookrightarrow VR_s(X)$.

A persistence module is a functor $[0, \infty) \to \boldsymbol{F}$-Vec, to the category of vector spaces over a fixed field $\boldsymbol{F}$. Explicitly, a persistence module M assigns to every $r \in [0, \infty)$ an $\boldsymbol{F}$-vector space M and to every pair $r \leq s$ in $[0, \infty)$ a map $M_{sr} : M_r \to M_s$, such that $M_{rr} = \mathrm{id}$ and $M_{sr} M_{ts} = M_{tr}$ for all $r \leq s \leq t$ in $[0, \infty)$.

If M is a persistence module such that $\dim M_r < \infty$ for all r, then there is an essentially unique decomposition $M \cong \bigoplus_{i \in I} \boldsymbol{F}_*^{[b_i, d_i)}$ of M into a direct sum of *interval modules* [11], defined as follows. For $0 \leq b < d \leq \infty$, the *interval module* $\boldsymbol{F}^{[b,d)}$ supported on $[b, d)$ is the persistence module with

$$\boldsymbol{F}_r^{[b,d)} = \begin{cases} \boldsymbol{F} & \text{if } b \leq r < d \\ 0 & \text{otherwise,} \end{cases} \qquad \boldsymbol{F}_{sr}^{[b,d)} = \begin{cases} \mathrm{id} & \text{if } b \leq r \leq s < d \\ 0 & \text{otherwise} \end{cases}$$

for all r, s. Taking the q-th simplicial homology $H_q(X, \boldsymbol{F}) := H_q(VR_*(X), \boldsymbol{F})$ of $VR_*(X)$ with coefficients in $\boldsymbol{F}$ defines a persistence module, called the q-th *persistent homology* of $VR_*(X)$. Informally, the intervals $[b_i, d_i)$ in the decomposition $H_q(X, \boldsymbol{F}) \cong \bigoplus_{i \in I} \boldsymbol{F}_*^{[b_i, d_i)}$ of $H_q(X, \boldsymbol{F})$ capture the ranges of parameters r for which a specific homological feature persists. Note that $H_q(X, \boldsymbol{F})$ depends on the field $\boldsymbol{F}$.

The intervals $[b_i, d_i)$ occurring in the decomposition $H_q(X, \boldsymbol{F}) \cong \bigoplus_{i \in I} \boldsymbol{F}_*^{[b_i, d_i)}$ are commonly drawn as dots in a *persistence diagram*, where the axes correspond to the endpoints of the intervals. In practice, one cuts off the Vietoris–Rips complex at some threshold r_∞. Points with $d_i \geq r_\infty$ are drawn on the upper boundary.

For instance, Fig. 14.3 shows the persistent homology of $\mathit{Cfg}(K)$ with respect to the two metrics $d_\angle$ and $d_\|$ with coefficients in the finite fields $\boldsymbol{F}_2$ and $\boldsymbol{F}_3$.

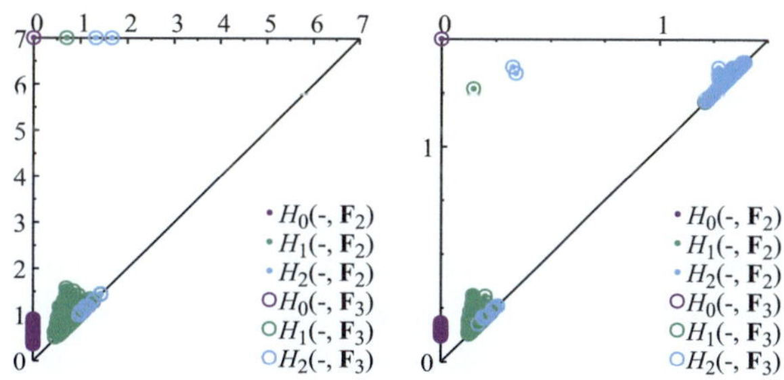

(a) $d_\angle$ (b) $d_{\|}$

Fig. 14.3 Persistent homology of $\mathit{Cfg}(K)$ with coefficients in $\mathbf{F}_2$ and $\mathbf{F}_3$ w.r.t. $d_\angle$ and $d_\|$. Each point (b, d) corresponds to one homology interval $[b, d)$. For this sample, the diagrams for $H_\bullet(-, \boldsymbol{F}_2)$ and $H_\bullet(-, \boldsymbol{F}_3)$ coincide; this is not the case for all subspaces we consider

The applicability of persistent homology is justified by the following statement: If M is a manifold and $X \subseteq M$ is a sufficiently large i.i.d. sample from M, then the long living intervals in $H_\bullet(VR_*(X))$ capture the actual singular homology $H_\bullet(M)$ of M with high probability [33, Thm. 3.1][1, Thm. 10]. See [13, 15, 34] for more details on persistent homology.

Several implementations are available for computing persistent homology of Vietoris–Rips complexes from a given distance matrix (see [34] for an overview). Because of its computational efficiency, we use the freely available software Ripser [3] which can compute the interval endpoints with respect to coefficients in finite fields.

For example, we learn the following about $Cfg(K)$ from Fig. 14.3: In the bottom left corner of the diagrams, near the diagonal, we see many short-lived topological features, while the four points in the top left represent long-lived features. In fact, the diagram shows that there is a long range of filtration parameters (from 2 to at least 7 for $d_\angle$) during which the homology type of the Vietoris–Rips complex does not change. One may thus argue that with high probability, the four points in the top left represent the true homology of $Cfg(K)$, so that $Cfg(K)$ has the $\boldsymbol{F}_2$- and $\boldsymbol{F}_3$-Betti numbers $(1, 1, 2)$ in dimensions $(0, 1, 2)$, respectively.

Computation of persistent homology depends on the field $\boldsymbol{F}$. We draw all our diagrams both for $\boldsymbol{F}_2$ and $\boldsymbol{F}_3$-valued persistent homology. Although both coincide in Fig. 14.3, we will see subspaces of $Cfg(K)$ where they do not. We do so because a closed surface (a two-dimensional compact manifold without boundary) is characterized uniquely up to homeomorphism by its homology with coefficients in $\boldsymbol{F}_2$ and $\boldsymbol{F}_3$, since closed surfaces are classified by their orientability and their genus, and homology over $\boldsymbol{F}_3$ detects orientability, while homology over $\boldsymbol{F}_2$ identifies the (orientable or non-orientable) genus.

14.2.3 Isomap

Using the coordinates of the vertices, the space $Cfg(K)$ can be embedded into $\mathbf{R}^{24}$ (eight vertices, with three spatial coordinates each). Analogously, using the torsion angles, it can be embedded into the 8-fold product $S^1 \times \cdots \times S^1$.

To obtain projections onto lower-dimensional Euclidean spaces, we apply a method called *Isomap* [38]. Given a finite subset $X = \{x_1, \ldots, x_k\} \subseteq \boldsymbol{R}^N$ of supposedly low intrinsic dimension embedded in a high-dimensional Euclidean space, this method first estimates an *intrinsic metric* d_X from the Euclidean metric $d_{\boldsymbol{R}^N}$ and then applies *multidimensional scaling* to compute a (non-linear) projection $\tilde{X} = \{\tilde{x}_1, \ldots, \tilde{x}_k\} \subset \boldsymbol{R}^n$ of X embedded in a low dimensional Euclidean space with $n \leq N$, such that the restriction of the Euclidean metric $d_{\boldsymbol{R}^n}$ to $\tilde{X}$ approximates d_X. With our Isomap projections, we also calculate the *metric distortion* $\log \frac{d_{\boldsymbol{R}^n}(\tilde{x}_i, \tilde{x}_j)}{d_X(x_i, x_j)}$ of each pair $(x_i, x_j) \in G(X)$. This quantity is large for points that are placed close together in $\tilde{X}$ that are actually distant in X.

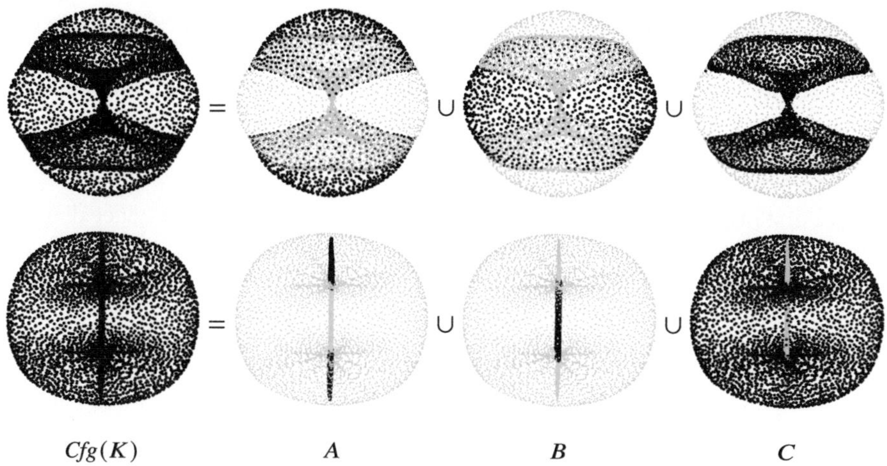

Fig. 14.4 Isomap projection (top: w.r.t. $d_{\|}$, bottom: $d_{\angle}$) of $Cfg(K)$ and its subspaces A, B, C

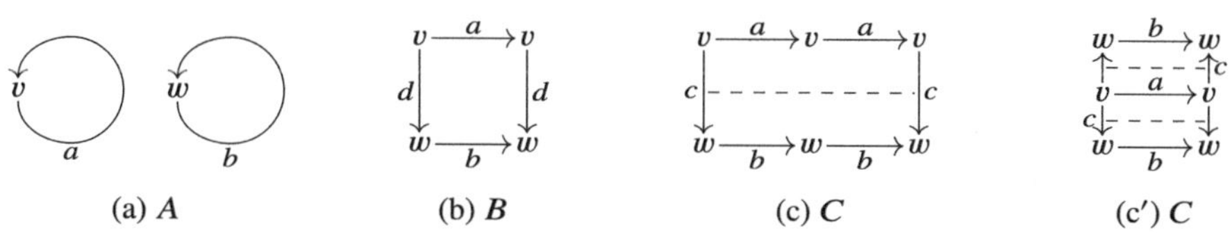

Fig. 14.5 Cellular model of the labeled configuration space $Cfg(K)$, and its subspaces A, B and C. The subspace A consists of two disjoint disks, B is a cylinder, and C is a model for a Klein bottle sewed together by two Möbius strips (both with the dashed line as boundary; one with core curve a, one with core curve b; see (c′) for a different drawing of the same model). The three subspaces are glued together along the 1-cells a and b. Thus, the whole space is the union of the sphere $S^2 \cong A \cup B$ the Klein bottle C, where $A \cup B$ and C intersect in the two disjoint circles a and b

Isomap has already been applied to the cyclooctane data set in [30]. In our case, we use the distances $d_{\angle}$ and $d_{\|}$ as input for Isomap and obtain projections to $\boldsymbol{R}^3$ of $Cfg(K)$ and various of its subsets. An Isomap projection of $Cfg(K)$ w.r.t. $d_{\angle}$ and $d_{\|}$ is shown in Fig. 14.4.

We observe that the projections exhibit regions of high metric distortion around the two circles in which the outer sphere and the inscribed hourglass intersect.

14.3 A Cell Structure for the Labeled Configuration Space

We will argue in the following that the labeled configuration space $Cfg(K)$ this space is the union of three subspaces, denoted by A, B and C, such that A consists of two disjoint 2-disks, B is a cylinder (without the caps) and C is a Klein bottle. Thus, $Cfg(K)$ is the union of the sphere $A \cup B$ and a Klein bottle C that intersect in two disjoint circles (the 1-dimensional cells a and b in Fig. 14.5). These are glued together as in Fig. 14.5.

This description has been previously provided in [29, 30]. We will provide additional evidence for this description. More importantly, we will show that this cell complex arises naturally from the data. Namely, we will see that the symmetry types (Sect. 14.3.1) endow $Cfg(K)$ with a cell structure refining Fig. 14.5.

As a first justification of the proposed model, we see from the persistence diagram (Fig. 14.3) that $Cfg(K)$ has the Betti numbers $\beta_\bullet(Cfg(K), \boldsymbol{F}_2) = \beta_\bullet(Cfg(K), \boldsymbol{F}_3) = (1, 1, 2)$ in dimensions $(0, 1, 2)$, respectively, which coincides with the Betti numbers of the proposed cell complex.

The Isomap projections of $Cfg(K)$ with respect to the two metrics $d_{\|}$ and $d_{\angle}$ (Fig. 14.4, left) serve as a starting point for our analysis. Note that these projections are not isometric embeddings, and do not necessarily reflect the topology of $Cfg(K)$.

14.3.1 Symmetry Types

We have not explained yet how the three subspaces A, B and C that partition $Cfg(K)$ are defined. A key ingredient to this end is the observation that the sequences of dihedral angles of the configurations in $Cfg(K)$ exhibit certain *symmetry types*. In the following, we define certain symmetry types, whose definitions have been chosen after scrutiny of the data, that we will use in our analysis, and give numbers to them for easy reference; see Table 14.1.

These symmetry types partition $Cfg(K)$ into various subspaces, which we analyze separately in the following. Note that a point can have more than one symmetry type. It follows from the definition that every point has either type 1 or type 2.

Two-dimensional strata First, we say a configuration has symmetry type 1 if its dihedral angles form a sequence of the form $(\alpha, \beta, \gamma, \delta, \alpha, \beta, \gamma, \delta)$, and we say that it has type 2 otherwise; see Table 14.1a. A configuration of type 1 is said to have also symmetry type 3 if its dihedral angles have signs of the form $(+, -, +, -, \dots)$ or $(-, +, -, +, \dots)$, with $+$ standing for non-negative and $-$ standing for non-positive, and of symmetry type 4 if any circular shift $(\alpha, \beta, \gamma, \delta, \dots)$ of its dihedral angle sequence either has signs $(+, -, -, -, \dots)$ and satisfies $\beta \leq \gamma \geq \delta$, or has signs $(-, +, +, +, \dots)$ and satisfies $\beta \geq \gamma \leq \delta$.

We let $A \subseteq Cfg(K)$ be the subspace of all configurations that have symmetry type 3 or 4; we let $B \subseteq Cfg(K)$ be the subspace of all configurations that have symmetry type 1 but not 3 or 4, and we let $C \subseteq Cfg(K)$ be the closure of the subspace of all configurations that have symmetry type 2. Then A, B and C are two-dimensional subspaces of $Cfg(K)$. Because type 2 is the complement of type 1, they cover $Cfg(K)$. Furthermore, their interiors are pairwise disjoint. Table 14.1a lists further symmetry types (5–12) that partition $Cfg(K)$ even further. All these types define two-dimensional subspaces of $Cfg(K)$.

Table 14.1 Dihedral angle patterns of the symmetry types. A configuration belongs to a certain symmetry type if its dihedral angle sequence or any circular shift of it satisfies the condition. Conditions with signs $\pm$ or $\lesseqgtr$ are meant as either the condition with all upper signs, or the condition with all lower signs. In types 4, 5, 6 and 16, the magnitude of α compared to β, γ or δ is not relevant

Type	Subspace	Condition
1	A, B	$(\alpha, \beta, \gamma, \delta, \alpha, \beta, \gamma, \delta)$
2	C	not $(\alpha, \beta, \gamma, \delta, \alpha, \beta, \gamma, \delta)$
3	A	$\begin{pmatrix} \alpha, & \beta, & \gamma, & \delta, & \alpha, & \beta, & \gamma, & \delta \\ + & - & + & - & + & - & + & - \end{pmatrix}$
4		$\begin{pmatrix} \alpha, & \beta & \gamma & \delta, & \alpha, & \beta & \gamma & \delta \\ \pm & \mp & \mp & \mp & \pm & \mp & \mp & \mp \end{pmatrix}$
5		$\begin{pmatrix} \alpha, & \beta & \lesseqgtr & \gamma & \lesseqgtr & \delta, & \alpha, & \beta & \lesseqgtr & \gamma & \lesseqgtr & \delta \\ \pm & \mp & & \mp & & \mp & \pm & \mp & & \mp & & \mp \end{pmatrix}$
	B	or $\begin{pmatrix} \alpha, & \beta & \lesseqgtr & \gamma & \lesseqgtr & \delta, & \alpha, & \beta & \lesseqgtr & \gamma & \lesseqgtr & \delta \\ \mp & \pm & & \pm & & \pm & \mp & \pm & & \pm & & \pm \end{pmatrix}$
6		$\begin{pmatrix} \alpha, & \beta & \gtreqless & \gamma & \lesseqgtr & \delta, & \alpha, & \beta & \gtreqless & \gamma & \lesseqgtr & \delta \\ \pm & \mp & & \mp & & \mp & \pm & \mp & & \mp & & \mp \end{pmatrix}$
7		$\begin{pmatrix} \alpha, & \beta, & \gamma, & \delta, & \alpha, & \beta, & \gamma, & \delta \\ + & + & - & - & + & + & - & - \end{pmatrix}$
8		$(++-++-+-)$
9		$(--+--+-+)$
10	C	$(--++-+-+)$
11		$(++--+-+-)$
12		$(--+-++-+)$

(a) Two-dimensional strata, and the subspace A, B or C they belong to

Type	Condition
13	$(\alpha, \beta, \gamma, \delta, -\delta, -\gamma, -\beta, -\alpha)$
14	$(\alpha, \beta, \gamma, \beta, \alpha, \delta, \epsilon, \zeta)$
15	$\begin{pmatrix} \alpha, & \beta, & -\alpha, & -\beta, & \alpha, & \beta, & -\alpha, & -\beta \\ + & + & - & - & + & + & - & - \end{pmatrix}$
16	$\begin{pmatrix} \alpha, & \beta & \lesseqgtr & \gamma & = & \delta, & \alpha, & \beta & \lesseqgtr & \gamma & = & \delta \\ \pm & \mp & & \mp & & \mp & \pm & \mp & & \mp & & \mp \end{pmatrix}$
	or $\begin{pmatrix} \alpha, & \beta & = & \gamma & \lesseqgtr & \delta, & \alpha, & \beta & = & \gamma & \lesseqgtr & \delta \\ \mp & \pm & & \pm & & \pm & \mp & \pm & & \pm & & \pm \end{pmatrix}$
17	$(\alpha, \beta, \gamma, \alpha, \epsilon, -\gamma, -\beta, \epsilon)$
18	$(\alpha, \beta, \gamma, \delta, -\alpha, -\beta, -\gamma, -\delta)$

(b) One-dimensional strata

Intersecting types	Cardinality	Pattern
3,13,14	2	$(\alpha, -\alpha, \alpha, -\alpha, \alpha, -\alpha, \alpha, -\alpha,)$
13,15	4	$(\alpha, 0, -\alpha, 0, \alpha, 0, -\alpha, 0)$
14,15	4	$(\alpha, \alpha, -\alpha, -\alpha, \alpha, \alpha, -\alpha, -\alpha)$
13,16	8	$(\alpha, -\alpha, 0, 0, \alpha, -\alpha, 0, 0)$
14,16	8	$(\alpha, \beta, \beta, \beta, \alpha, \beta, \beta, \beta)$
13,18	8	$(\alpha, -\alpha, \beta, \beta, -\alpha, \alpha, -\beta, -\beta)$
14,18	8	$(\alpha, -\beta, 0, \beta, -\alpha, \beta, 0, -\beta)$

(c) Zero-dimensional strata. For examples of these configurations, see Fig. 14.8

Figures 14.6 and 14.7 show a configuration of each of these (and the following) symmetry type, together with an indication where these symmetry types occur in the Isomap projection from Fig. 14.4.

One-dimensional strata Additionally, we define the symmetry types defined by the conditions in Table 14.1b. All of these define one-dimensional subspaces of $Cfg(K)$. Symmetry type 16 is the intersection of types 4 and 5 and will turn out later to be the pairwise intersection of A, B and C. Because of their appearance in Figs. 14.6 and 14.7, we will refer to the types 13, 14 and 17 as *meridians*, and to type 15 (which lies in B) and type 18 (which lies in C) as *equators*. Of course, intersections of the closures of the one-dimensional strata defines an additional abundance of one-dimensional strata; however, we refrain from giving names to all of them.
Zero-dimensional strata Finally, the intersections of the (closures of) strata listed in Table 14.1c contain only finitely many elements that are particularly symmetric; see Fig. 14.8.

For $S \subseteq \mathbf{N}$, we denote by $Cfg(K, S)$ the subspace of $Cfg(K)$ of configurations that have a symmetry type in S. In the following, we analyze the subspaces $A = Cfg(K, \{3, 4\})$, $B = Cfg(K, \{5, 6, 7\})$ and $C = Cfg(K, \{2\}) = Cfg(K) \setminus Cfg(K, \{1\})$ separately.

Note: If $Cfg(K, S)$ defines a zero- or one-dimensional stratum, we cannot expect to find configurations for which the defining conditions hold on the nose. Instead, we test the equations defining these symmetry types up to a certain tolerance ($\pm 5°$ for one-dimensional and $\pm 2.5°$ for two-dimensional strata). In particular, points lying close to the boundary of a component will also be assigned the symmetry type of the neighboring component.

14.3.2 The Spherical Subspace $A \cup B$

Figure 14.9 shows an Isomap projection of the subspace $A \cup B$ alone, together with the symmetry types of the samples. These projections suggest that $A \cup B$ is homeomorphic to a 2-sphere S^2. Since there are no visible singularities in the projection, we assume that $A \cup B$ is a surface. Consider the persistence diagram of $A \cup B$ in Fig. 14.9. It shows that the $\boldsymbol{F}_2$- and $\boldsymbol{F}_3$-Betti numbers are (1,0,1) in dimensions (0,1,2), respectively. In particular, they coincide, hence $A \cup B$ is orientable. By the classification of surfaces (see Sect. 14.2.2), it follows that $A \cup B$ is homeomorphic to S^2.

According to Fig. 14.9, the sphere $A \cup B$ decomposes further into the following subspaces: $A = Cfg(K, \{3, 4\})$ forms two opposing "pole caps" on the sphere, and $B = Cfg(K, \{5, 6, 7\})$ forms a cylinder between the two discs. The 1-dimensional subspace $Cfg(K, \{13\})$ forms four "meridians" on the two discs and the cylinder, which partition B and each of the two connected subspaces of A into four subspaces, giving the cell structure shown in Fig. 14.13a, b. The colors in Fig. 14.13 correspond to the symmetry types in Fig. 14.9.

We see that the Isomap projection in Fig. 14.9 maps $A \cup B$ to a sphere, such that the two connected components of A can be viewed as the "pole caps" of this sphere. The two poles then are the two configurations $P_1 := (\alpha, -\alpha, \alpha, -\alpha, \ldots)$ and $P_2 := (-\alpha, \alpha, -\alpha, \alpha, \ldots)$, with $\alpha \approx 89°$. The sphere is divided into two hemi-

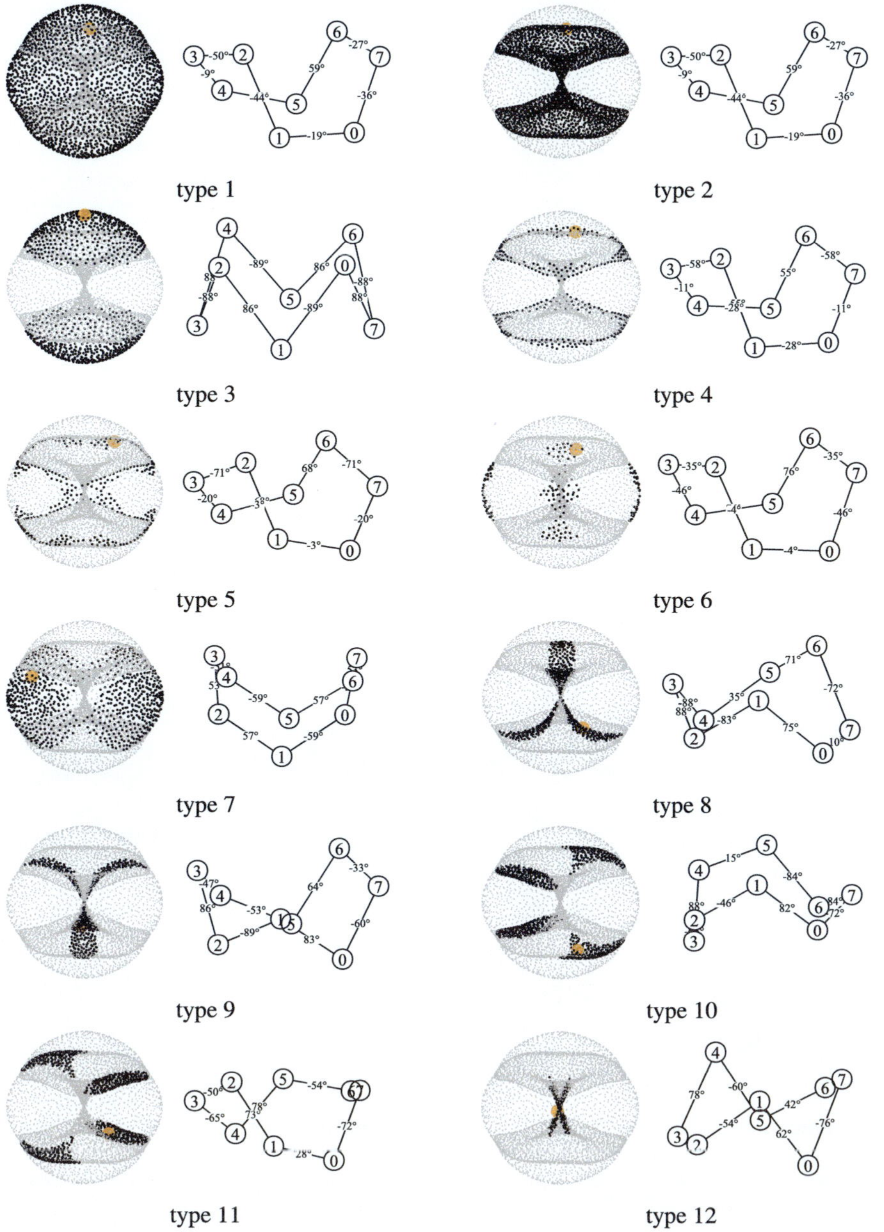

Fig. 14.6 Location of symmetry types in the Isomap projection w.r.t. $d_{\|}$ from Fig. 14.4. The right picture shows a typical standard representatives, whose location in the projection on the left is highlighted. See also Fig. 14.7 for a different projection of the symmetry types on the "hourglass" subspace A

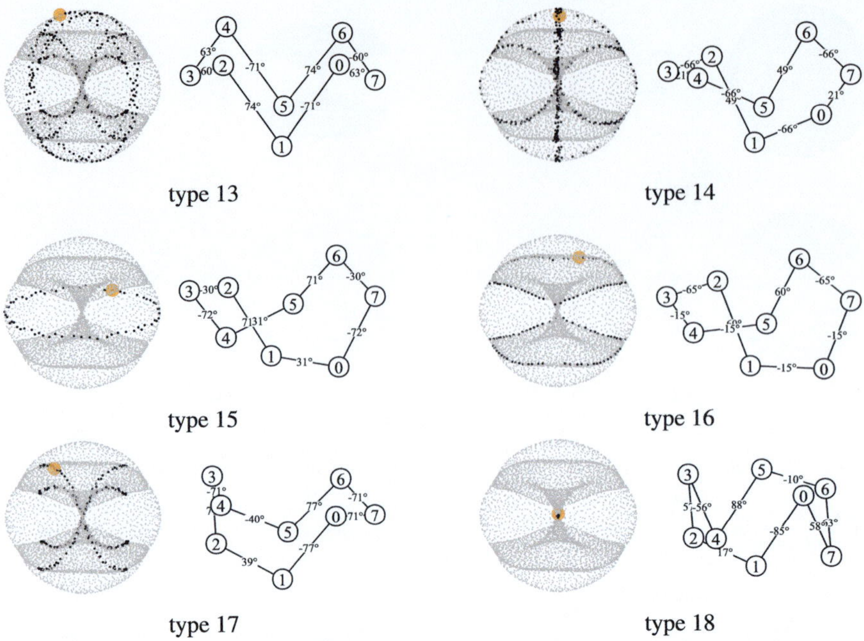

Fig. 14.6 Location of symmetry types (continued)

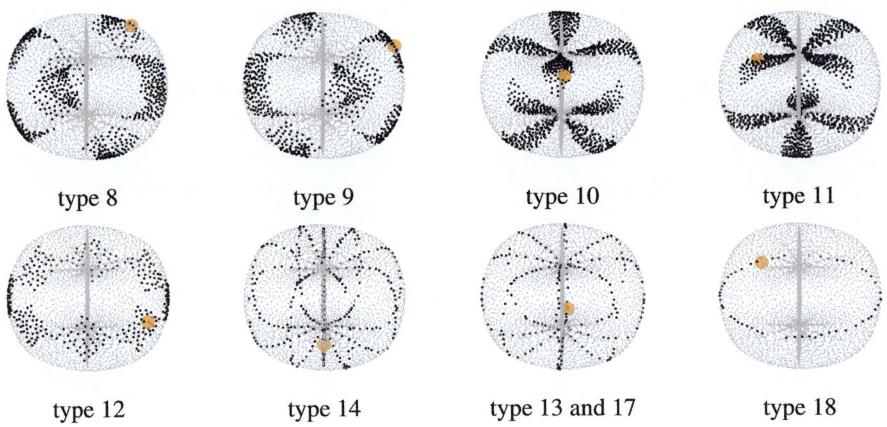

Fig. 14.7 Location of symmetry types existing on the Klein bottle subspace C, in the Isomap projection w.r.t. $d_\angle$ from Fig. 14.4

spheres H_1 and H_2, where H_1 (resp. H_2) contains the configurations in $A \cup B$ whose angle sequence $(\alpha_1, \alpha_2, \dots)$ satisfies $\alpha_1 + \alpha_3 + \dots \geq \alpha_2 + \alpha_4 + \cdots$ (resp. $\leq$). In particular, $P_i \in H_i$ for $i = 1, 2$, and $H_1 \cap H_2 = Cfg(K, \{15\})$.

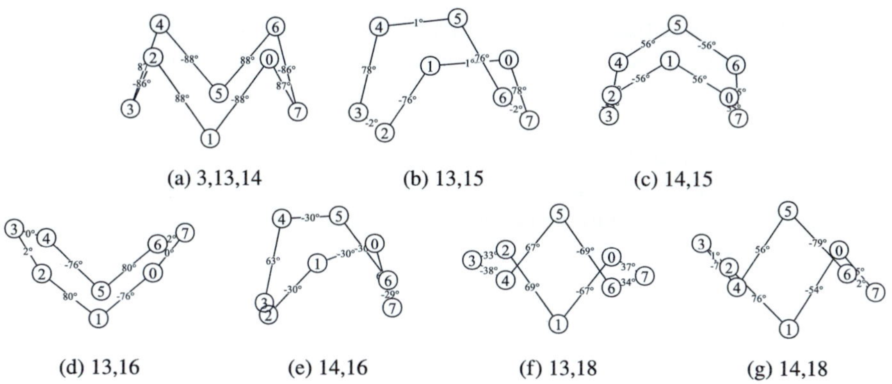

Fig. 14.8 Examples for the zero-dimensional strata from Table 14.1c

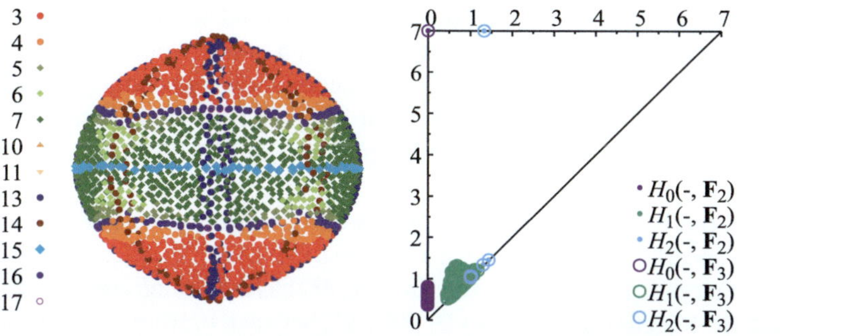

Fig. 14.9 Isomap projection and persistence diagram (w.r.t. $d_\angle$) the component $A \cup B$, defined as the set of points with symmetry types 3–7. Note that a point in $A \cup B$ can *additionally* have one ore more of the symmetry types 13–18 that define one-dimensional strata. Furthermore, because we test the conditions from Table 14.1 only up to a tolerance, we also see points that have symmetry types 10 and 11, in addition to one of the types 3–7

14.3.3 The Klein Bottle Subspace C

This subspace has been referred to in the literature [30, 42] as the "interior" or "hourglass" subspace, in reference to its appearance as the hourglass inscribed into the sphere in Fig. 14.4. It has been suggested that this subspace can be described as a Klein bottle [30]. We provide additional evidence for this observation, extending the results from [42].

The Isomap projection in Fig. 14.4 exhibits high metric distortion around the two circles in which $A \cup B$ and C intersect. Figure 14.10 shows the Isomap projection of C alone. In this sphere-like projection, this intersection is mapped to the two "poles" of the sphere and is subject to large metric distortion; i.e., Isomap places points close to each other in the projection that are distant in the actual space. We suspect that this region is in fact a 1-cell to which the 2-cell forming C is attached by a degree-2-

map, yielding a space that is not isometrically embeddable in $\boldsymbol{R}^3$. This would align well with the fact that as an algebraic surface, $Cfg(K)$ cannot have one dimensional singularities with an uneven number of leaves attached to it.

Therefore, we propose that C can be represented by the cell complex shown in Fig. 14.13c, which refines the complex in Fig. 14.5c. Again, all edges with the same endpoints denote the same 1-cell. This cell complex is a model for a Klein bottle; see also Fig. 14.5c–c′. As we will see in Sect. 14.3.4, the vertices v_j and w_i in Fig. 14.13c correspond to the vertices in Fig. 14.13a, b.

To verify that Fig. 14.13c models C, consider the persistence diagram of C in Fig. 14.10. It suggests that C has Betti numbers $(1, 2, 1)$ over $\boldsymbol{F}_2$ and (1,1,0) over $\boldsymbol{F}_3$ in dimensions 0, 1 and 2, respectively. Assuming that C is a manifold, the classification of closed surfaces gives that C is a Klein bottle.

To provide additional evidence for this claim, we have selected only the points from C whose angle sequence $(\alpha_1, \alpha_2, \ldots)$ satisfies $\alpha_1 + \alpha_3 + \cdots \geq \alpha_2 + \alpha_4 + \cdots$. As for the sphere component $A \cup B$, this condition (and the analogous condition with $\leq$) define two subspaces M_1 and M_2 (the upper and lower "hemisphere" in Fig. 14.10) that cover C and satisfy $Cfg(K, \{18\})$. This corresponds to cutting the cell complex in Figs. 14.5c–14.13c along the horizontal line. If C is a Klein bottle, then M_1 and M_2 will be two Möbius strips with configurations of type 16 as core curve and type 18 as boundary.

Indeed, the Isomap projection of M_1 in Fig. 14.11a resembles a Plücker conoid, which is a specific immersion of the Möbius strip into $\boldsymbol{R}^3$. Furthermore, according to the persistence diagram of M_1 in Fig. 14.11a, M_1 has the homology type of a 1-sphere. Through the bijection $H^1(M_1, \mathbf{Z}) \cong [M_1, S^1]$ between the first integral cohomology of M_1 and homotopy classes of maps from M_1 to S^1, one can associate a representative map $M_1 \to S^1$ to each class in the persistent cohomology $H^1(VR_*(M), \mathbf{Z})$. For details on the method, see [41].

The color-coding in Fig. 14.11a shows such a map $f\colon M_1 \to S^1$ corresponding to the prominent 1-cohomology class in the persistence diagram in Fig. 14.11a. We see that f winds two times around the boundary if M. An Isomap projection of the core curve $M_1 \cap Cfg(K, \{16\})$ of M_1, together with the restriction of f, is shown

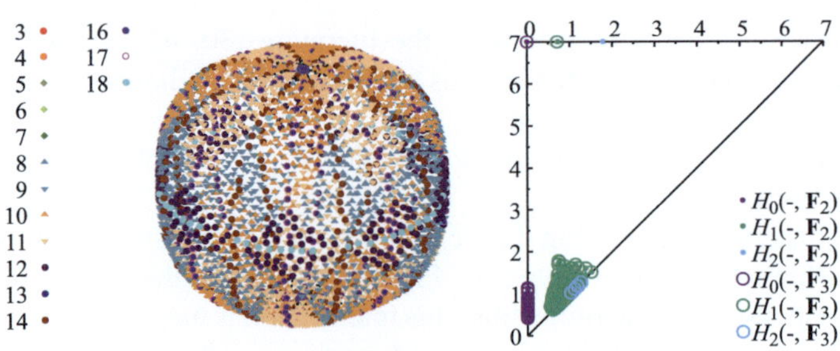

Fig. 14.10 Isomap projection and persistence diagram (w.r.t. $d_{\angle}$) of C. See Fig. 14.9 for an explanation of the appearance of symmetry types 4–7 in this plot

in Fig. 14.11b. We see that f winds around this core curve once, as expected. We therefore argue that M_1 (and analogously, M_2) are two Möbius strips that are glued together along their boundaries $Cfg(K, \{18\})$. Therefore, C is a Klein bottle.

14.3.4 Gluing Both Subspaces

It remains to understand how the subspaces A, B and C and are glued together. We have already seen that A and B intersect in two circles given by $Cfg(K, \{16\})$. In the previous section, we have also seen that the core curves of the two Möbius strips M_1, M_2 are also $Cfg(K, \{16\})$. Thus, A, B and C commonly intersect in two circles given by $Cfg(K, \{16\})$.

To make clear how this gluing works, consider the aforementioned type 13 "meridians" on $Cfg(K)$; see Figs. 14.6 and 14.7. These meridians bifurcate in the points where $A \cup B$ and C intersect. We can thus track where these bifurcation points lie in the projections of Fig. 14.4.

To this end, we use the following observation: consider the points in $Cfg(K, \{13\}) \cap C$. A point of type 13 has the dihedral angle pattern $(\alpha, \beta, \gamma, \delta, -\delta, -\gamma, -\beta, -\alpha)$ or one of its circular shifts $(\alpha, \beta, \gamma, \delta, \dots)$, $(\beta, \gamma, \delta, -\delta, \dots)$, ... There are four distinct circular shifts before one gets back the original pattern. If one imposes the additional condition $\delta \geq 0$, then one gets eight distinguishable shifts of this pattern, In Fig. 14.12, we have highlighted these eight distinct shifts of this pattern on $Cfg(K, \{13\}) \cap C$ in a different color.

In the Isomap projection of C alone (Fig. 14.12a), we see that this approach indeed distinguishes all the eight meridians on the Klein bottle component. When applying the same color-scheme in the Isomap projection of the entire space (see Fig. 14.12b)

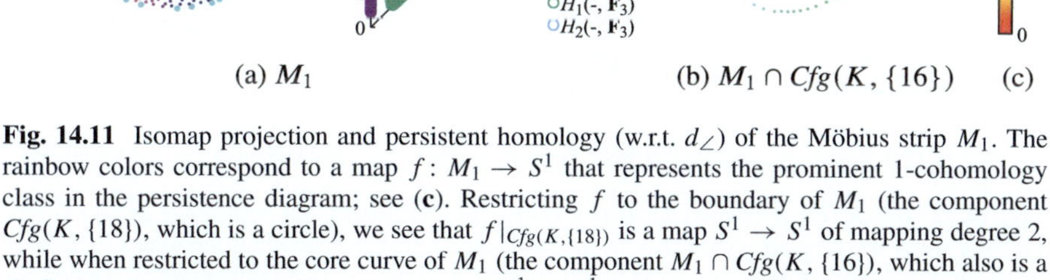

(a) M_1 (b) $M_1 \cap Cfg(K, \{16\})$ (c)

Fig. 14.11 Isomap projection and persistent homology (w.r.t. $d_\angle$) of the Möbius strip M_1. The rainbow colors correspond to a map $f\colon M_1 \to S^1$ that represents the prominent 1-cohomology class in the persistence diagram; see (**c**). Restricting f to the boundary of M_1 (the component $Cfg(K, \{18\})$, which is a circle), we see that $f|_{Cfg(K,\{18\})}$ is a map $S^1 \to S^1$ of mapping degree 2, while when restricted to the core curve of M_1 (the component $M_1 \cap Cfg(K, \{16\})$, which also is a circle, see (**a**)) yields a map $f|_{M_1 \cap Cfg(K,\{16\})}\colon S^1 \to S^1$ of degree 1. This supports the hypothesis that M_1 is a Möbius strip in which C is projected to the "hourglass" sitting inside the sphere (which represents $A \cup B$), we see that each two antipodal meridians of C are projected to the same image in Fig. 14.12b. This happens because samples of $A \cup B$ have a dihedral angle pattern of the form $(\alpha, \beta, \gamma, \delta, \alpha, \beta, \gamma, \delta)$ and therefore only have four distinct circular shifts. Furthermore, we see that any pair of antipodal meridians on C (e.g., the two blue ones) are attached to two antipodal meridians on $A \cup B$; see Fig. 14.12b; see also the Isomap projection of $Cfg(K, \{13\})$ alone in Fig. 14.12c.

From this, we obtain the gluing shown in Fig. 14.13. As mentioned before, this cell complex has the $\mathbf{F}_2$- and $\mathbf{F}_3$-Betti numbers (1,1,2) in dimensions (0,1,2), respectively, which matches the observed persistent homology from Fig. 14.3.

14.4 A Cell Structure for the Quotient Space $Cfg(K)/C_8$

Recall that $Cfg(K)$ is the *labeled* configuration space. We are also interested in a cell structure of the *unlabeled* configuration space. In this space, two configurations are considered equivalent if they differ only by a cyclic renumbering of the vertices, or a reversal of the numbering. Mathematically, this corresponds to taking the quotient space $Cfg(K)/D_8$ with respect to the *dihedral group* D_8 (the symmetry group of the regular octagon), which is the semidirect product $D_8 = C_8 \rtimes C_2$ of two cyclic groups, where C_8 acts by cyclic shifting of the numbering, and C_2 acts by reversing the numbering. Thus, our goal is to find a cell structure of $Cfg(K)/D_8$, and to describe the quotient map $Cfg(K) \to Cfg(K)/D_8$.

The quotient $Cfg(K) \to Cfg(K)/D_8$ factors through the quotient $Cfg(K)/C_8$ with respect to the normal subgroup $C_8 \triangleleft D_8$. Therefore, in this section, we investigate the quotient space $Cfg(K)/C_8$ as an intermediate step, together with the quotient map $Cfg(K) \to Cfg(K)/C_8$. We then describe $Cfg(K)/D_8$ and the quotient map $Cfg(K)/C_8 \to Cfg(K)/D_8$ in Sect. 14.5.

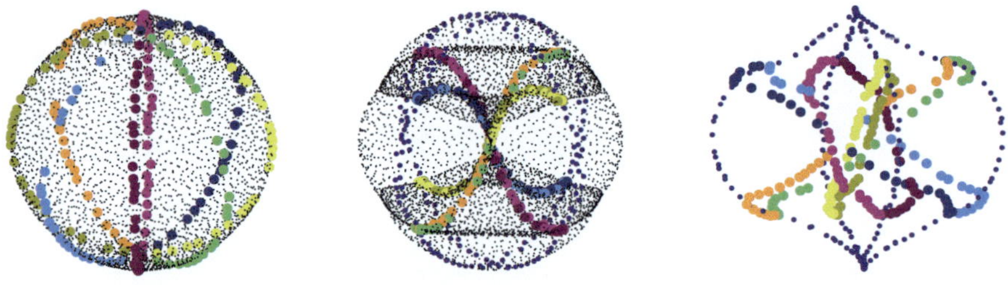

(a) C (w.r.t. $d_\angle$) (b) $Cfg(K)$ (w.r.t. $d_\parallel$) (c) $Cfg(K, \{13\})$ (w.r.t. $d_\parallel$)

Fig. 14.12 $Cfg(K, \{13\})$ endows C with a system of "meridians"; cf. Figs. 14.6 and 14.7. The pictures show the Isomap projections from Fig. 14.4, where the individual meridians of C have been equipped with colors, which allows to understand how the two components C (a and the "hourglass" in b) and $A \cup B$ (the sphere in Fig. 14.12b) are glued together

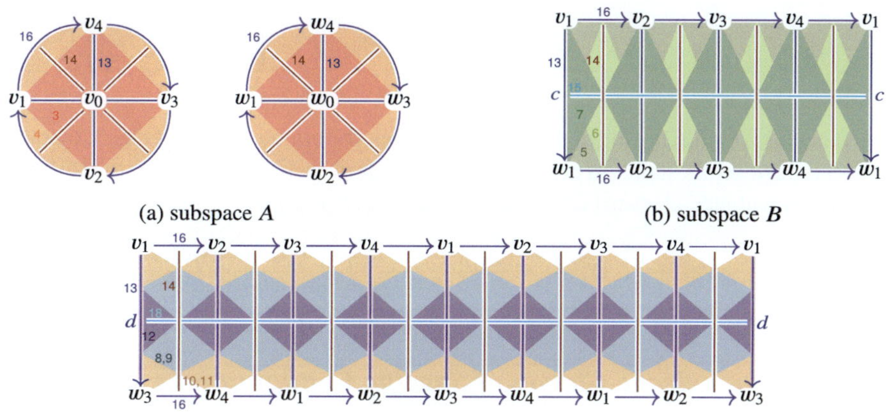

(a) subspace *A*

(b) subspace *B*

(c) subspace *C*

Fig. 14.13 Proposed cell complex representing $Cfg(K)$. The colors represent the symmetry types (small numbers, same colors as in Figs. 14.9 and 14.10). The domains of the different symmetry types are obtained from the Isomap projections. For clarity, not all symmetry types listed in Table 14.1 are explicitly drawn or labeled. Areas left white have only symmetry type 2 assigned. This cell complex is a refinement of Fig. 14.5

The following definition turns $Cfg(K)/C_8$ into a metric space.

Definition Let X be a metric space and $Y = X/\sim$ be a quotient of X w.r.t. an equivalence relation $\sim$. The *quotient (pseudo-)metric* is the finest metric d_Y on Y such that $d_Y([x],[y]) \leq d_X(x, y)$ for all $x, y \in X$. If all equivalence classes of X w.r.t. $\sim$ are finite, d_Y is a metric.

14.4.1 Actions

Let t denote the generator of C_8, which acts on the vertices by shifting the indexing by one. On the torsion angles $(\sigma_i(\bar{x}))_i$ of a configuration $\bar{x} \in Cfg(K)$, the generator t acts by

$$t(\sigma_0(\bar{x}), \ldots, \sigma_7(\bar{x})) = (\sigma_1(\bar{x}), \ldots, \sigma_7(\bar{x}), \sigma_0(\bar{x})).$$

The angular metric $d_\angle$ on $Cfg(K)/C_8$ then is given by

$$d_\angle([\bar{x}], [\bar{y}]) = \min\{d_\angle\big(\sigma(\bar{x}), t^n\sigma(\bar{y})\big); n = 0, \ldots, 7\}.$$

For the Euclidean metric, the situation is a little more intricate. Recall from Sect. 14.2.1 that we have defined the Euclidean metric $d_\parallel$ on $Cfg(K)$ in terms of standard realizations x^{std} that satisfy the Eckart condition. On standard realizations, t acts by

$$(x_0, \ldots, x_7) \mapsto (Tx_1, \ldots, Tx_7, Tx_0) \quad \text{for } T = \begin{pmatrix} \sqrt{2}/2 & \sqrt{2}/2 & 0 \\ -\sqrt{2}/2 & \sqrt{2}/2 & 0 \\ 0 & 0 & 1 \end{pmatrix},$$

i.e., by cyclic reindexing and a rotation by $\frac{\pi}{4}$ about the z-axis. One may check that this action indeed sends standard realizations to standard realizations. Thus, the distance of two equivalence classes $[x^{\mathrm{std}}]$ and $[y^{\mathrm{std}}]$ in $Cfg(K)/C_8$ is

$$d_{\|}([x^{\mathrm{std}}], [y^{\mathrm{std}}]) = \min\{d_{\|}(t^n x, y); n = 1, \ldots, 7\}.$$

14.4.2 Equivalence Classes of Specific Configurations

We start our analysis with the zero dimensional strata from Table 14.1c. There are two configurations jointly of types 1, 6 and 8 (pattern: $(\alpha, -\alpha, \alpha, -\alpha, \alpha, -\alpha, \alpha, -\alpha)$), corresponding to the two possible signs for α. Each of the two configuration sits in the middle of the two connected components of subspace A (see Fig. 14.4). These belong to the same equivalence class mod C_8, which we denote by v_0'.

Analogously, all four configurations jointly of type 13 and 15 (pattern: $(\alpha, 0, -\alpha, 0, \alpha, 0, -\alpha, 0)$) (each one sits in the middle of one face of B in Fig. 14.13) belong to the same C_8-orbit, and similarly for the eight configurations of type 13 and 18 (pattern: $(\alpha, -\alpha, \beta, \beta, -\alpha, \alpha, -\beta, -\beta)$). This suggests that each of the 2-cells in Fig. 14.13b, c is a fundamental domain for the C_8-action on B and C, respectively. As a justification, recall that the four 2-cells of A and B (see Fig. 14.13) correspond to the four possible shifts of the dihedral angle pattern $(\alpha, \beta, \gamma, \delta, \alpha, \beta, \gamma, \delta)$. An analogous justification holds for configurations in C.

The four configurations jointly of type 13 and 16 (pattern: $(\alpha, -\alpha, 0, 0, \alpha, -\alpha, 0, 0,))$ that lie in the intersections of A, B and C belong to *two* C_8 orbits, one for each sign of α. We denote these by v_1' and v_2'.

14.4.3 Cell Structure

Based on these observations, we arrive at the following description of the cell structure of $Cfg(K)/C_8$ and the quotient map $Cfg(K) \to Cfg(K)/C_8$: In Fig. 14.14, we have replaced the labels of the vertices by the C_8-equivalence classes they belong to. Edges with identical labels are identified. The resulting quotient cell complex is shown in Fig. 14.15. The subcomplex $A' := A/C_8$ is a disc; the subcomplex $B' := B/C_8$ and $C' := C/C_8$ both are isomorphic to a Möbius strip *MS*. This cell complex represents a gluing $D^2 \cup_{S_1} MS \cup_{S_1} MS$ of a disc with two Möbius strips along their boundaries.

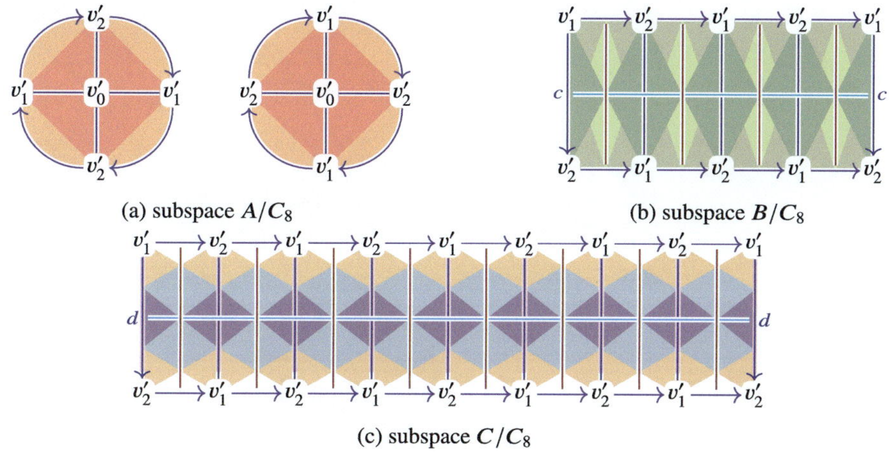

Fig. 14.14 The cell structure for $Cfg(K)$ from Fig. 14.13, with vertices replaced by the equivalence classes w.r.t. C_8 they belong to (See Fig. 14.13 for the meaning of the colors)

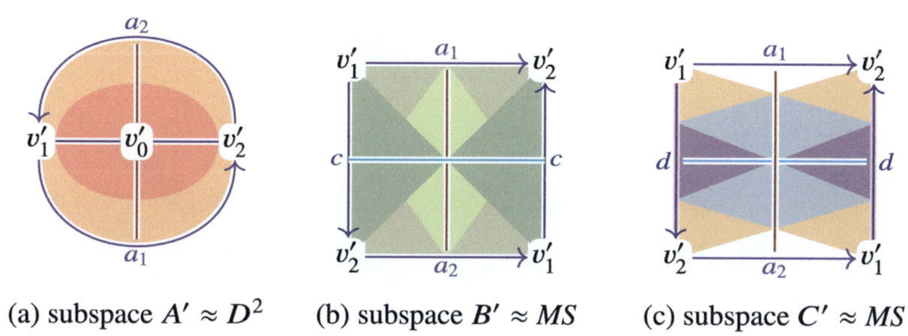

Fig. 14.15 Cell complex representing $Cfg(K)/C_8$

Table 14.2 Expected Betti numbers of various subspaces of $Cfg(K)/C_8$, according to the cell complex from Fig. 14.15

Subspace		$\beta_\bullet(-, \boldsymbol{F}_2)$	$\beta_\bullet(-, \boldsymbol{F}_3)$
A'	$\approx D^2$	(1,0,0)	(1,0,0)
B', C'	$\approx MS$	(1,0,0)	(1,0,0)
$A' \cup B', A' \cup C'$	$\approx \boldsymbol{P}^2$	(1,1,1)	(1,0,0)
$B' \cup C'$	$\approx KB$	(1,2,1)	(1,1,0)
$A' \cup B' \cup C'$	$\approx D^2 \cup_{S_1} MS \cup_{S_1} MS$	(1,1,2)	(1,0,0)

14.4.4 Plausibility Check

If the model is correct, then both $A' \cup B'$ and $A' \cup C'$ are homeomorphic to a real projective plane $\boldsymbol{P}^2 \cong D^2 \cup_{S_1} MS$. Similarly, $B' \cup C'$ is homeomorphic to a Klein bottle $KB \cong MS \cup_{S_1} MS$. We should thus expect to find the Betti numbers listed in

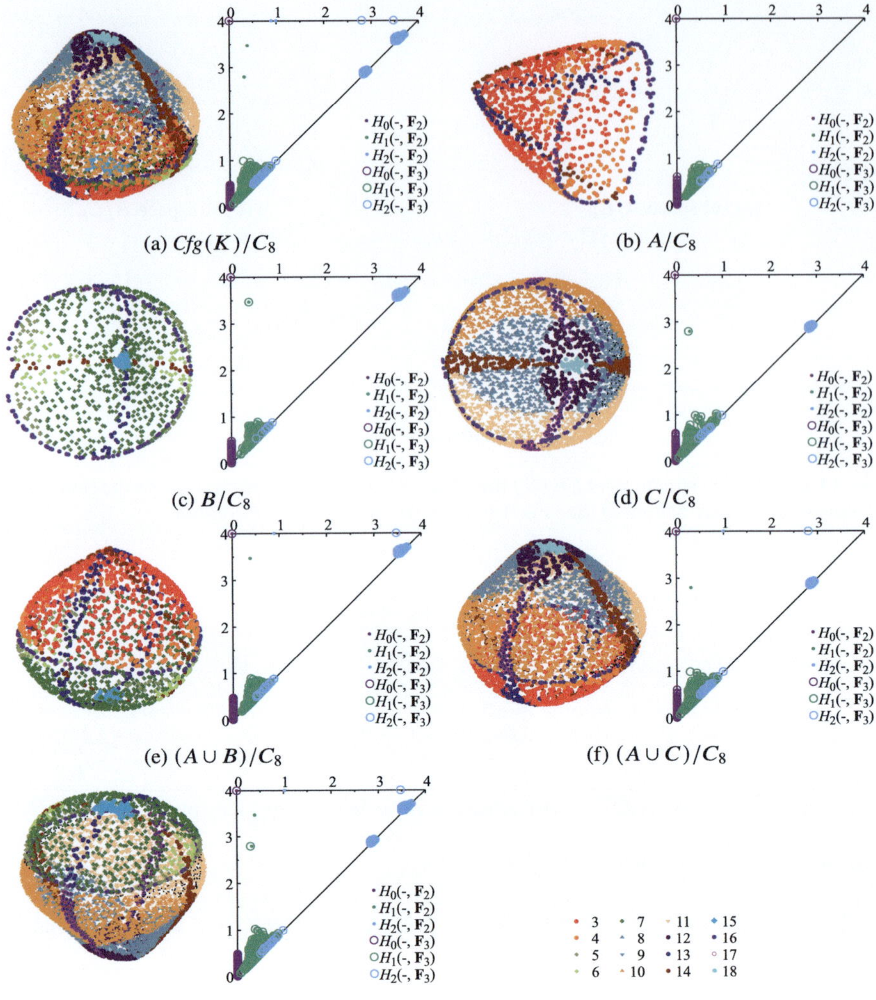

(a) $Cfg(K)/C_8$ (b) A/C_8 (c) B/C_8 (d) C/C_8 (e) $(A \cup B)/C_8$ (f) $(A \cup C)/C_8$ (g) $(B \cup C)/C_8$

Fig. 14.16 Isomap projections and persistence diagram (w.r.t. $d_\angle$) of different subspaces of $Cfg(K)/C_8$

Table 14.2. In Fig. 14.16, we see the persistence diagrams and Isomap projections of these subspaces. Indeed, we observe the expected Betti numbers.

To produce further evidence supporting our model, we have computed maps $B/C_8, C/C_8 \to S^1$ of the presumable Möbius strips B' and C' that correspond to the unique most prominent 1-cohomology class in Fig. 14.16c–d, respective; see Fig. 14.17. We see that the computed maps to S^1 wind twice around the common boundary curve ($Cfg(K, \{16\})/C_8$) and once about the respective core curves

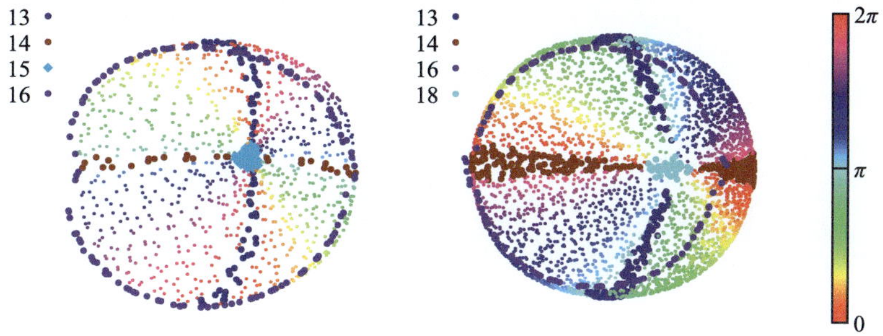

(a) B/C_8 (b) C/C_8

Fig. 14.17 Color-coded maps $(-) \to S^1$ representing the respective unique most prominent 1-cohomology class in Figs. 14.16b–14.16. See Sect. 14.3.3 for a description of the method

$(Cfg(K, \{14\})/C_8$ and $Cfg(K, \{18\})/C_8)$ of B/C_8 and C/C_8, which supports our hypothesis.

14.5 A Cell Structure for the Quotient Space $Cfg(K)/D_8$

Finally, we want to understand the unlabeled configuration space $Cfg(K)/D_8$; this is the space of all possible configurations of a chemical molecule with indistinguishable atoms. As $D_8 = C_8 \rtimes C_2$, where C_2 acts on C_8 by inversion. We have to describe the action of C_2 on configurations.

14.5.1 Actions

The generator s of C_2 acts on dihedral angles of a $\bar{x} \in Cfg(K)$ by reversing the order; that is,

$$s(\sigma(\bar{x}))_i = \sigma(\bar{x})_{1-i} \tag{14.2}$$

with indices taken modulo 8. Analogously, s acts on standard realizations x^{std} by

$$(sx)_i = Sx_{1-i} \quad \text{for} \quad S = \begin{pmatrix} 0 & 1 & 0 \\ 1 & 0 & 0 \\ 0 & 0 & -1 \end{pmatrix}; \tag{14.3}$$

i.e., by rotating them by π about the axis $(1, 1, 0)^{\text{T}}$. The reindexing and the rotation matrix S are chosen to preserve the Eckart condition Eq. 14.1. We obtain a Euclidean and an angular quotient metric on $Cfg(K)/D_8$ by minimizing over actions of s and t analogously to the description in Sect. 14.4.1.

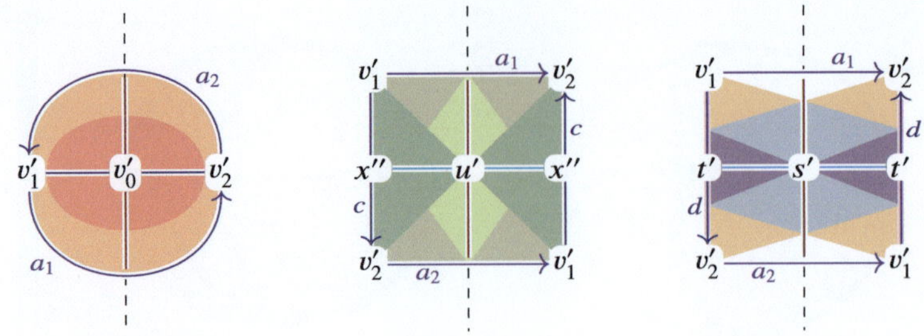

(a) subspace $A' \approx D^2$ of $Cfg(K)/C_8$

(b) subspace $B' \approx MS$ of $Cfg(K)/C_8$

(c) subspace $C' \approx MS$ of $Cfg(K)/C_8$

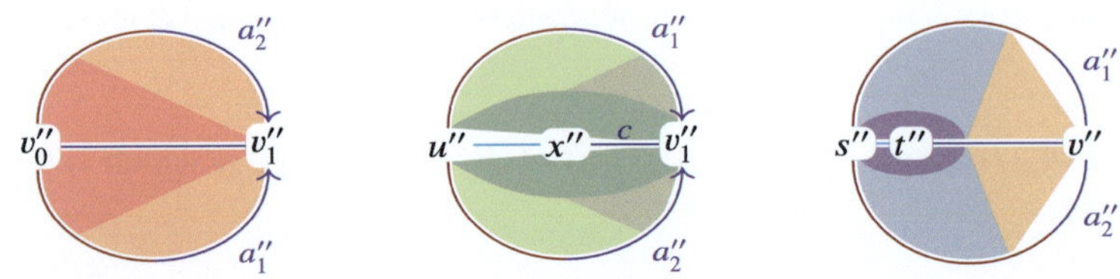

(d) subspace A'' of $Cfg(K)/D_8$

(e) subspace B'' of $Cfg(K)/D_8$

(f) subspace C'' of $Cfg(K)/D_8$

Fig. 14.18 Upper row: cell $Cfg(K)/C_8$ from Fig. 14.15. The D_8-action takes mirror images across the dashed vertical line. With respect to Fig. 14.15, additional nodes and sub-edges have been added. Lower row: quotient cell complex structure $Cfg(K)/D_8$. The dashed lines are free edges that are the images of the dashed lines in the upper row. The three subspaces are glued along the two lines a''_1, a''_2, which are the image of the edges a'_1 and a'_2 under the projection

14.5.2 Cell Structure

Recall that the two equivalence classes v'_1 and v'_2 in $Cfg(K)/C_8$ correspond to the two torsion angle patterns $(\beta, 0, 0, -\beta, \beta, 0, 0, -\beta)$ for $\beta \gtrless 0$. Since the C_2 action reverses the order of the angles, v'_1 and v'_2 belong to the same class in $Cfg(K)/D_8$. Furthermore, the samples that are both of types 1 and 14 have the dihedral angle pattern $(\alpha, \beta, \alpha, \gamma, \dots)$, which is invariant under the action of C_2. Therefore, the quotient map $Cfg(K)/C_8 \twoheadrightarrow Cfg(K)/D_8$ identifies points as indicated in Fig. 14.18. The upper row of the figure shows the cell complex representing $Cfg(K)/C_8$ from Fig. 14.15. The quotient map $Cfg(K)/C_8 \twoheadrightarrow Cfg(K)/D_8$ identifies mirror images across the dashed line (type 14). The lower row contains the image of that cell structure and thus represents a cell complex for $Cfg(K)/D_8$. The space $Cfg(K)/D_8$ thus consists of three sheets glued together along a common line (type 16). This line is the image of the intersection $A \cap B \cap C$, which is the set of configurations of symmetry type 16. In particular, $Cfg(K)/D_8$ is a contractible space. This is confirmed by the persistence diagrams in Fig. 14.19.

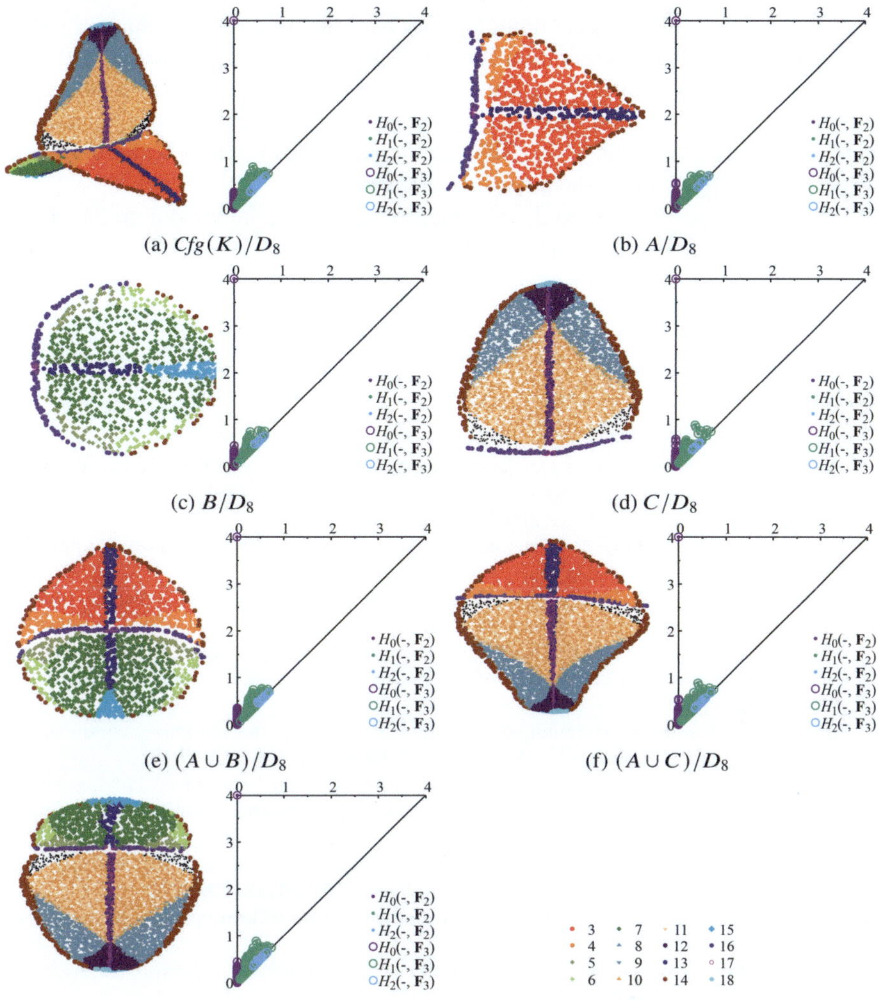

Fig. 14.19 Isomap projections and persistence diagram (w.r.t. $d_\angle$) of different subspaces of $Cfg(K)/D_8$

14.6 Summary and Outlook

We have investigated the topology of the configuration space of the cyclic eight-vertex linkage that corresponds to the chemical cyclooctane molecule, based on the two metrics $d_\parallel$ and $d_\angle$. Only the latter actually descends to a metric on the linkage configurations as is invariant under global translations and rotations of the configuration.

Using Isomap and persistent homology we have identified the topology of various subsets of the labeled configuration space $Cfg(K)$ determined by the symmetries of their elements. These subsets suggest a cell structure on the space that is compatible with the quotient maps and hence also descends to a cell structure of the unlabeled configuration space $Cfg(K)/D_8$. The latter space turned out to be a contractible surface consisting of three sheets glued together.

We have not yet considered the configurational energy functional on the conformation space of cyclooctane. It is an interesting question to study the critical points of this function in light of the cell structure proposed here.

References

1. Attali, D., Lieutier, A., Salinas, D.: Vietoris–Rips complexes also provide topologically correct reconstructions of sampled shapes. Comput. Geom. **46**, 448–465 (2013)
2. Bauer, U.: Ripser (2019). https://github.com/Ripser/ripser
3. Bauer, U.: Ripser: efficient computation of Vietoris–Rips persistence barcodes. J. Appl. Comput. Topol. **5**, 391–423 (2021)
4. Bauer, U., Lange, C., Wardetzky, M.: Optimal topological simplification of discrete functions on surfaces. Discret. Comput. Geom. **47**, 347–377 (2012)
5. Blumenthal, L.: Theory and Applications of Distance Geometry (Chelsea Pub. Co., 1970)
6. Boeyens, J., Evans, D.: Group theory of ring pucker. Acta Crystallogr. Sect. B Struct. Sci. **45**, 577–581 (1989)
7. Breiding, P., Timme, S.: HomotopyContinuation.Jl: a package for homotopy continuation in Julia. Mathematical Software – ICMS 2018, vol. 10931, pp. 458–465 (2018)
8. Breiding, P., Timme, S.: An Introduction to the Numerical Solution of Polynomial Systems. (HomotopyContinuation.jl,2019). https://www.juliahomotopycontinuation.org/guides/introduction
9. Brown, W., Martin, S., Pollock, S., Coutsias, E., Watson, J.: Algorithmic dimensionality reduction for molecular structure analysis. J. Chem. Phys. **129**, 064118 (2008)
10. Coutsias, E., Seok, C., Wester, M., Dill, K.: Resultants and loop closure. Int. J. Quantum Chem. **106**, 176–189 (2006)
11. Crawley-Boevey, W.: Decomposition of pointwise finite-dimensional persistence modules. J. Algebr. Appl. **14**, 1550066 (2015)
12. Cremer, D., Pople, J.: General definition of ring puckering coordinates. J. Am. Chem. Soc. **97**, 1354–1358 (1975)
13. Edelsbrunner, H., Harer, J.: Persistent homology—a survey. Contemp. Math. **453**, 257–282 (2008)
14. Edelsbrunner, H., Harer, J., Zomorodian, A.: Hierarchical Morse-Smale complexes for piecewise linear 2-manifolds. Discret. Comput. Geom. **30**, 87–107 (2003)
15. Edelsbrunner, H., Morozov, D.: Persistent Homology. Handbook of Discrete and Computational Geometry, 3rd ed. Chapman and Hall/CRC. ISBN: 978-1-4987-1139-5 (2017)
16. Evans, D., Boeyens, J.: Conformational analysis of ring pucker. Acta Crystallogr. Sect. B Struct. Sci. **45**, 581–590 (1989)
17. Forman, R.: Morse theory for cell complexes. Adv. Math. **134**, 90–145 (1998)
18. Freedman, D.: An incremental algorithm for reconstruction of surfaces of arbitrary codimension. Comput. Geom. **36**, 106–116 (2007)
19. Ghrist, R.: Barcodes: the persistent topology of data. Bull. Am. Math. Soc. **45**, 61–76 (2007)
20. Gibson, K., Scheraga, H.: Energy minimization of rigid-geometry polypeptides with exactly closed disulfide loops. J. Comput. Chem. **18**, 403–415 (1997)

21. Günther, D., Reininghaus, J., Seidel, H., Weinkauf, T.: Notes on the simplification of the Morse-Smale complex. In: Topological Methods in Data Analysis and Visualization, Vol. III, pp. 135–150. Springer, Cham (2014)
22. Gyulassy, A., Bremer, P., Hamann, B., Pascucci, V.: A practical approach to Morse-Smale complex computation: scalability and generality. IEEE Trans. Vis. Comput. Graph. **14**, 1619–1626 (2008)
23. Gyulassy, A., Bremer, P., Hamann, B., Pascucci, V.: Practical considerations in Morse-Smale complex computation. Topol. Methods Data Anal. Vis. 67–78 (2011)
24. Hendrickson, J.: Molecular geometry. V. Evaluation of functions and conformations of medium rings. J. Am. Chem. Soc. **89**, 7036–7043 (1967)
25. Hendrickson, J.: Molecular geometry. VII. Modes of interconversion in the medium rings. J. Am. Chem. Soc. **89**, 7047–7061 (1967)
26. Jordan, D., Steiner, M.: Configuration spaces of mechanical linkages. Discret. Comput. Geom. **22**, 297–315 (1999)
27. Kempe, A.: How to Draw a Straight Line: A Lecture on Linkages (MacMillan, 1877)
28. Martin, S.: NMTRI: Non-Manifold Surface Reconstruction. (Software by Shawn Martin, 2012). http://www.cs.otago.ac.nz/homepages/smartin/software.php
29. Martin, S., Watson, J.: Non-manifold surface reconstruction from high dimensional point cloud data. Comput. Geom. **44**, 427–441 (2011)
30. Martin, S., Thompson, A., Coutsias, E., Watson, J.: Topology of cyclo-octane energy landscape. J. Chem. Phys. **132**, 234115 (2010)
31. Meiboom, S., Hewitt, R., Luz, Z.: The conformation of cyclooctane: an experimental determination by NMR in an oriented solvent. J. Chem. Phys. **66**, 4041–4051 (1977)
32. Membrillo-Solis, I., Pirashvili, M., Steinberg, L., Brodzki, J., Frey, J.: Topology and Geometry of Molecular Conformational Spaces and Energy Landscapes (2019). arXiv:1907.07770
33. Niyogi, P., Smale, S., Weinberger, S.: Finding the homology of submanifolds with high confidence from random samples. Discret. Comput. Geom. **39**, 419–441 (2008)
34. Otter, N., Porter, M., Tillmann, U., Grindrod, P., Harrington, H.: A roadmap for the computation of persistent homology. EPJ Data Sci. **6**, 17 (2017)
35. Pickett, H., Strauss, H.: Symmetry and conformation of the cycloalkanes. J. Chem. Phys. **55**, 324–334 (1971)
36. Porta, J., Ros, L., Thomas, F., Corcho, F., Cantó, J., Pérez, J.: Complete maps of molecular-loop conformational spaces. J. Comput. Chem. **28**, 2170–2189 (2007)
37. Sommese, A., Wampler, C.: The Numerical Solution of Systems of Polynomials Arising in Engineering and Science (Hackensack, NJ World Scientific, 2005). http://openlibrary.org/books/OL3438210M
38. Tenenbaum, J.: A global geometric framework for nonlinear dimensionality reduction. Science **290**, 2319–2323 (2000)
39. Thomas, F., Porta, J., Ros, L.: Distance constraints solved geometrically. On Advances In Robot Kinematics, pp. 31–40 (2004)
40. Zomorodian, A., Carlsson, G.: Computing persistent homology. Discret. Comput. Geom. **33**, 249–274 (2005)
41. Silva, V., Morozov, D., Vejdemo-Johansson, M.: Dualities in persistent (co)homology. Inverse Probl. **27**, 124003 (2011)
42. Adams, H., Moy, M.: Topology applied to machine learning: from global to local. Front. Artif. Intell. **4**, 668302 (2021). https://doi.org/10.3389/frai.2021.668302/full
43. Eckart, C.: Some studies soncerning rotating axes and polyatomic molecules. Phys. Rev. **47**, 552–558 (1935)
44. Pólya, G.: Kombinatorische Anzahlbestimmungen für Gruppen, Graphen und chemische Verbindungen. Acta Math. **68**, 145–254 (1937)

MIX
Papier aus verantwortungsvollen Quellen
Paper from responsible sources
FSC® C105338

FSC
www.fsc.org

If you have any concerns about our products,
you can contact us on
ProductSafety@springernature.com

In case Publisher is established outside the EU,
the EU authorized representative is:
Springer Nature Customer Service Center GmbH
Europaplatz 3, 69115 Heidelberg, Germany

Printed by Libri Plureos GmbH
in Hamburg, Germany